Computational Modeling and Simulation of Solids and Structures: Recent Advances and Practical Applications

Computational Modeling and Simulation of Solids and Structures: Recent Advances and Practical Applications

Editors

Jin-Gyun Kim

Jae Hyuk Lim

Peter Persson

MDPI • Basel • Beijing • Wuhan • Barcelona • Belgrade • Manchester • Tokyo • Cluj • Tianjin

Editors
Jin-Gyun Kim
Kyung Hee University
Republic of Korea

Jae Hyuk Lim
Jeonju-si
Republic of Korea

Peter Persson
Lund University
Sweden

Editorial Office
MDPI
St. Alban-Anlage 66
4052 Basel, Switzerland

This is a reprint of articles from the Special Issue published online in the open access journal *Applied Sciences* (ISSN 2076-3417) (available at: https://www.mdpi.com/journal/applsci/special_issues/Modeling_Simulation_of_Solids_and_Structures).

For citation purposes, cite each article independently as indicated on the article page online and as indicated below:

LastName, A.A.; LastName, B.B.; LastName, C.C. Article Title. *Journal Name* **Year**, *Volume Number*, Page Range.

ISBN 978-3-0365-7748-7 (Hbk)
ISBN 978-3-0365-7749-4 (PDF)

Contents

applied sciences

MDPI

Editorial

Special Issue on "Computational Modeling and Simulation of Solids and Structures: Recent Advances and Practical Applications"

Jin-Gyun Kim [1,*], Jae Hyuk Lim [2,*] and Peter Persson [3,*]

[1] Department of Mechanical Engineering, Kyung Hee University, Yongin-si 17104, Korea
[2] Department of Mechanical Engineering, Jeonbuk National University, Jeonju-si 54896, Korea
[3] Department of Construction Sciences, Lund University, 220 02 Lund, Sweden
* Correspondence: jingyun.kim@khu.ac.kr (J.-G.K.); jaehyuklim@jbnu.ac.kr (J.H.L.); peter.persson@construction.lth.se (P.P.)

Citation: Kim, J.-G.; Lim, J.H.; Persson, P. Special Issue on "Computational Modeling and Simulation of Solids and Structures: Recent Advances and Practical Applications". *Appl. Sci.* **2022**, *12*, 3660. https://doi.org/10.3390/app12073660

Received: 29 March 2022
Accepted: 1 April 2022
Published: 5 April 2022

1. Introduction

Computational modeling and simulation are essential to solid and structural mechanics. They have not only covered entire engineering fields (civil, aerospace, mechanical, etc.) but also various scales (from nano to macro) and physics (mono- and multi-physics). Recently, they have been found to be able to offer theoretical backgrounds of digital transformation. Society at large is increasingly enthusiastic about data-driven modeling and simulation, and the possibilities they offer. The aim of this Special Issue is to provide a forum for researchers to discuss recent advanced computational modeling and simulation techniques of solids and structures, and applications to solve challenging engineering problems. Innovative and novel modeling approaches, numerical methods, and industrial applications are of special interest. The industrial applications should include a strong connection to computational modeling and simulation.

2. Special Issues

The call for papers in the Special Issue "Computational Modeling and Simulation of Solids and Structures: Recent Advances and Practical Applications" in *Applied Sciences* was open from 1 March 2021, to 28 February 2022, and received 16 manuscripts, of which 9 were selected to be published, giving a 56% approval rate. These manuscripts cover the wide range of the topics introduced above and are listed here in order of publication date.

Kim et al. [1] studied piezoresistive characteristics of NCSS (nanocarbon composite strain sensor) and a simple design to improve the NCSS sensitivity by using its geometric pattern at a macro scale.

Macho et al. [2] describe features of the GIM software that are frequently used to support and exemplify the theoretical concepts taught in lectures. GIM integrated into different learning activities is also introduced to show its potential as a tool for learning and self-evaluation.

Tian et al. [3] inversely identify the pre tightening torque dependent parameters for empirical modeling of bolted joints with reference to the modal test results. To consider the contact performance of the joint structure of the bolt lap joint, the thin element method with linear constitutive relation was employed, which makes the simulation result more accurate.

Hoffer et al. [4] provide a framework for selecting mesh-free surrogate methods. Their evaluations show that surrogate modeling can be a competitive tool in engineering applications of various complexity.

Jeong [5] provides a scheme for topology optimization of deformable bodies with linear dynamic impact and frictionless contact conditions. When the nonmatching mesh occurred during sliding contact, a mortar method was employed, which was combined with the solid isotropic method with penalization (SIMP) method for the topology optimization. From the results, it was shown that the proposed scheme was quite efficient and general to solve topology optimization problems considering frictionless contact conditions.

Manconi et al. [6] used the Wave Finite Element method to study both free and forced wave propagation in structures with radial periodicities. In relation to a standard Finite Element model using Perfectly Matched Layers, the developed approach was proved accurate and provided significant savings of computational time.

Choi et al. [7] present modeling and validation of a passive truss-link mechanism for deployable structures considering friction compensation with response surface methods. To make an excellent correlation between test analysis, an inverse analysis to fit the parameters of the empirical friction model with reference to the test results was conducted using the response surface methods.

Hoffer et al. [8] present a surrogate modeling technique considering both aleatoric and epistemic uncertainties. Concerning the engineering application of hot metal forming, the developed approach was shown computationally efficient as compared to the use of standard Finite Element simulation.

Bugaru and Vasile [9] propose a modeling technique of the forced bending vibrating (FBV) movements for the elements of an automotive driveshaft using a perturbation technique, the asymptotic method approach (AMA), in the region of principal parametric resonance (PPR).

Funding: This research received no external funding.

Acknowledgments: We would like to thank the authors of the manuscripts submitted to this Special Issue. Further, we thank the reviewers and the *Applied Sciences* team for your support.

Conflicts of Interest: The authors declare no conflict of interest.

References

1. Kim, S.; Choi, B.; Oh, G.; Kim, C.; Jung, Y.; Jang, J.; Joung, K.; Suh, J.; Kang, I. Piezoresistive Characteristics of Nanocarbon Composite Strain Sensor by Its Longitudinal Pattern Design. *Appl. Sci.* **2021**, *11*, 5760. [CrossRef]
2. Macho, E.; Urízar, M.; Petuya, V.; Hernández, A. Improving Skills in Mechanism and Machine Science Using GIM Software. *Appl. Sci.* **2021**, *11*, 7850. [CrossRef]
3. Tian, Y.; Qian, H.; Cao, Z.; Zhang, D.; Jiang, D. Identification of Pre-Tightening Torque Dependent Parameters for Empirical Modeling of Bolted Joints. *Appl. Sci.* **2021**, *11*, 9134. [CrossRef]
4. Hoffer, J.; Geiger, B.; Ofner, P.; Kern, R. Mesh-Free Surrogate Models for Structural Mechanic FEM Simulation: A Comparative Study of Approaches. *Appl. Sci.* **2021**, *11*, 9411. [CrossRef]
5. Jeong, G. Topology Optimization of Deformable Bodies with Linear Dynamic Impact and Frictionless Contact Condition. *Appl. Sci.* **2021**, *11*, 10518. [CrossRef]
6. Manconi, E.; Sorokin, S.; Garziera, R.; Quartaroli, M. Free and Forced Wave Motion in a Two-Dimensional Plate with Radial Periodicity. *Appl. Sci.* **2021**, *11*, 10948. [CrossRef]
7. Choi, H.; Kim, D.; Park, J.; Lim, J.; Jang, T. Modeling and Validation of a Passive Truss-Link Mechanism for Deployable Structures Considering Friction Compensation with Response Surface Methods. *Appl. Sci.* **2022**, *12*, 451. [CrossRef]
8. Hoffer, J.; Geiger, B.; Kern, R. Gaussian Process Surrogates for Modeling Uncertainties in a Use Case of Forging Superalloys. *Appl. Sci.* **2022**, *12*, 1089. [CrossRef]
9. Bugaru, M.; Vasile, O. Modeling and Analysis of FBV Movements for Automotive Driveshafts in the PPR Region. *Appl. Sci.* **2022**, *12*, 3237. [CrossRef]

applied sciences

Article

Piezoresistive Characteristics of Nanocarbon Composite Strain Sensor by Its Longitudinal Pattern Design

Sung-Yong Kim [1,†], Baek-Gyu Choi [1,†], Gwang-Won Oh [1], Chan-Jung Kim [1], Young-Seok Jung [1], Jin-Seok Jang [2], Kwan-Young Joung [3], Jun-Ho Suh [4] and Inpil Kang [1,*]

1 Department of Mechanical and Design Engineering, Pukyong National University, Busan 48513, Korea; ksy1357@pukyong.ac.kr (S.-Y.K.); cbg8901@pukyong.ac.kr (B.-G.C.); 201955212@pukyong.ac.kr (G.-W.O.); cjkim@pknu.ac.kr (C.-J.K.); yousjung@pknu.ac.kr (Y.-S.J.)
2 Mechanical Components and Materials R&D Group, Korea Institute of Industrial Technology (KITECH), Daegu 42994, Korea; jsjang@kitech.re.kr
3 Innovative Smart Manufacturing R&D Department, Korea Institute of Industrial Technology (KITECH), Cheonan 31056, Korea; j6044@kitech.re.kr
4 Department of Mechanical Engineering, Pusan National University, Busan 46241, Korea; junhosuh@pusan.ac.kr
* Correspondence: ipkang@pknu.ac.kr; Tel.: +82-51-629-6971
† These authors contributed equally as a first author to this work.

Abstract: For an engineering feasibility study, we studied a simple design to improve NCSS (nanocarbon composite strain sensor) sensitivity by using its geometric pattern at a macro scale. We fabricated bulk- and grid-type sensors with different filler content weights (wt.%) and different sensor lengths and investigated their sensitivity characteristics. We also proposed a unit gauge factor model of NCSS to find a correlation between sensor length and its sensitivity. NCSS sensitivity was improved proportional to its length incremental ratio and we were able to achieve better linear and consistent data from the grid type than the bulk type one. We conclude that the longer sensor length results in a larger change of resistance due to its piezoresistive unit summation and that sensor geometric pattern design is one of the important issues for axial load and deformation measurement.

Keywords: carbon nanotube; strain sensor; piezoresistive mechanism; sensor pattern design

Citation: Kim, S.-Y.; Choi, B.-G.; Oh, G.-W.; Kim, C.-J.; Jung, Y.-S.; Jang, J.-S.; Joung, K.-Y.; Suh, J.-H.; Kang, I. Piezoresistive Characteristics of Nanocarbon Composite Strain Sensor by Its Longitudinal Pattern Design. *Appl. Sci.* **2021**, *11*, 5760. https://doi.org/10.3390/app11135760

Academic Editor: David Charles Barton

Received: 11 May 2021
Accepted: 18 June 2021
Published: 22 June 2021

1. Introduction

Due to their versatility and exceptional mechanical and electrical properties, nanocarbon materials have been studied for use in reinforced composites as well as in transducers [1–3]. The incorporation of nanocarbon materials, with their pertinent electrical properties, allows for applications as sensors in a wide range of fields [4–11].

A nanocarbon composite can be used as a sensory material due to the electrically conductive fillers in the matrix. Having electrical conductivity, the nanocarbon composite has a piezoresistive feature that can be used like a commercial foil strain gauge. We reported the comparable strain sensing performance of a nanocarbon composite strain sensor (NCSS) to a commercial foil strain gauge [12]. Although the strain sensing performances of the two sensors are similar, the piezoresistive mechanism of the NCSS is more complicated than that of the metallic type gauge [13]. The piezoresistive mechanism of the NCSS is related to various factors, such as the properties of the matrix, the content and arrangement of the fillers, and the geometric shape of the composite.

In particular, the piezoresistivity is directly related to the NCSS sensitivity, and it may be controllable via the fabrication process variables. During the sensor fabrication process, we believe that sensor irregularities generally occur due to filler randomness, including quality and dispersion, which dominate a sensor's electrical properties. Such irregularities may hamper the ability to achieve uniform performance of an individual sensor. To predict nanoscale internal filler behavior, researchers have studied physical models [14–17], computer-based molecular dynamics, and other types of simulations [18–21].

Some researchers have studied variations in NCSS sensitivity based on fabrication methodology. Lee et al. fabricated a single-walled carbon nanotube (SWCNT) film strain sensor using the spray method in a grid form and studied the sensor resistance and sensitivity characteristics according to the number of grids and the thickness of the sensor [22]. In addition, Huang et al. fabricated a SWCNT nano strain sensor by alcohol catalytic chemical vapor deposition (ACCVD) and studied the sensor sensitivity characteristics according to the SWCNT growth time and the number of sensor grids [23]. Wang et al. used the spray deposition modeling (SDM) method to fabricate carbon nanotube (CNT) sensors with varying thicknesses and investigated their piezoresistive properties [24]. Kong et al. fabricated a composite sensor using carbon black and PDMS (Polydimethylsiloxane), and proposed an application case with different patterns and shapes for the sensor [25].

There have been other similar macroscopic approaches based on sensor pattern studies. Li et al. fabricated multi-walled carbon nanotube (MWCNT) film sensors using a solution/filtration method, and they tested piezoresistivity with an optimized aspect ratio to obtain a proper signal from the sensor [26]. Xu and Allen fabricated a MWCNT/PDMS strain sensor. They showed the correlation between the initial resistance according to sensor thickness and sensor sensitivity [27]. Other studies have investigated improvements to NCSS sensitivity by using hybrid nanocarbon composites or functionalized nanocarbon fillers [28–30].

Since the quality control of most nanocomposites remains a challenge at the mass production level, the electrical characteristics of nanocarbon composites are also hard to consistently control. Consequently, most fabricated NCSSs may not have identical piezoresistive properties in terms of the nanocomposite process at the nano or micro scales. However, a commercial strain sensor yields uniform performance according to specifications, and a customer can expect reliable output from the sensor to meet the desired purpose. Unlike conventional strain sensors, NCSSs tend to involve many complexities at micro and macro scales and, furthermore, they are difficult to fabricate.

Therefore, for an engineering feasibility study, we used a simple design to improve NCSS sensitivity by using its macroscale geometric pattern. We fabricated bulk- and grid-type sensors with different filler contents (wt.%) and different sensor lengths, and we investigated their sensitivity characteristics. We also proposed a unit gauge factor model of NCSS to determine the correlation between sensor length and its sensitivity.

2. Experimental Program

2.1. NCSS Fabrication Process

To examine the percolation threshold and sensitivity of nanocarbon composite sensors, we fabricated the samples with five weight ratios (0.125, 0.25, 0.35, 0.5, and 1 wt.%) via the process shown in Figure 1. We used MWCNTs (Hanhwa Chemical Co., Korea, CM-280) as internal electrical fillers, epoxy (Kukdo Chemical Co., Korea, YD-128) as base material, and methylene chloride (Samchun Pure Chemical Co., Korea, purity 99.8%) as a solvent. The MWCNTs have a length of 180~200 μm, a diameter of 10~15 nm, and an aspect ratio of 12,000~20,000.

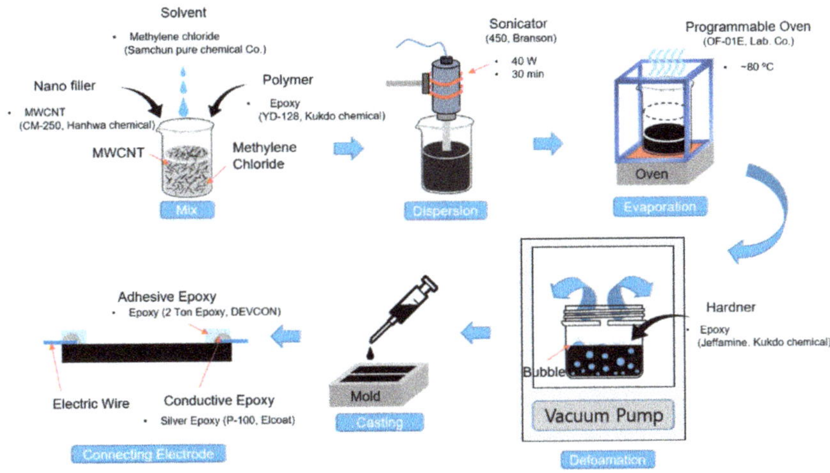

Figure 1. Schematic diagram of the nanocarbon composite fabrication process. MWCNT: multi-walled carbon nanotube.

To obtain the effective electrical performance of the manufactured composite material, the solution was dispersed at about 40 W for 30 min using a sonicator (Branson Co., Brookfield, CT, USA, 450). To control the mixing viscosity with the hardener (Kukdo Chemical Co., Korea, Jeffamine), the dispersed solution was evaporated in a programmable oven (Lab Companion, Daejeon, Korea, OF-01E) at 80 °C. After that, the mixture was degassed in a vacuum oven at 0 atm and 50 °C for about 10 min to remove residual bubbles. The mixture was injected into silicone molds using a syringe. Silicone molds were prepared in various casting widths (8, 10, 12 mm), thicknesses (0.6, 1.2, 2.4, 3.6 mm), and lengths (30, 40, 50 mm) to fabricate different sizes of sensors. The degassed mixture in the mold was cured at 80 °C using the programable oven. To complete the sensor samples, the casted samples were separated from the molds, and electrical wires were connected to the samples with conductive epoxy (CANS, Japan, Elcoat-P-100).

2.2. Test Setup

To investigate the electrical characteristics and sensitivity of the NCSS, the experimental apparatus shown in Figure 2 was constructed. The NCSS was tightly bonded with an adhesive epoxy on top of a thick steel cantilever (300 mm × 25 mm × 2 mm). The center of all sensor samples was located 50 mm from the fixed end of the cantilever. We installed a pair of NCSSs on the cantilever to double-check the outputs. The sensor-installed cantilever was mounted on an optical table by fixtures. We manually bent the free end of cantilever at steps of 20 mm from −140 mm to 140 mm, and the deflection was measured with a laser sensor (KEYENCE Co., Seongnam-si, Korea, IL-300). When a displacement is applied to the free end of the cantilever, strain is generated on the sensor attached to the upper part of the cantilever. The induced strain changes into sensor resistance change due to its piezoresistive characteristics. The sensor resistance was measured by a multi-meter (KEYSIGHT technologies Co., Santa Rosa, CA, USA, 34465A) with the two-probes method. The induced strain was later calculated in terms of the deflection by strain and bending relation equation.

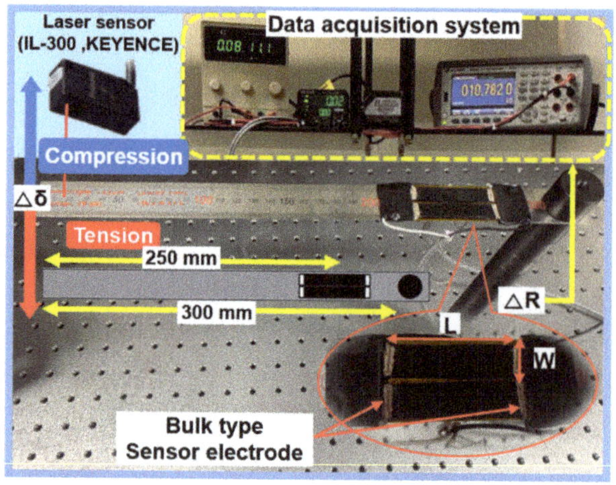

Figure 2. Experimental setup for the electrical and sensitivity characteristics of the nanocarbon composite strain sensor (NCSS).

3. Results

3.1. Percolation Threshold and Sensor Sensitivity Characteristics

We varied the percolation threshold to find an appropriate piezoresistivity boundary and to check the uniformity of the samples fabricated in each batch. We fabricated the samples at five different MWCNT weight fractions and measured their resistances, as shown in Figure 3a. The fabrication procedure was repeated two times to obtain reliable data. The resistances changed rapidly below 0.35 wt.%, and we decided on which piezoresistive design to use based on these results.

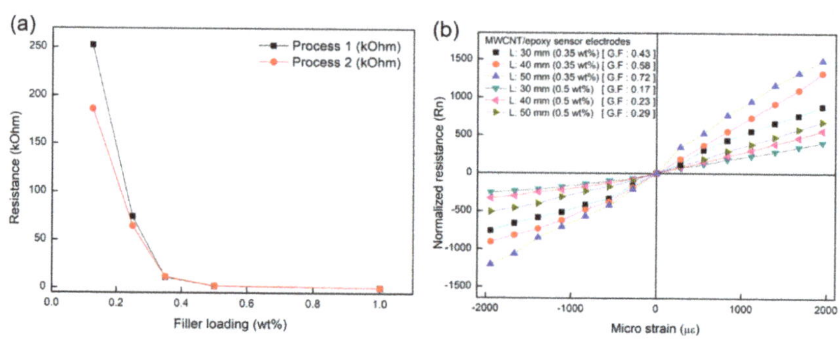

Figure 3. (**a**) Percolation threshold of the samples and (**b**) sensitivity characteristics (according to wt.%) of the NCSS electrodes.

When the samples were fabricated using a different process with the same recipe, their electrical properties were not identical under 0.35 wt.%, as shown in Figure 3a. This may be due to the relationship between the electrical conducting path and the filler-loading fraction in composites. In the case of higher filler content, we might be able to obtain similar electrical conductivity from the samples due to their high electrical filler density, as shown for 0.5 wt.% in Table 1. However, in the case of lower filler content, the electrical conducting path may vary widely due to their lower filler-loading fraction, as shown for 0.35 wt.% in Table 1. We also fabricated sensor samples with different lengths for cases with both 0.35 and 0.5 wt.% filler to test the piezoresistivity in terms of the geometrical factors of the sensors. We prepared samples with lengths of 30, 40, and 50 mm and fixed their width (10 mm) and thickness (1.2 mm), as shown in Table 1. The resistance change ratio (normalized resistance, Rn) of the sensors were also increased according to their length, as shown in Figure 3b. This indicates that the sensor length may correlate with sensitivity, and the length should therefore be one of the factors considered in sensor design.

Table 1. The resistance of the sensor according to the length and wt.% at no load.

Sensor	0.35 wt.%			0.5 wt.%		
T = 1.2, W = 10 (mm)	L = 30	L = 40	L = 50	L= 30	L = 40	L = 50
Sample 1 (kΩ)	9.47	11.37	14.33	1.53	1.87	2.33
Sample 2 (kΩ)	8.63	10.89	14.10	1.44	1.87	2.50

3.2. Length and Sensitivity Correlation Based on Piezoresistive Effect

To determine the correlation between sensor length and sensitivity, we proposed a unit gauge factor model of NCSS. The gauge factor (G. F.) is defined in Equation (1), and the strain (ε) of the cantilever is given in Equation (2).

$$\text{G.F.} = \frac{R_n}{\varepsilon} = \frac{\frac{\Delta R}{R}}{\varepsilon} \tag{1}$$

$$\varepsilon = \frac{3c(L-a)}{L^3} y(L) \tag{2}$$

Here, c is the height from the center of the cantilever, L is the length of the cantilever, a is the distance from the fixed end of the cantilever to the center of the sensor, and y (L) is the displacement applied to the end of the cantilever. Substituting the above equation into the general gauge factor equation, we yield the following.

$$\text{G.F.} = \frac{\frac{\Delta R}{R}}{\frac{3c(L-a)}{L^3} y(L)} = \frac{L^3 \Delta R}{3c(L-a)y(L)R} \tag{3}$$

Because the fillers dispersed in the matrix construct the electrical conductive paths with contact resistance, its conductivity is less than that of a metal. Therefore, we assumed that the NCSS is an electrical conductive series based on an electrical conductivity model of nanocomposites and tried to determine its piezoresistive sensitivity mechanism based on this assumption. The NCSS is considered as a chain series of piezoresistive units having the same geometric size (Figure 4a). The electrical resistance of the sensor can be expressed as a linear summation of each unit, as shown in Figure 4b.

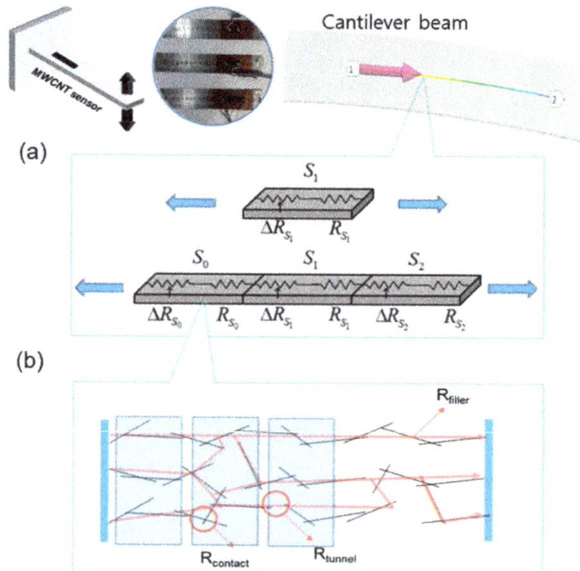

Figure 4. Unit gauge factor model of NCSS: (**a**) Single piezoresistive unit (top) and chain series of the units with linear summation (bottom) under uniaxial load; and (**b**) an electrical conductivity schematic concept model of nanocomposite.

As strains happen across all of the units of the NCSS, the gauge factor with respect to the strain can be expressed as follows.

$$\text{G.F.} = \frac{\frac{\Delta R_{S_0}}{R_{S_0}}}{\varepsilon_{S_0}} + \frac{\frac{\Delta R_{S_1}}{R_{S_1}}}{\varepsilon_{S_1}} + \frac{\frac{\Delta R_{S_2}}{R_{S_2}}}{\varepsilon_{S_2}} \tag{4}$$

In addition, the electrical resistance model of the nanocarbon composite, shown in Figure 4b, can be expressed as follows:

$$R_{CNT-composites} = R_{filler} + R_{tunnel} + R_{contact} \tag{5}$$

where R_{filler} is the electrical resistance of the fillers in the composite, R_{tunnel} is the resistance between the fillers inside the composite, and $R_{contact}$ is the contact resistance between fillers. Substituting the above equation into the gauge factor equation, the gauge factor of the nanocarbon composite sensor can be expressed as follows.

$$\text{G.F.}_{CNT-composites} = \frac{R_N}{\varepsilon} = \frac{\frac{\Delta R}{R}}{\varepsilon} = \frac{\frac{\Delta R_{filler} + \Delta R_{tunnel} + \Delta R_{contact}}{R_{filler} + R_{tunnel} + R_{contact}}}{\varepsilon} \tag{6}$$

Thus, the electrical resistance change of the composite is dominated by ΔR_{tunnel} [12], and the resistance due to the tunneling effect can be expressed as follows in Equation (7) [31]:

$$R_{tunnel} = \frac{h^2 d}{Ae^2\sqrt{2m\lambda}} exp\left(\frac{4\pi d}{h}\sqrt{2m\lambda}\right) \tag{7}$$

where h is Planck's constant, d is the spacing between fillers, A is the cross-sectional area of the tunnel, e is the quantum of electricity, m is the mass of electrons, and λ is the height of the matrix barrier. The amount of change in the tunneling resistance is generated exponentially by the applied strain of each single unit part. Accordingly, the amount of change in the tunneling resistance with respect to the strain of the entire sensor line, that is, the sensitivity to the length of the nanocarbon composite, may be expressed as follows:

$$\begin{aligned} G.F._{CNT-composites} &= \frac{\frac{\Delta R_{tunnel-S_0}}{R_{filler-S_0}+R_{tunnel-S_0}+R_{contact-S_0}}}{\varepsilon_{S_0}} + \frac{\frac{\Delta R_{tunnel-S_1}}{R_{S_1}}}{\varepsilon_{S_1}} + \frac{\frac{\Delta R_{tunnel-S_2}}{R_{S_2}}}{\varepsilon_{S_2}} + \cdots \\ &= \sum_{n=S_0}^{S_S} \frac{\frac{\Delta R_{tunnel-n}}{R_n}}{\varepsilon_n} \quad (; R_{S_0} = R_{S_1} = \cdots = R_{S_S}) \end{aligned} \tag{8}$$

From the above hypothesis, in this study, we supposed that the NCSS is a chain series of individual piezoresistive units, and the whole piezoresistive change is a linear summation of the units.

3.3. Improvement of Piezoresistive Characteristics by Using Sensor Pattern

Based on the first experiment described in Section 3.2, we hypothesized that the sensitivity of the NCSS is closely related to its length because structural deformation brings whole piezoresistive change from entire units of the sensor. We designed three grid sensors via a thin polyimide film mask. The mask pattern was designed by CAD (Computer Aided Design) program and was fabricated by laser cutting process. The CNT ink was manually printed through the mask on the cantilever, as shown in the lower right-hand corner of Figure 5. To secure sensor electrical stability with better conductivity, we conservatively used the 0.5 wt.% to fabricate the grid sensor samples.

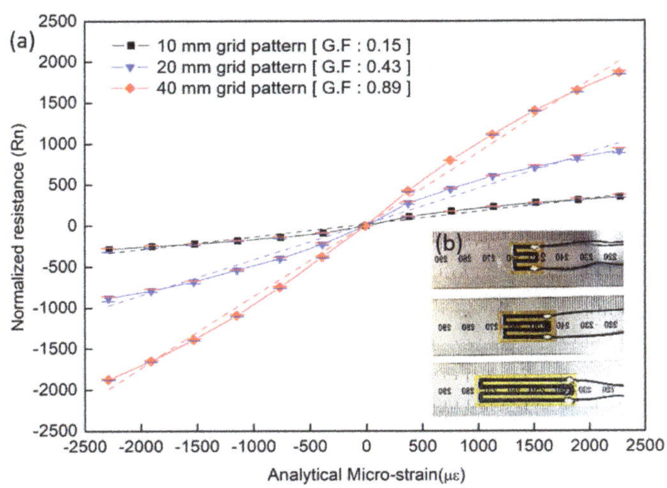

Figure 5. Grid-type NCSS experiment: (**a**) grid-type NCSS samples and (**b**) sensitivity characteristic of the grid-type NCSS with respect to length pattern change.

We measured each resistance change with respect to beam bending five times and averaged the measured values after eliminating the maximum and minimum data points. Figure 5 shows the results. When comparing the amount of change in the normalized resistance with respect to the strain of the sensor samples shown in Figures 3 and 5, it can be seen that the 40-mm pattern type (G.F.: 0.87) has a larger change in the normalized resistance for the same strain than the 40-mm bulk type (G.F.: 0.22).

As we expected from the above, the NCSS sensitivity improvement is exactly proportional to its incremental length ratio. We obtained higher sensitivity in cases where the sensor had a longer length or denser pattern per unit area. We also achieved more linear, higher sensitivity and consistent data from the grid-type NCSS than from the bulk-type sensor. Additionally, we deduced that NCSS strain output may be expressed by the summation of the entire sensor covered surface. According to our literature survey, the strain sensitivity of a conventional foil strain gauge is only related to its material properties and is not much affected by its length [32], which is a notable piezoresistive feature of the NCSS. We conclude that since NCSSs are affected by the overall strain change in a given direction, the geometric pattern design is one of the most important variables in axial load and deformation measurements.

4. Conclusions

In this study, the piezoresistive characteristics of NCSSs were experimentally studied in terms of the sensor's geometric length to determine its sensitivity to the design. We fabricated bulk- and grid-type sensors with different filler contents (wt.%) and different sensor lengths.

In the first experiment, we used silicone molds to prepare samples with three different lengths (30, 40, 50 mm) and two different filler compositions (0.35 and 0.5 wt.%). We measured the sensor resistance change with respect to the bending strain variation of a cantilever. For both compositions, we obtained higher sensitivity at longer sensor lengths, which suggested the possibility of using the geometric design to control the sensitivity.

We deduced that sensor length may correlate with sensitivity, and we proposed a unit gauge factor model of NCSSs to explain the proportional relationship between their length and sensitivity. We supposed that the NCSS is a chain series of individual piezoresistive units and that the whole piezoresistive change can be a linear summation of these units. To verify our hypothesis, we performed a second experiment with more sophisticated samples. We designed grid-type sensors by using thin polyimide film masks. We printed the grid-type NCSS (0.5 wt.% filler) on a cantilever and repeated the same test. Results indicated that the improvement in NCSS sensitivity was exactly proportional to its incremental length ratio (because structural deformation brings piezoresistive change from whole units of the sensor). We were able to obtain higher sensitivity in the cases where the sensor had a longer length and denser pattern per unit area. We also achieved more linear and consistent data from the grid-type NCSS than from the bulk-type sensor.

From the analysis, we concluded that the greater sensor length brings a greater change in resistance due to its piezoresistive unit summation. Eventually, the sensitivity of NCSSs directly relates to their length, and the patternized length can be used to easily control the sensor sensitivity. We also found that the fine sensor pattern can improve its performance, allowing better results to be achieved. Sensor geometric pattern design is one of the most important aspects of axial load and deformation measurements. For further study, we are studying advanced sensitivity design considering the three-dimensional pattern variables of the NCSS as well.

Author Contributions: Conceptualization: S.-Y.K., B.-G.C., and I.K.; methodology: S.-Y.K., B.-G.C., C.-J.K., Y.-S.J., J.-S.J., J.-H.S., and I.K.; formal analysis: S.-Y.K. and B.-G.C.; investigation: S.-Y.K. and B.-G.C.; data curation: S.-Y.K., B.-G.C., and G.-W.O.; writing—original draft preparation: S.-Y.K. and B.-G.C.; writing—review and editing: C.-J.K., Y.-S.J., J.-S.J., K.-Y.J., and J.-H.S.; visualization: S.-Y.K., B.-G.C., and G.-W.O.; project administration: I.K.; funding acquisition: I.K. All authors have read and agreed to the published version of the manuscript.

Funding: This work was supported by a Research Grant of Pukyong National University (2019).

Institutional Review Board Statement: Not applicable.

Informed Consent Statement: Not applicable.

Data Availability Statement: Data available in a publicly accessible repository.

Conflicts of Interest: The authors declare no conflict of interest.

References

1. Salvado, R.; Lopes, C.; Araujo, P.; Gorski, M.; Velez, F.J.; Gomez, J.C.; Krzywon, R. Carbon fiber epoxy composites for both strengthening and health monitoring of structures. *Sensors* **2015**, *15*, 10753–10770. [CrossRef]
2. Arash, B.; Park, H.S.; Rabczuk, T. Mechanical properties of carbon nanotube reinforced polymer nanocomposites: A coarse-grained model. *Compos. B. Eng.* **2015**, *80*, 92–100. [CrossRef]
3. Cho, B.G.; Hwang, S.H.; Park, Y.B. Fabrication and characterization of carbon nanotube/carbon fiber/polycarbonate multiscale hybrid composites. *Compos. Res.* **2016**, *29*, 269–275. [CrossRef]
4. Kang, I.; Schulz, M.J.; Kim, J.H.; Shanov, V.; Shi, D. A carbon nanotube strain sensor for structural health monitoring. *Smart Mater. Struct.* **2006**, *15*, 737–748. [CrossRef]
5. Loyola, B.R.; Saponara, V.L.; Loh, K.J. In situ strain monitoring of fiber-reinforced polymers using embedded piezoresistive nanocomposites. *J. Mater. Sci.* **2010**, *45*, 6786–6798. [CrossRef]
6. Hu, N.; Itoi, T.; Akagi, T.; Kojima, T.; Xue, J.; Yan, C.; Atobe, S.; Fukunaga, H.; Yuan, W.; Ning, H.; et al. Alamusi Ultrasensitive strain sensors made from metal-coated carbon nanofiller/epoxy composites. *Carbon* **2013**, *51*, 202–212. [CrossRef]
7. Dinh, N.T.; Kanoun, O. Temperature-compensated force/pressure sensor based on multi-walled carbon nanotube epoxy composites. *Carbon* **2015**, *15*, 11133–11150. [CrossRef]
8. Ferreira, A.; Mendez, S.L. Piezoresistive response of spray-printed carbon nanotube/poly(vinylidene fluoride) composites. *Compos. B. Eng.* **2016**, *96*, 242–247. [CrossRef]
9. Kim, S.Y.; Kang, I. A study on the development of a novel pressure sensor based on nano carbon piezoresistive composite by using 3D printing. *Trans. Korean Soc. Mech. Eng. A* **2017**, *41*, 187–192.
10. Kim, S.Y.; Park, S.H.; Choi, B.G.; Kang, I.; Park, S.W.; Shin, J.W.; Kim, J.H.; Baek, W.K.; Lim, K.T.; Kim, Y.; et al. Development of a spoke type torque sensor using painting carbon nanotube strain sensors. *J. Nanosci. Nanotechnol.* **2018**, *18*, 1782–1786. [CrossRef]
11. Le, T.H.; Lee, D.H.; Kim, J.H.; Park, S.J. Polypyrrole/graphene quantum dot composites as a sensor media for epinephrine. *J. Nanosci. Nanotechnol.* **2020**, *20*, 4005–4010. [CrossRef] [PubMed]
12. Choi, G.; Lee, J.W.; Cha, J.Y.; Kim, Y.; Choi, Y.; Schulz, M.J.; Moon, C.K.; Lim, K.T.; Kim, S.Y.; Kang, I. A Spray-On Carbon Nanotube Artificial Neuron Strain Sensor for Composite Structural Health Monitoring. *Sensors* **2016**, *16*, 1171. [CrossRef]
13. Kim, S.Y.; Choi, B.G.; Baek, W.K.; Park, S.H.; Park, S.W.; Shin, J.W.; Kang, I. Impact paint sensor based on polymer/multi-dimension carbon nano isotopes composites. *Smart Mater. Struct.* **2019**, *28*, 035025. [CrossRef]
14. Hu, N.; Karube, Y.; Yan, C.; Masuda, Z.; Fukunaga, H. Tunneling effect in a polymer/carbon nanotube nanocomposite strain sensor. *Acta Mater.* **2008**, *56*, 2929–2936. [CrossRef]
15. Ram, R.; Rahaman, M.; Khastgir, D. Electrical properties of polyvinylidene fluoride (PVDF)/multi-walled carbon nanotube (MWCNT) semi-transparent composites: Modelling of DC conductivity. *Compos. Part A Appl. Sci. Manuf.* **2015**, *69*, 30–39. [CrossRef]
16. Bonavolonta, C.; Camerlingo, C.; Carotenuto, G.; Nicola, S.; Longo, A.; Meola, C.; Boccardi, S.; Palomba, M.; Pepe, G.P.; Valentino, M. Characterization of piezoresistive properties of graphene-supported polymer coating for strain sensor application. *Sens. Actuator A Phys.* **2016**, *252*, 26–34. [CrossRef]
17. Liu, R.; Xu, Z.; Yin, J.; Huang, J.; Liu, D.; Xie, G. A coupled mechanical and electrical model concerning piezoresistve effect of CFRP materials. *Compos. B. Eng.* **2016**, *96*, 125–135. [CrossRef]
18. Hu, N.; Karube, Y.; Arai, M.; Watanabe, T.; Yan, C.; Li, Y.; Liu, Y.; Fukunaga, H. Investigation on sensitivity of a polymer/carbon nanotube composite strain sensor. *Carbon* **2010**, *48*, 680–687. [CrossRef]
19. Rahman, R.; Soltanian, S.; Servati, P. Coupled effects of film thickness and filler length on conductivity and strain sensitivity of carbon nanotube/polymer composite thin film. *IEEE Sens. J.* **2016**, *16*, 77–87. [CrossRef]
20. Micaela, C.; Massimo, R.; Imran, S.M.; Alberto, T. Conductivity in carbon nanotube polymer composites: A comparison between model and experiment. *Compos. Part A Appl. Sci. Manuf.* **2016**, *87*, 237–242.
21. Lee, B.M.; Loh, K.J. Carbon nanotube thin film strain sensors: Comparison between experimental tests and numerical simulations. *Nanotechnology* **2017**, *28*, 155502. [CrossRef] [PubMed]
22. Lee, D.; Hong, H.P.; Lee, C.J.; Park, C.W.; Min, N.K. Microfabrication and characterization of spray-coated single-wall carbon nanotube film strain gauges. *Nanotechnology* **2011**, *22*, 455301. [CrossRef]
23. Huang, Y.; Huang, S.; Hsu, C.; Chao, R.; Vu, T.K. Design and fabrication of single-walled carbon nanonet flexible strain sensors. *Sensors* **2012**, *12*, 3269–3280. [CrossRef] [PubMed]
24. Wang, X.; Sparkman, J.; Gou, J. Strain sensing of printed carbon nanotube sensors on polyurethane substrate with spray deposition modeling. *Compos. Commun.* **2017**, *3*, 1–6. [CrossRef]
25. Kong, J.H.; Jang, N.S.; Kim, S.H.; Kim, J.M. Simple and rapid micropatterning of conductive carbon composites and its application to elastic strain sensors. *Carbon* **2014**, *77*, 199–207. [CrossRef]
26. Li, X.; Levy, C.; Elaadil, L. Multiwalled carbon nanotube film for strain sensing. *Nanotechnology* **2008**, *19*, 045501. [CrossRef] [PubMed]
27. Xu, W.; Allen, M.G. Deformable strain sensors based on patterned MWCNTs/polydimethylsiloxane composites. *J. Polym. Sci. B. Polym. Phys.* **2013**, *51*, 1505–1512. [CrossRef]
28. Chun, K.Y.; Oh, Y.; Rho, J.; Ahn, J.H.; Kim, Y.J.; Choi, H.R.; Baik, S. Highly conductive, printable and stretchable composite films of carbon nanotubes and silver. *Nat. Nanotechnol.* **2010**, *5*, 853–857. [CrossRef]

29. Wei, Y.; Chen, S.; Dong, X.; Lin, Y.; Liu, L. Flexible piezoresistive sensor based on "dynamic bridging effect" of silver nanowired toward graphene. *Carbon* **2017**, *113*, 395–403. [CrossRef]
30. Pei, Z.; Liu, Y.; Zhang, Q.; Zhao, D.; Wang, J.; Yuan, Z.; Zhang, W.; Sang, S. Highly sensitive, stretchable strain sensor based on Ag@COOH- functionalized CNTs for stroke and pronunciation recognition. *Adv. Electron. Mater.* **2019**, *5*, 1900227. [CrossRef]
31. Esmaeli, A.; Sbarufatti, C.; Casati, R.; Jiménez-Suárez, A.; Ureña, A.; Hamouda, A.M.S. Effective addition of nanoclay in enhancement of mechanical and electromechanical properties of SWCNT reinforced epoxy: Strain sensing and crack-induced piezoresistivity. *Theor. Appl. Fract. Mech.* **2020**, *110*, 102831. [CrossRef]
32. Hoffmann, K. An Introduction to Stress Analysis and Transducer Design using Strain Gauges. *HBM* **2012**, 34–41.

applied sciences

MDPI

Article

Improving Skills in Mechanism and Machine Science Using GIM Software

Erik Macho *, Mónica Urízar, Víctor Petuya and Alfonso Hernández

Faculty of Engineering in Bilbao, University of the Basque Country (UPV/EHU), Plaza Ingeniero Torres Quevedo, 48013 Bilbao, Spain; monica.urizar@ehu.es (M.U.); victor.petuya@ehu.es (V.P.); a.hernandez@ehu.es (A.H.)
* Correspondence: erik.macho@ehu.eus; Tel.: +34-94-601-7441

Abstract: The field of education has evolved significantly in recent years as it has incorporated new pedagogical methodologies. Many of these methodologies are designed to encourage students' participation in the learning process. The traditional role of the student as a passive receiver of content is no longer considered valid. Teaching in mechanical engineering is no stranger to these changes either, where new learning activities have been designed to complement theory-heavy lectures. These activities take place in both physical and virtual laboratories. In case of the latter, the use of the GIM software (developed at the Department of Mechanical Engineering of the University of the Basque Country UPV/EHU, Spain) is a promising option. In this paper, features of the GIM that are most frequently used to support and exemplify the theoretical concepts taught in lectures are described using a case study. In addition, GIM is integrated into different learning activities to show its potential as a tool for learning and self-evaluation.

Keywords: learning tools; teaching methodology; educational software; mechanism science; problem-based learning

Citation: Macho, E.; Urízar, M.; Petuya, V.; Hernández, A. Improving Skills in Mechanism and Machine Science Using GIM Software. *Appl. Sci.* **2021**, *11*, 7850. https://doi.org/10.3390/app11177850

Academic Editors: Jae Hyuk Lim, Jin-Gyun Kim and Peter Persson

Received: 27 July 2021
Accepted: 24 August 2021
Published: 26 August 2021

1. Introduction

Programs to simulate a variety of mechanisms were initially developed to solve mathematical equations involved in the interconnection among elements of such mechanisms. The solving of systems of non-linear algebraic equations (or algebraic–differential equations in the case of dynamics) was a constant "burden" every time a mechanism had to be analyzed. General-purpose multibody dynamics programs subsequently emerged with the same target as general-purpose finite element programs. Among the softwares available at the time, the following general-purpose multibody dynamics programs were prominent: ADAMS [1,2], AUTOLEV [3], COMPAMM [4], DADS [5], DYMAC [6], and MESA [7]. This paper focuses on programs intended for mechanism simulations related to education, and few of them have any didactic capacity. It is challenging to find references in the literature where general-purpose programs are used for teaching. Some specific cases can be found, such as the Virtual Lab based on ADAMS presented in [8]. The authors claimed, however, that the exercises to be solved by students need to be carefully selected due to the complexity of the program. On the contrary, they remarked the advantage of employing a program that is commonly used in the industry so that students can learn it in college.

Another group of more specific programs for the analysis and synthesis of mechanisms was developed in the context of mechanisms and machine theory (MMT). The following are noteworthy: KINSYN [9], LINCAGES [10], and RECSYN [11]. University lecturers in machines and mechanisms have participated in the development of these programs, because of which they focus on a didactic approach. Recent softwares in this group are MechDev [12] and WinMecC [13,14], both dedicated to the analysis and synthesis of planar mechanisms. The most remarkable characteristic of WinMecC is its dimensional synthesis module based on evolutionary algorithms. MechDev has a special architecture based on plug-ins, including an algorithm that combines analytical with numerical computation.

Another powerful program similar to them is Working Model 2D [15], which is capable of modeling mechanisms in an interactive way.

Dynamic geometric environments (DGEs) are software tools that are being commonly used nowadays to teach MMT. They allow for the building of parametrized geometric entities. GeoGebra [16] is a representative DGE. Mathematical libraries that can be used to support the teaching of many subjects related to mechanism and machine science are another type of technological tools. For example, the graphical programming environment Simulink of MATLAB is a good choice for implementing and solving the dynamics of the mathematical models of mechanical systems [17]. The MATLAB framework has proven to be useful for the development of active teaching–learning methodologies, such as the one designed to teach kinematic and dynamic analyses of 3D multibody systems proposed in [18]. Finally, there are some cases in which CAD systems have been used to simulate mechanisms for teaching. However, this does not appear to be the most adequate option as they are not specific programs for mechanisms, are usually expensive, and do not focus on didactic approaches.

The software presented in this paper (GIM), the initial steps of which were introduced in [19], is a general-purpose software that can handle planar and spatial systems of any number of degrees of freedom. It was developed by the COMPMECH research group of the Department of Mechanical Engineering of the University of the Basque Country UPV/EHU in Spain. It is designed for teaching and learning activities related to important subjects in mechanical engineering, such as applied mechanics, mechanism and machine theory, computational kinematics and dynamics, mechanical design, and robotics. GIM has also been used as a tool for the development of several doctoral theses, and their results have served as important feedback to develop and improve its computational modules.

2. Capabilities and Potential of GIM

One of the aims of using GIM is to help students better understand the theoretical concepts explained in lectures in class, and to motivate them to work independently with the software to develop their skills on it. The software is available for free.

This section provides a general idea to the reader of the tools provided by GIM related to the learning process. A planar example is first developed. The case study chosen is the quick return mechanism, and is posed in the same way that it would be to students. The goal is to obtain the value of the actuating torque required to achieve a specified motion when some resisting loads are applied as shown in Figure 1.

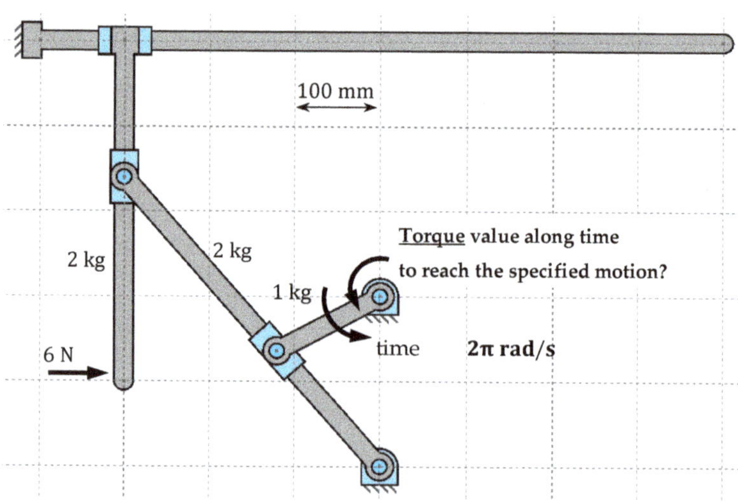

Figure 1. Proposed case study.

To solve this exercise, students have to follow basic steps:

- Create the geometrical model of the mechanism.
- Define the function of input motion and solve the kinematics.
- Define existing resistant loads and solve the inverse dynamics (kinetostatic problem).

Since simply an example is not enough to show the potential of this software and all options that the students can exploit to improve their learning skills, additional aspects are explained.

2.1. Geometry Module—Case Study

A kinematic sketch is a geometrical model of the mechanism in which ideal joints are considered and elements are modeled as perfectly rigid bodies. The student can build the geometrical model to make a kinematics or dynamics simulation simply and quickly. A trained user can have any model ready for simulation in a few minutes.

For analysis, the mechanism is defined directly in an assembled position, but not necessarily with empirical dimensions. The first step consists of defining the points of interest of the mechanical system, called nodes. To do this, the students can directly type the coordinates of the nodes or set their positions using the mouse pointer. In the second step, using these nodes, elements of the mechanism are built. The user simply selects the nodes belonging to the same element. Finally, the kinematic joints between elements are set from a list.

Even in the process of geometrical definition, students can compare the results with their knowledge, because each time a geometrical change is made, the program computes the number of degrees of freedom and the number of redundant constraints of the model in its given state.

Once the topology of the mechanism has been defined, the user can still modify the position of any node and orientation of any joint, or—what constitutes a more practical capability—can edit such values of the geometrical constraints as lengths of elements or angles between them. This is useful because the empirical geometrical data of a mechanism is often given by the sizes of elements and certain positional constraints, but does not include the coordinates of the moving nodes.

Figure 2 shows the process of generating the geometrical model of the Quick Return mechanism, which is used as an example to illustrate the teaching/self-learning capabilities of the program. This paper is not intended to be a user manual, and accordingly, practical instructions are not provided. The images of the software provided here should be sufficient for the reader to understand the main idea of GIM.

Geometry Module–Additional Considerations

The GIM is a general-purpose software that can to deal with planar and spatial systems of any number of degrees of freedom. This is why the geometry module implements a wide collection of kinematic joints including all the most commonly used ones in practice (Figures 3 and 4).

Some of them require the definition of additional geometric information—for example, the orientation of axes of the joint for spatial revolution. In these cases, once more, the student can use the keyboard or the mouse (with or without a grid option). Any joint type can be set as a fixed joint by connecting elements with a fixed frame. Moreover, the most convenient system of coordinates (Cartesian or polar) can be used in the geometric function (Figure 5). The spatial joints and links are modeled as easily as the planar ones.

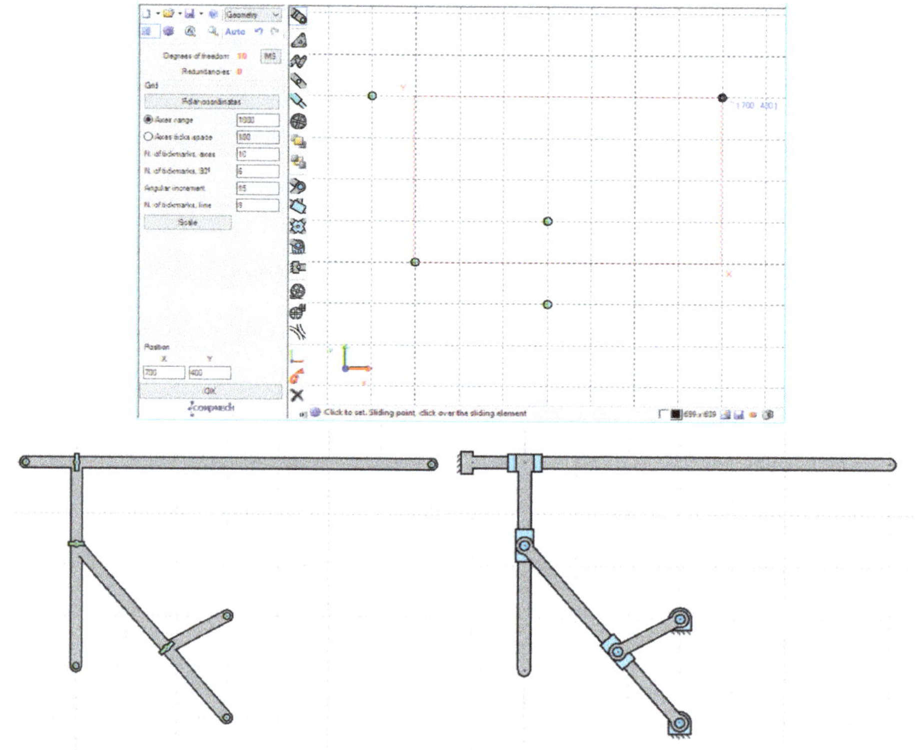

Figure 2. Main steps of geometric definition: 1—Nodes, 2—Elements, 3—Joints.

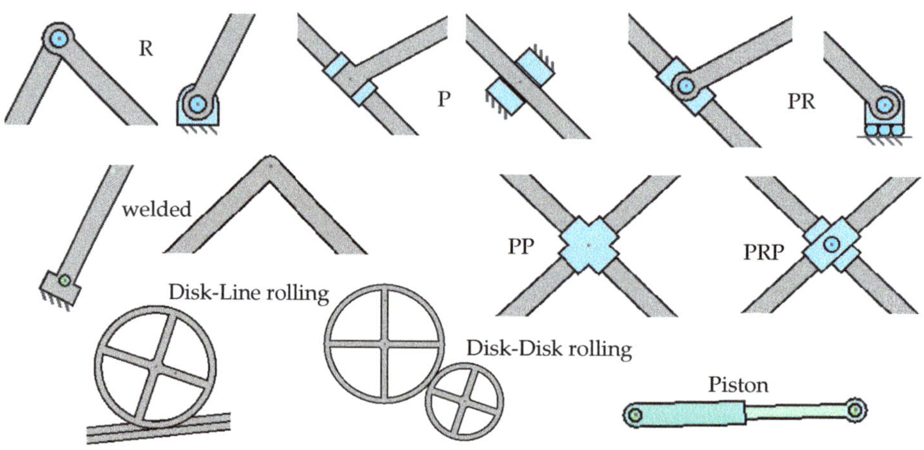

Figure 3. Planar joints implemented in GIM.

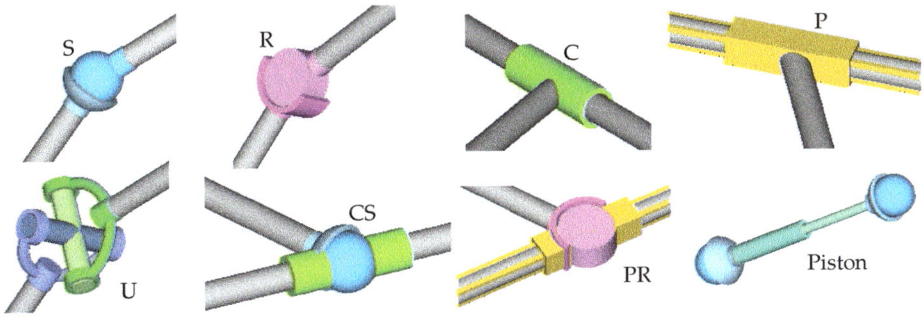

Figure 4. Spatial joints implemented in GIM.

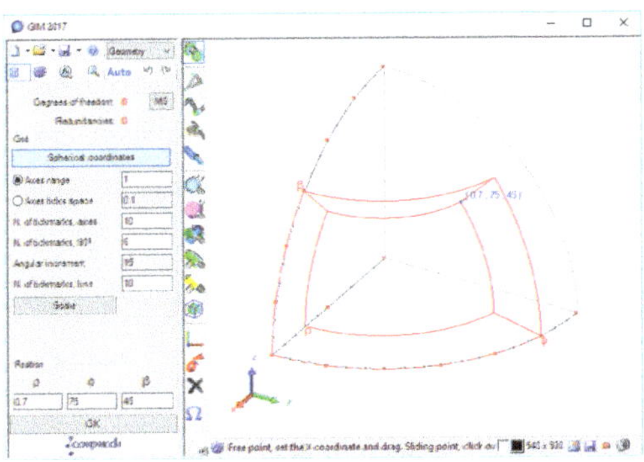

Figure 5. Position of a point in spherical coordinates.

2.2. Kinematics Module—Case Study

The three main kinematic problems, i.e., those of position, velocity, and acceleration, can be solved in GIM. For this, the student needs to define as many actuators as the number of degrees of freedom of the mechanism. Each actuator is established as a motion function that specifies the value of its position variable, and the first and second derivatives along time. The most common types of actuators (fixed rotations, relative rotations, pistons, and sliders) as well as the most common types of motion functions (constant velocity or acceleration, polynomial, sinusoidal) are provided. The multibody approach based on natural coordinates is implemented to obtain the primary results: the simulation of motion of the mechanism (trajectories of all nodes, and velocities and accelerations at all nodes in each position).

As shown in Figure 6, all results can be graphically plotted along with the mechanism at any position, which makes it possible to visually analyze their evolution during motion. It also becomes possible to evaluate the variation of any parameter along the time tracing the corresponding function.

Using this computation module, apart from having at hand the numerical results of the proposed problem (as a mechanism calculator), students can corroborate the relevant theoretical concepts studied during lectures. Some examples are shown in Figure 7:

- They can observe how the velocity is always tangential to the trajectory while watching the motion of the mechanism.
- They verify the decomposition of acceleration into a tangential component and a normal component pointing to the center of curvature of the trajectory. The ways in which intrinsic components are related to changes in the module and direction of the velocity vector can be graphically verified.

- In the context of relative motion, they realize how this decomposition is made in relative and frame (and the Coriolis acceleration) components. They can verify that the Coriolis acceleration is always perpendicular to relative velocity.

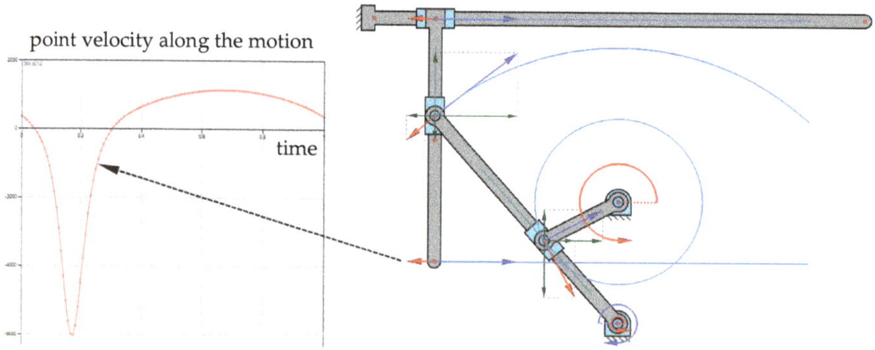

Figure 6. Main kinematic results: Trajectories, velocities, and accelerations for given inputs.

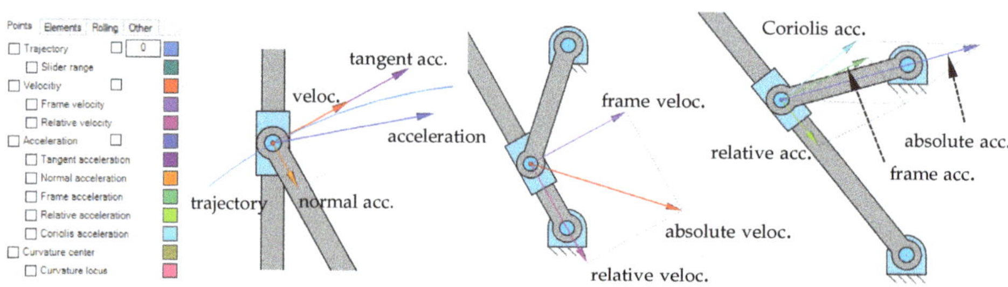

Figure 7. Visualizing decomposition of velocity and acceleration.

Once the analysis has been concluded, the student can modify any kinematic or geometric parameter to see how this affects the results, which are recomputed in real time.

Kinematics Module—Additional Considerations

Apart from the main results, additional derived results are provided, such as the centers of curvature of the points, instantaneous centers of rotation of the elements, and fixed and moving centrodes. In addition, all results can be computed in terms of absolute motion or that relative to any moving element. Some of these additional capacities are shown in Figures 8–11.

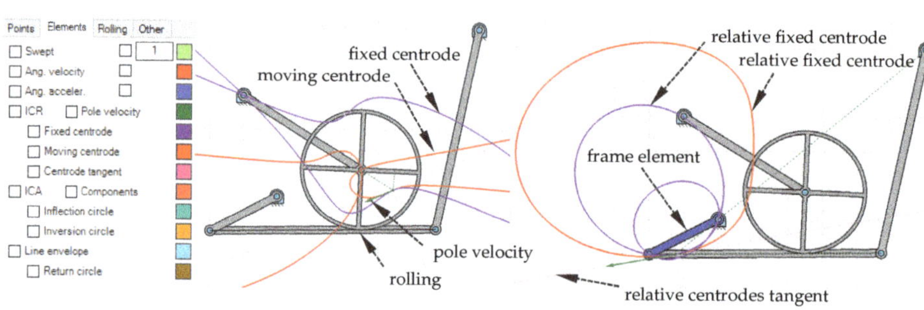

Figure 8. Disk element fixed and moving centrodes in absolute and relative motions. Velocities of poles.

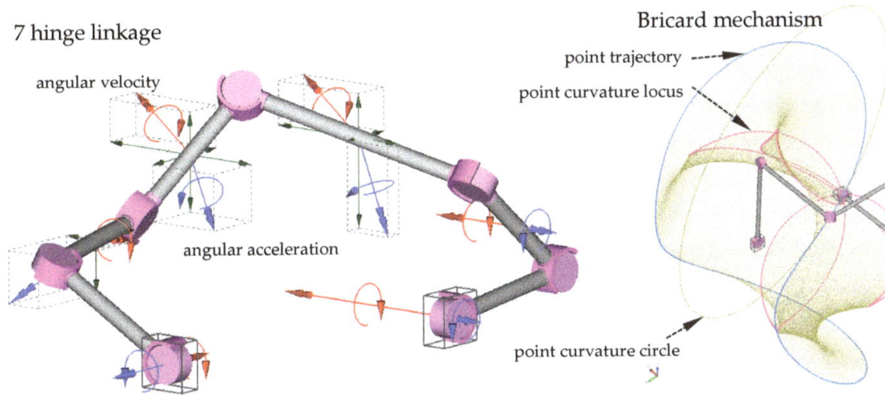

Figure 9. Angular velocities and accelerations. Loci of center of curvature of a trajectory.

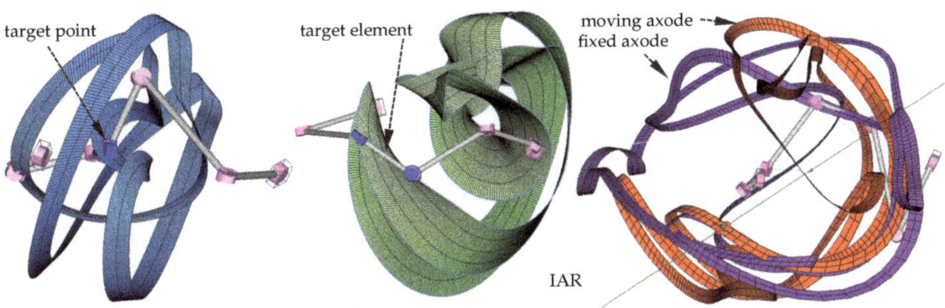

Figure 10. Swept of revolute joint axis swept. Element's swept. Fixed and moving axodes with the instantaneous axes of rotation and sliding.

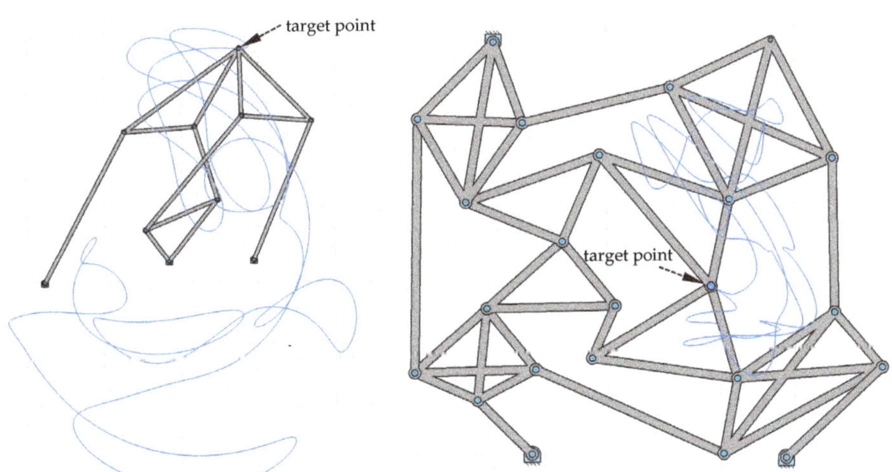

Figure 11. Complex trajectories of one degree-of-freedom mechanisms computed automatically (see Videos S1 and S2 of the Supplementary Material to visualize the motions of both mechanisms).

2.3. Dynamics Module–Case Study

Dynamic problems consider simultaneously motion and forces. GIM can solve two main types of dynamics problems, i.e., problems in inverse dynamics (kinetostatic problem) and direct dynamics. If all actuators are controlled in terms of position, velocity, and acceleration, as in the kinematics module, the motion of the mechanism is defined and can be computed independently from the existing forces. The unknowns of this problem are the values of the driving loads required to achieve such motion (considering resistant applied loads), and are computed by means of a kinetostatic approach. On the contrary, when all applied load values are known, the resulting motion is dependent on such values, i.e., the motion cannot be computed using a purely kinematic approach, and the direct dynamics problem has to be simulated.

Apart from driving loads (known values in direct dynamics but unknowns in inverse dynamics), the student can define as many external resistant loads as desired. The available loads are punctual as well as linearly distributed forces and torques. In this module, to compute inertial loads, the properties of mass of each element need to be specified. The default values of the center of gravity and moment of inertia are automatically computed depending on the shape of the element, but because this shape is sometimes just a kinematic sketch (the real element may have a different shape), these defaults can be substituted by custom values. Elemental weights are also considered if a value of gravitational acceleration is defined. Figure 12 shows the proposed case study, and the value of the torque required over time to achieve a specific constant rotational velocity in the actuator under certain external loads.

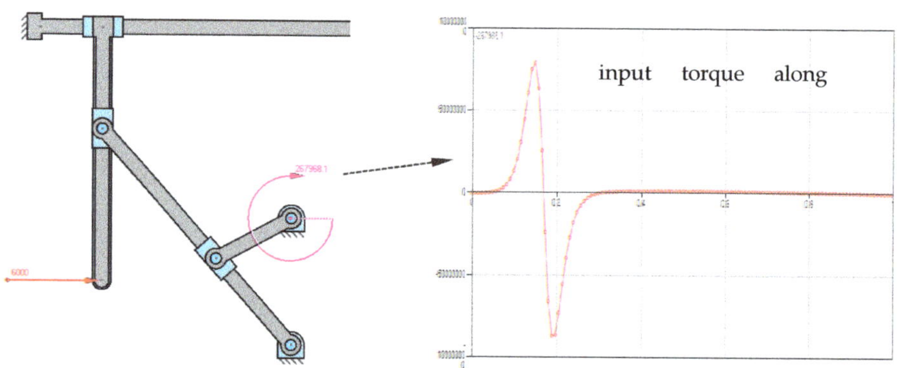

Figure 12. Value of driving load for the computation of inverse dynamics.

A number of additional results can be visualized apart from the above. In any dynamics problem, if the mechanism is non-redundant, a complete analysis of the force of the mechanical system is performed to compute the reaction forces and torques at the joints between elements. In addition, for linear elements, a diagram of internal efforts can be computed. As shown in Figure 13, the real distribution of weight and inertial forces along the element are considered. This force analysis is conducted for each position of the simulation, and thus the evolution of any of these results can be represented along the motion.

The student has the option of visualizing the values of any internal effort as a color map for the entire mechanical system at a position. Any internal effort diagram of an element can be also traced along time (motion), as shown in Figure 14.

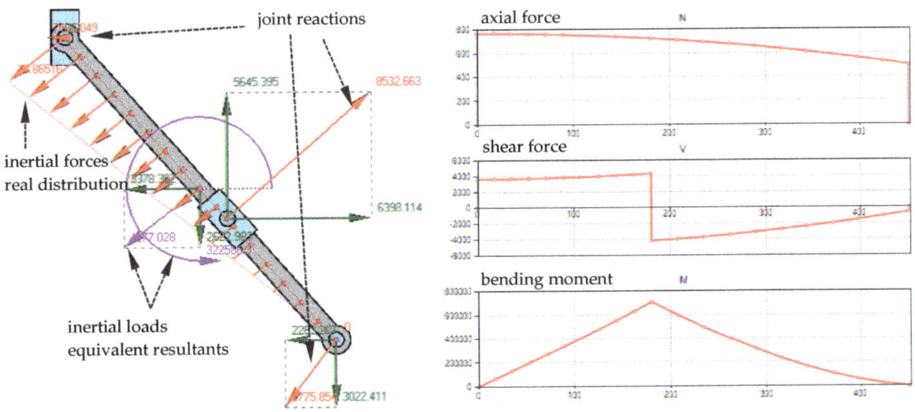

Figure 13. Free solid and internal efforts' diagrams.

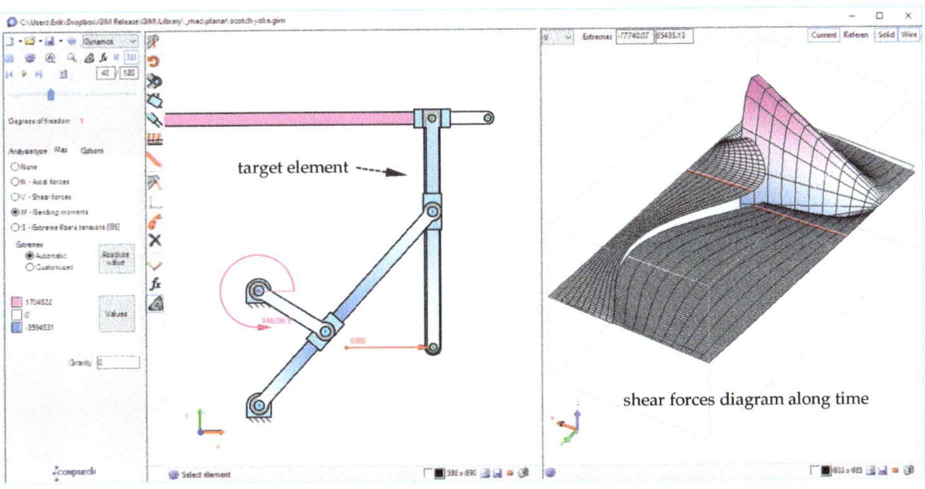

Figure 14. Color map of internal effort for all systems and evolution over time for an element.

Using the tools provided in the computation module, the students can check/verify in a very clear way some of the concepts imparted to them theoretically, such as the following:

- Although they had the same kinematics, the inertial properties of elements affected their dynamical results.
- Discontinuities in the internal efforts' diagrams are related to the existence of punctual loads at the relevant points.
- The type of reaction load depends on joint type, but the action–reaction principle is always verified.

Other characteristics, such as the fact that when a body has only forces at two points, both are in the direction of the line connecting the points, and are in the opposite direction when a body has forces only at three points and their action lines intersect at a point.

Dynamics Module—Additional Considerations

Other standard elements commonly used in dynamic analyses, like springs and dampers, are also implemented in the software to use to model mechanical systems. Because the methodology implemented has a general purpose, it is valid for a system with any number of degrees of freedom, including isostatic structures (mechanical systems with zero degrees of freedom and zero redundancies), such as the one shown in Figure 15.

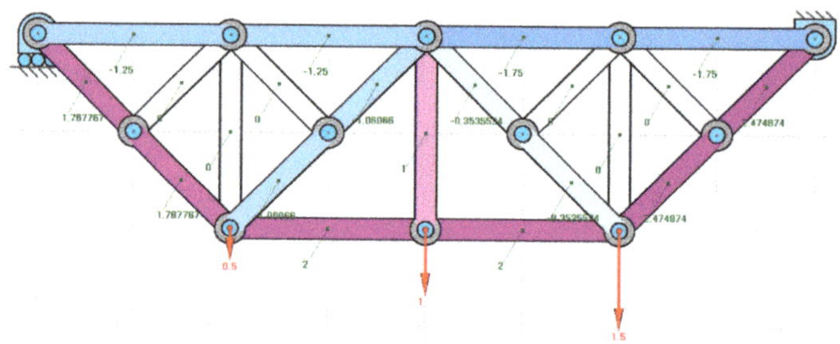

Figure 15. Axial forces of isostatic structure.

When a direct dynamics problem is computed, the motion is unknown, and simulating it often requires the numerical integration of a system of differential equations because, due to the complexity inherent to the problem, it does not have an analytical solution. This task is programmed in the software, thus students can observe the result (Figure 16). The motion of the mechanism shown in Figure 16 can be visualized in the third video included in the Supplementary Material.

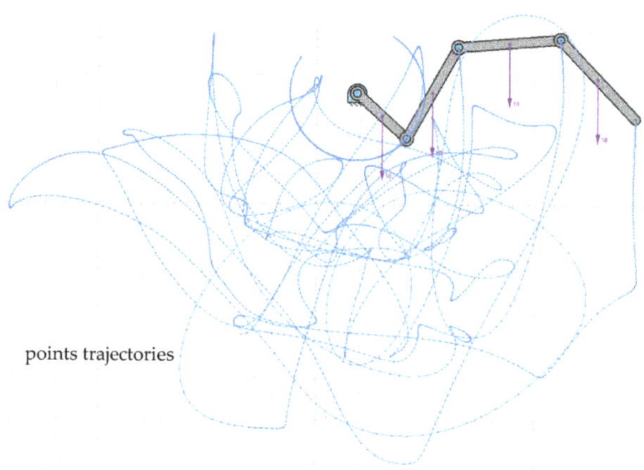

Figure 16. Simulating the motion of a chain released to the effect of gravity from repose.

2.4. Synthesis Module

The synthesis module focuses on an approach to teaching based on four-bar linkage. It covers the main types of synthesis, i.e., path generation, rigid body motion, and function generation. Path generation synthesis establishes some precision points that must belong to the trajectory of the four-bar linkage coupler, as shown in Figure 17. Rigid body motion synthesis establishes poses in which the coupler element must be placed during its motion, as shown in Figure 18a. Function generation synthesis establishes relations between the angles of input and output that need to be satisfied as shown in Figure 18b.

For each type of synthesis problem, the interface of the program shows to the student all the graphical constructions required to solve the problem and displays all possible solutions to it, as in the case of a non-linear problem. Students can easily identify such graphical constructions as represented in their handbooks.

In the software, the student can easily change the position of any precision point or any desired posture of the output element, and the impact of this modification in the lengths of the resulting bars is shown in real time.

In any synthesis problem, some additional results are provided, such as the line envelope, cubic of stationary curvature, pivot point curve, the Ball point, or cognate mechanisms and their derivations. These are shown in Figure 19.

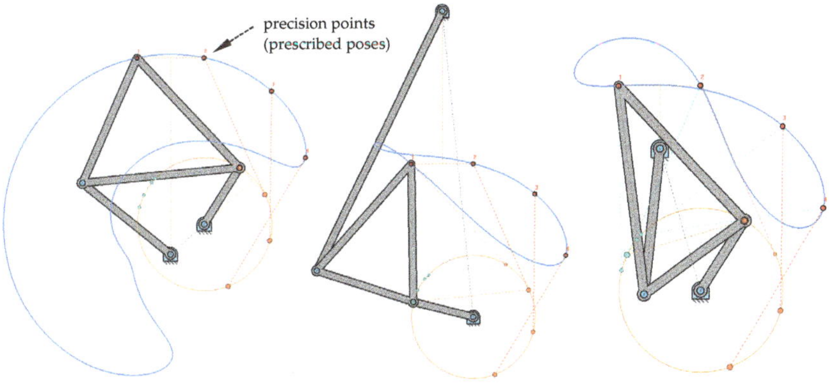

Figure 17. Path planning synthesis; multiple solutions.

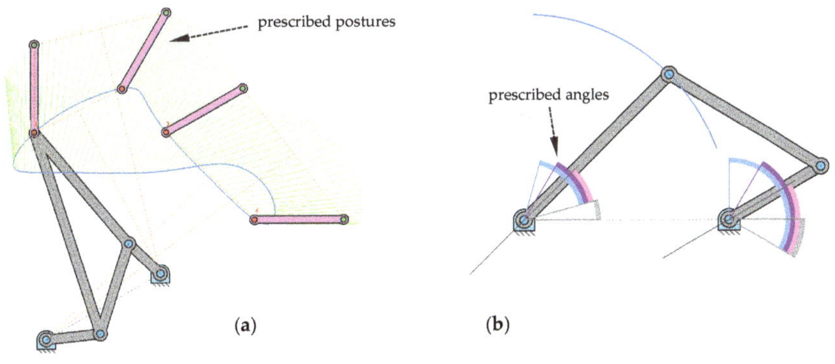

Figure 18. Rigid body motion (**a**) and function generation (**b**) synthesis.

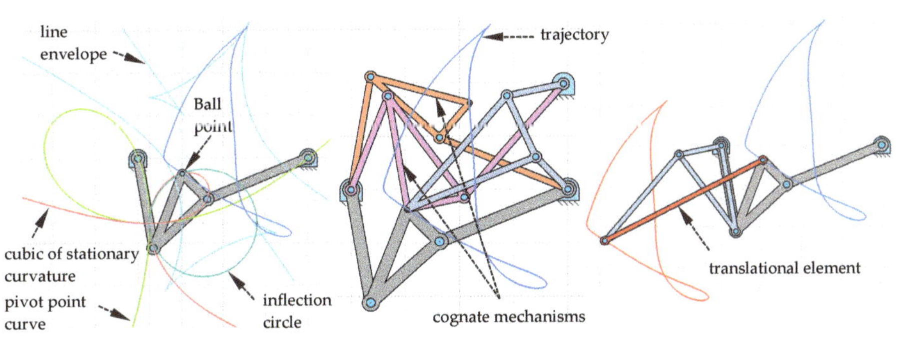

Figure 19. Line envelope, cubic of stationary curvature, pivot point curve, and Ball point. Cognates and derived translational mechanism.

3. Active Learning Activities

To highlight the potential of GIM for teaching, different activities performed using it are described in this section. A software can be used to complement and reinforce the complex theoretical concepts of subjects in mechanism and machine science. On the one hand, during academic courses, students have practical sessions in which different exercises are solved using the software. On the other hand, some active learning activities are developed during the seminars for students to enhance their ability to design and analyze different mechanical systems or structures. These activities, commonly known as problem-based learning (PBL), focus on applications. In this way, the students learn to approach problems in engineering.

3.1. Practical Teaching Support Sessions

During an academic course, practical sessions using GIM software are offered to students. In these sessions, lab groups of a maximum of 25 people are created. Initially, the teacher explains in a step-by-step manner the main modules of the GIM software and different options that are implemented. Then, a report template containing some proposed exercises is given to the students. They work independently on their computers, solve the exercises using GIM, and update the report in the Moodle platform for the course once they finish it. Depending on the subject, several problems can be proposed. Some examples are as follows:

- Modeling mechanical systems in GIM by first defining the structural diagram of the mechanism being studied and representing it in the Geometry module.
- Performing the kinematic analysis of planar or spatial mechanisms to obtain the velocities, accelerations, fixed and moving centrodes, such main circles as inflection, inversion, and cuspidal (or return), and blocking postures. This is done using the Kinematics module.
- Making use of the Synthesis module to assess options for the analysis of the four-bar mechanism, such as function generation, path generation, and rigid body motion, to obtain the cognates, and elements with permanent translational motion.
- Solving dynamic problems in planar mechanisms to obtain free solid diagrams of elements of the mechanism as well as inner forces and moments.

During these practical sessions, the teacher answers the students' questions. Once the students have updated their reports to the virtual platform, the teacher reviews them to provide feedback to each student.

3.2. Problem-Based Learning

It is sometimes challenging to capture the interest of students when introducing complex concepts during teaching. Thus, to show them the relation between theory and practice, some case studies based on applications have been developed during seminars. This has helped motivate them to explore different ideas and exercise their creativity [20].

In general, the methodology followed is as follows: The teacher, at the beginning of the session, explains the case study (the initial data, design criteria, and relevant hints); then, the students work in groups to solve the proposed problem by combining the necessary theoretical developments with the tools offered by GIM. Depending on the difficulty of the case study, this task can be completed during the seminar or finished in the students' study time. They can then update the report to the Moodle platform within an established period.

A wide variety of practical examples can thus be approached: designing mechanisms to perform certain specific tasks, analyzing the motion of robots and parallel manipulators common in the industry, designing and computing efforts in structures based on trusses, and vibration analysis of simple mechanical systems.

To illustrate the example of a case study proposed during these PBL activities, Figure 20 shows the rooftop of the San Mamés football stadium in Bilbao (Spain). The stadium is located in front of our university and is very well known to students. The objective of this PBL, which is among the activities designed through an educational innovation project in our Department of Mechanical Engineering (Ref: PIE2012/14), was to tackle the static problem of the truss that conformed to the structure of the rooftop. Figure 20a shows a floor plan of the truss of the rooftop, and Figure 20b shows a result obtained by a student. This consisted of reactions under specific loads.

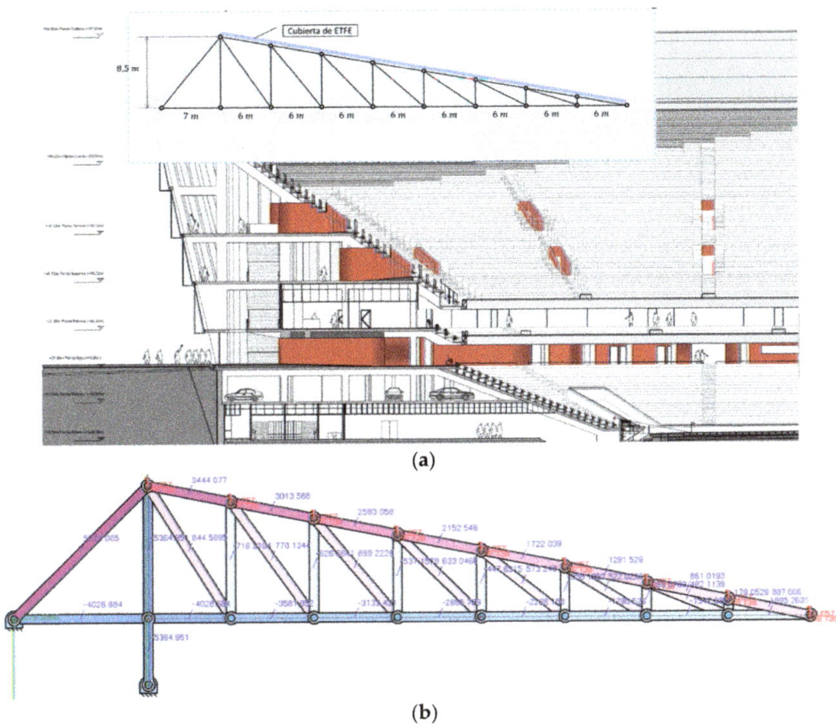

Figure 20. Example of PBL: (**a**) Rooftop of the football stadium in Bilbao (Spain), (**b**) Results.

3.3. *Self-Learning and Self-Checking*

The GIM software is used in practical sessions or seminars that teachers have developed as part of courses that they teach, and any student can install it for free. The website of our research group, COMPMECH Research Group, features a simple user manual to start using the software and many video tutorials to solve different problems with GIM.

Students are encouraged to practice with GIM and make the most of it. Note that the software can act as a self-checking assistant for students. Indeed, students usually take advantage of the software by solving many of the exercises from exams from previous years. They can verify whether their results, obtained by applying the theoretical procedure, coincide with those generated by GIM. In this way, they can identify errors that they made and enhance their skills.

4. GIM in Universities and Companies

The main channel for the dissemination of GIM is the direct download from our COMPMECH research group webpage. In Table 1, detailed data for downloads of the software to date are presented. The number of institutions adopting GIM has increased over the last few years to more than 500 per year. This software can have a significant positive impact not only on universities and educational centers all over the world, but also on companies and research centers related with innovation and research activities. GIM has also been cited and used in publications related to education and the development of virtual labs [21–25].

Table 1. GIM download data.

Academic Year	Institutions	Countries	Universities or Other Educational Institutions	Companies or Research Centers
2015/2016	341	66	248	93
2016/2017	551	90	417	134
2017/2018	560	88	395	165
2018/2019	502	86	376	126
2019/2020	660	92	516	144

Figure 21 shows the use of the GIM software throughout the world. The countries marked in red are those in which downloads of the software have been registered since 2015.

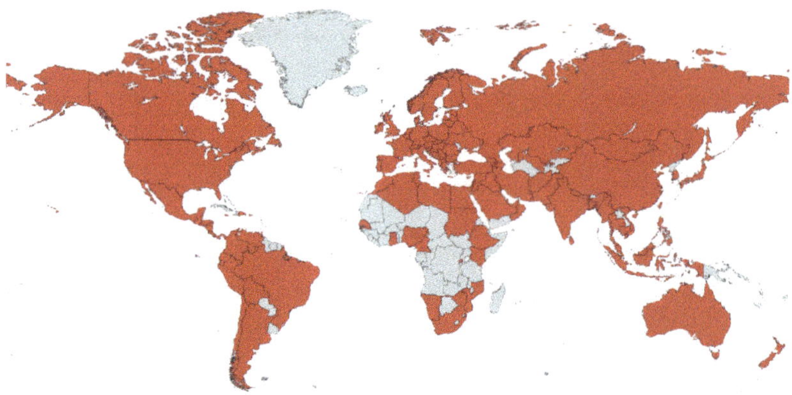

Figure 21. Countries using GIM software.

Tutorials and lectures have been organized to educate the academic community about GIM. In the scope of the IFToMM (International Federation for the Promotion of Mechanism and Machine Science) community, we can cite as instance the tutorial held during the 14th IFToMM World Congress in 2015 in Taipei, and lectures scheduled during IFToMM summer schools (Timisoara 2014 and Palermo 2016). Moreover, in the contexts of the Erasmus + internships and visits to universities, the capabilities of GIM have been presented. This is exemplified by lectures given at the Odessa National Polytechnic University (Ucrania) in 2017, and the presentation given at the Tokyo Institute of Technology in 2019. GIM has been also presented in educational conferences, such as ISEMMS (International Symposium on the Education in Mechanism and Machine Science) [26] and INTED (International Technology, Education and Development Conference) [27].

However, a key factor exhibiting the usefulness and interest of the software presented in this paper is the feedback from students that have used it. In an anonymous survey conducted via Google Forms at the end of the academic year 2017–2018 featuring 176 students of the Faculty of Engineering in Bilbao, after having received two lectures (4 h) on the use of the program, 90% stated that the software had been useful for them to better understand the subjects studied (machine theory and applied mechanics). They also highlighted its simplicity of use (91%) and assigned an average score of 3.7/5 to the entire *GIM experience*.

Recently, in January 2020, an online voluntary and anonymous Google Forms survey was sent to the people that have downloaded the latest version of GIM since September

2019. This survey collected 81 responses. The users' profile is shown in Figure 22. As it can be seen, the highest percentage of users is linked to the educational sector as a teacher or a student. This confirms the interest of GIM software as an educational tool. In any case, GIM is also used in companies and for research activities.

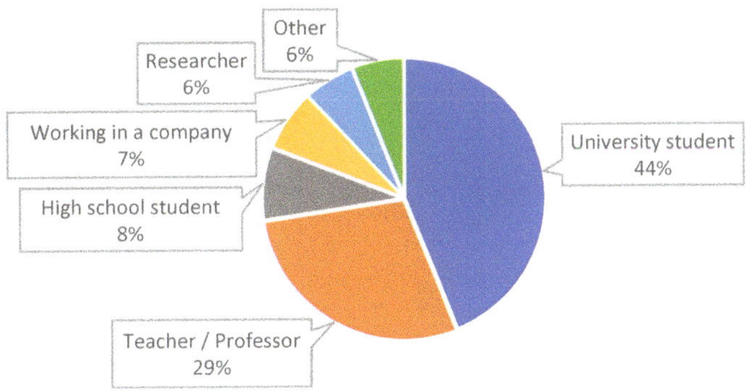

Figure 22. GIM users' profile.

In relation with the use of GIM as a complementary tool for students, the main results are depicted in Figure 23. It is shown that they use the software not only during the lectures, but also in their self-study time. The students rate the usefulness of GIM as an educational tool with an average rate of 3.9 (scale 0–5).

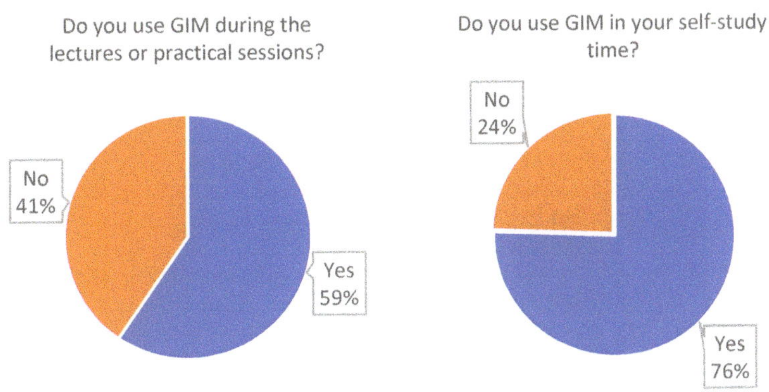

Figure 23. Use of GIM in the learning process.

The use of GIM by professors/teachers is shown in Figure 24. As expected, it has more impact during the practical sessions than in the theoretical lectures. The professors/teachers score the usefulness of GIM as a complement to the lectures with 4.2 (scale 0–5).

The ease of use of GIM is rated by all GIM users with 3.4 (scale 0–5). The score of the overall GIM experience is 3.9 (scale 0–5). Finally, only 6% of the users state that they do not intend to download a future version of GIM. The 68% of the users are sure they will download it and 26% are not sure.

Regarding the specific comments of the users, the negative comments focus on the lack of detailed tutorials for learning how to use all the features offered by the software: "Could really benefit from a comprehensive tutorial", "Lacking resources to fully utilize all features", and "Please provide a complete tutorial". We agree with these comments; the current tutorial is quite brief and the program has much more capabilities than the ones explained in it. We will enhance it by incorporating all the options of GIM.

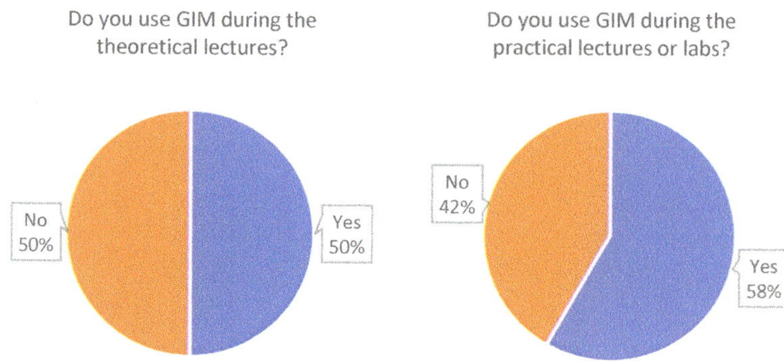

Figure 24. Use of GIM in the teaching process.

On the other side, the majority of the comments emphasizes the educational and design capabilities of GIM, such as: "Excellent app to help me understand my course", "Very interesting software that lets you think directly in the core of the mechanism", "Plan to have it as an integrated part in my lectures", and "Wonderful. It is a very effective tool for engineering students and designers".

5. Conclusions

This paper explored the impact of the GIM software as a supporting tool for teaching university courses on mechanism and machine science. The authors reviewed the potential of the software for modeling planar and spatial linkages, and for carrying out kinematic and dynamic analyses, as well as dimensional synthesis. Although GIM also offers additional, advanced features for PhD Students and other researchers, they are not presented here as they are beyond the scope of the paper.

GIM is being used in many universities as a powerful and valuable tool for boosting students' learning throughout the academic year. The learning activities presented in the paper show the versatility of GIM and consist of practical and problem-based learning sessions. It is also worth noting the positive effects of the use of the software observed in students in terms of increased motivation for working independently. The feedback received from the students from a survey on the use of GIM reinforced these conclusions.

The impact of GIM was quantified by monitoring direct downloads both from universities (or other educational institutions) and companies or research centers. These downloads have increased annually. In addition, dissemination activities have been organized to present GIM in summer schools, educational conferences, and during lectures in several universities. GIM is constantly evolving to incorporate more capabilities to respond to the demands of different users.

Supplementary Materials: The following are available online at https://www.mdpi.com/article/10.3390/app11177850/s1. The three videos referred to in the paper showing the kinematics and dynamics simulations are available.

Author Contributions: Conceptualization, V.P.; methodology, E.M.; software, E.M.; validation, M.U.; formal analysis, V.P.; investigation, E.M.; resources, V.P.; data curation, M.U.; writing—original draft preparation, E.M., M.U., V.P., and A.H.; writing—review and editing, M.U.; visualization, M.U.; supervision, A.H.; project administration, A.H.; funding acquisition, A.H. and V.P. All authors have read and agreed to the published version of the manuscript.

Funding: This research was funded by Ministerio de Economía y Competitividad, Spanish Government Project, MINECO/FEDER, UE (grant number DPI2015-67626-P), Departamento de Educación, Política Lingüística y Cultura, Regional Government of the Basque Country (grant number IT949-16) and University of the Basque Country UPV/EHU (grant number PIE2012/14).

Institutional Review Board Statement: Not applicable.

Informed Consent Statement: Not applicable.

Data Availability Statement: Not applicable.

Conflicts of Interest: The authors declare no conflict of interest.

References

1. Orlandea, N.; Chace, M.A.; Calahan, D.A. A Sparsity-Oriented Approach to the Dynamic Analysis and Design of Mechanical Systems–Part 1. *J. Eng. Ind.* **1977**, *99*, 773. [CrossRef]
2. Orlandea, N.; Chace, M.A.; Calahan, D.A. A Sparsity-Oriented Approach to the Dynamic Analysis and Design of Mechanical Systems–Part 2. *J. Eng. Ind.* **1977**, *99*, 780. [CrossRef]
3. Levinson, D. *AUTOLEV-User's Manual*; Online Dynamics Inc.: Sunnyvale, CA, USA, 1990.
4. Unda, J.; Avelló, A.; Jiménez, J.M.; García de Jalón, J. COMPAMM-A Program for the Dynamic Analysis of Multi-Rigid-Body Systems. In *Engineering Software IV*; Springer: Berlin/Heidelberg, Germany, 1985; pp. 983–996. [CrossRef]
5. Wehage, R.A.; Haug, E.J. Generalized Coordinate Partitioning for Dimension Reduction in Analysis of Constrained Dynamic Systems. *J. Mech. Des.* **1982**, *104*, 247. [CrossRef]
6. Paul, B. DYMAC (Dynamics of MAChinery). In *Multibody Systems Handbook*; Schiehlen, W., Ed.; Springer: Berlin/Heidelberg, Germany, 1990; pp. 305–322.
7. Wittenburg, J.; Wolz, U. MESA-VERDE a Symbolic Program for Nonlinear Articulated Rigid-Body Dynamics. In Proceedings of the ASME Design Engineering Technical Conference, Cincinnati, OH, USA, 10–13 September 1985; Volume 8.
8. Suñer, J.L.; Carballeira, J. Enhancing Mechanism and Machin Science Learning by Creating Virtual Labs with ADAMS. In *New Trends in Educational Activity in the Field of Mechanism and Machine Theory*; García-Prada, J.C., Castejón, C., Eds.; Springer: Berlin/Heidelberg, Germany, 2014; pp. 221–228.
9. Rubel, A.J.; Kaufman, R.E. KINSYN III: A new human-engineered system for interactive computer-aided design of planar linkages. *J. Eng. Ind. Trans. ASME* **1977**, *99*, 440–448. [CrossRef]
10. Erdman, A.G.; Gustafson, J. LINCAGES: A linkage interactive computer analysis and graphically enhanced synthesis package. In Proceedings of the ASME Design Engineering Technical Conferences, 77-DTC-5, Chicago, IL, USA, 26–30 September 1977.
11. Chuang, J.C.; Strong, R.T.; Waldron, K.J. Implementation of solution rectification techniques in an interactive linkage synthesis program. *J. Mech. Des.* **1981**, *103*, 657–661. [CrossRef]
12. Müller, M.; Mannheim, T.-P.; Hüsing, M.; Corves, B. MechDev-A new Software for Developing Planar Mechanisms. In *Interdisciplinary Applications of Kinematics. Mechanisms and Machine Science*; Kecskeméthy, A., Geu Flores, F., Carrera, E., Elias, D., Eds.; Springer: Berlin/Heidelberg, Germany, 2018; Volume 71, pp. 167–175.
13. Bataller, A.; Ortiz, A.; Cabrera, J.A.; Nadal, F. WinMecC: Software for the Analysis and Synthesis of Planar Mechanisms. In *New Trends in Mechanism and Machine Science, Mechanism and Machine Science*; Wenger, P., Flores, P., Eds.; Springer: Berlin/Heidelberg, Germany, 2017; Volume 43, pp. 233–242.
14. Nadal, F.; Cabrera, J.A.; Bataller, A.; Castillo, J.J.; Ortiz, A. Turning functions in optimal synthesis of mechanisms. *J. Mech. Des.* **2015**, *137*, 6. [CrossRef]
15. Technologies, Design Simulation. Working Model. 2007–2012. Available online: https://www.design-simulation.com/ (accessed on 26 August 2021).
16. Hohenwarter, M.; Preiner, J. Dynamic mathematics with GeoGebra. *J. Online Math. Its Appl.* **2007**, *7*, 1448.
17. Calvo, J.A.; Álvarez-Caldas, C.; San Román, J.L. Analysis of Dynamic Systems Using Bond Graph and SIMULINK. In *New Trends in Educational Activity in the Field of Mechanism and Machine Theory*; García-Prada, J.C., Castejón, C., Eds.; Springer: Berlin/Heidelberg, Germany, 2014; pp. 155–162.
18. García de Jalón, J.; Callejo, A. A straight methodology to include multibody dynamics in graduate and undergraduate subjects. *Mech. Mach. Theory* **2011**, *46*, 168–182. [CrossRef]
19. Petuya, V.; Macho, E.; Altuzarra, O.; Pinto, C.; Hernández, A. Educational software tools for the kinematic analysis of mechanisms. *Comput. Appl. Eng. Educ.* **2014**, *22*, 72–86. [CrossRef]
20. Royalty, A. Design-base d Pedagogy: Investigating an emerging approach to teaching design to non-designers. *Mech. Mach. Theory* **2018**, *125*, 137–145. [CrossRef]
21. Hamon, C.L.; Green, M.G.; Dunlap, B.; Camburn, B.A.; Crawford, R.H.; Jensen, D.D. Virtual or physical prototypes? Development and testing of a Prototyping Planning Tool. In Proceedings of the 121st ASEE Annual Conference and Exposition, Indianapolis, IN, USA, 15–18 June 2014.
22. Wang, Y.; Ong, S.K.; Nee, A.Y.C. Enhancing mechanisms education through interaction with augmented reality simulation. *Comput. Appl. Eng. Educ.* **2018**, *26*, 1552–1564. [CrossRef]
23. Adrian, P.; Reinoso, O.; Gil, A. A virtual laboratory to simulate the control of parallel robots. In Proceedings of the 3rd IFAC Workshop on Internet Based Control Education (IBCE 2015), Brescia, Italy, 4–6 November 2015; Volume 48, pp. 19–24.
24. Hampali, S.; Chittawadigi, R.G.; Saha, S.K. MechAnalyzer. 3D Model Based Mechanism Learning Software. In Proceedings of the 14th IFToMM World Congress, Taipei, Taiwan, 25–30 October 2016; pp. 425–431.
25. Craifaleanu, A.; Dragomirescu, C.; Craifaleanu, I.G. Virtual Laboratory of Dynamics. In Proceedings of the 6th International Conference on Education and New Learning Technologies (EDULEARN), Barcelona, Spain, 7–9 July 2014; pp. 4612–4619.

26. Urízar, M.; Altuzarra, O.; Diez, M.; Campa, F.J.; Macho, E. Dynamics and Mechanical Vibrations. Complementing the Theory with Virtual Simulation and Experimental Analysis. In *New Trends in Educational Activity in the Field of Mechanism and Machine Theory*; García-Prada, J.C., Castejón, C., Eds.; Springer: Berlin/Heidelberg, Germany, 2019; pp. 64–71.
27. Petuya, V.; Altuzarra, O.; Pinto, C.; Hernández, A. A new educational software for the kinematic analysis of spatial mechanisms. In Proceedings of the International Technology, Education and Development Conference (INTED 09), Valencia, Spain, 9–11 March 2009; ISBN 978-84-612-7578-6.

applied sciences

MDPI

Article

Identification of Pre-Tightening Torque Dependent Parameters for Empirical Modeling of Bolted Joints

Yu Tian [1], Hui Qian [1], Zhifu Cao [2], Dahai Zhang [3] and Dong Jiang [1,*]

1 School of Mechanical and Electronic Engineering, Nanjing Forestry University, Nanjing 210037, China; tianyu@njfu.edu.cn (Y.T.); qianhui@njfu.edu.cn (H.Q.)
2 School of Aeronautics and Astronautics, Chongqing University, Chongqing 400044, China; aezfcao@cqu.edu.cn
3 Institute of Aerospace Machinery and Dynamics, Southeast University, Nanjing 211189, China; dzhang@seu.edu.cn
* Correspondence: jiangdong@njfu.edu.cn

Abstract: The vibration characteristics of bolted structures are crucially affected by the pre-tightening torque. An approach for identifying the pre-tightening torque dependent stiffness parameters of bolted joints is proposed in this paper. Firstly, the interface of the bolted joint is characterized by the thin layer element with the isotropic material property, and the parameter value of the property is assigned relative to the distance from the center of the bolt; the influence of the bolt is ignored. Secondly, the model updating method is adopted to identify the parameters of thin layer elements using experimental data, and modal data under different values of pre-tightening torque in the range of 2 N·m~22 N·m are obtained; the torque wrench is used to determine the pre-tightening torque in the modal test. Finally, after identifying the material parameters using partial experimental data on pre-tightening torque range, the empirical equation of the interface parameters with the pre-tightening torque parameter is obtained by curve fitting and the rest of the experimental data are used to verify the accuracy of the fitted empirical equations. It is concluded that this method can obtain all the parameters of the equivalent thin layer elements within a certain range of pre-tightening torque, which can provide a reference for the empirical modeling of bolted structures, improve modeling efficiency and reflect the characteristic performance of real structural dynamics.

Keywords: bolted joints; parameter identification; thin-layer element; pre-tightening torque; contact interface

Citation: Tian, Y.; Qian, H.; Cao, Z.; Zhang, D.; Jiang, D. Identification of Pre-Tightening Torque Dependent Parameters for Empirical Modeling of Bolted Joints. *Appl. Sci.* **2021**, *11*, 9134. https://doi.org/10.3390/app11199134

Academic Editors: Jae Hyuk Lim and Angelo Luongo

Received: 16 August 2021
Accepted: 23 September 2021
Published: 30 September 2021

1. Introduction

Bolt connection has been widely used in mechanical structures; the vibration behavior of the assembly structure is closely related to the mechanical performance of the contact interface of the bolted joints [1–4]. In a well-connected joint, nonlinearity is often neglected by structural dynamics analysis due to the sufficient and appropriate pre-tightening torque. The dynamic characteristics of bolted structures always depend on the pre-tightening torque.

High-precision modeling of bolt connection is a key problem in structural design. The bolt can be well simulated by the 3D solid refinement model [5]. This modeling method can accurately analyze the deformation and stress characteristics of the inner and surrounding structures, and the effect of surface slip and contact effect on the mechanical properties of the structure can be fully considered. However, this method is too time-consuming and not conducive to engineering structures when many bolt connections exist. Under the condition that simulation precision is satisfied, some simplified models are proposed to analyze the bolt connection. Li et al. [6] proposed a bolt-spring model, which replaces the bolt model with a set of spring models. This method can improve the efficiency of computing. Vilela et al. [7] proposed a unitary model which can consider the contact interfaces, friction and preload. Fukuoka [8] utilized the three-dimensional FE model to analyze the mechanical behavior of multi-bolted joints. Yan et al. [9] used the thin-layer element method to represent the nonlinear factors in this study and verify the accuracy of thin-layer element modeling. Ahmadian et al. [10] compared the measured responses with the predictions of the model containing a parametric generic joint element. The parameters of the joint interface model are successfully identified by minimizing the difference between the measured responses and the model predictions. Yao et al. [11] proved that the material parameters of the partitioned thin-layer elements can be expressed for modeling aero-engine bolted joints. These methods can provide the deformation and stress results in good agreement with experimental results, and the effects of surface slip and contact clearance on the mechanical properties of the bolted structure can also be fully considered. However, these methods do not pay enough attention to the pre-tightening force of the bolts.

The appropriate bolt pre-tightening force will make the whole structure have better performance. The clamping conditions also have been found to affect the performance of joints. In the area of experimental study of bolt pre-tightening force, Cooper and Turvey [12] studied the fatigue test of a single bolt lap and concluded that failure loads, critical end distances and critical widths increased as the bolt clamping torque increased. Zhao et al. [13] used case studies on the equal pre-tightening force and bending moment effect to accurately predict the dynamic characteristic of a bolted assembly. In the field of numerical modeling, the effect of pre-tightening force must be taken into account for the establishment of an accurate FE model of the bolt structure. Zhao et al. [14] represented interface contact stiffness by implementing thin layer elements into the FE model and obtained the regularity of the contact stiffness in bolted interface changes with bolt preload. Ultimately, these studies establish the viewpoint that the element method is better for simulating the pre-tightening force. On this basis, the parameter identification of the finite elements of bolted structures has been widely studied. Yang and Park [15] proposed a method for inversion identification of the structural stiffness and damping of joints using frequency response function (FRF). Furthermore, in order to handle errors in test measurement, they applied this method to eliminate the noise in the original FRFs. Jiang et al. [16] identified the mechanical parameters of the contact interface, considering the uncertainty in the bolted structure by adopting the thin layer element method. Ren and Beards [17] improved the techniques for identifying joints using experimental data; these techniques are insensitive to measurement noise. Tsai and Chou [18] presented a synthesis formula to predict the FRF of the two-bolt-joint structure where the FRF was used to identify the stiffness and damping of the subject investigated. Yang et al. [19] proposed an approach for identifying the rotational stiffness and translational stiffness of the joint model by combining substructure synthesis and FRFs. Xu et al. [20] proposed a high-fidelity modeling method for clamping boundary conditions. Most of these works focus on the mechanical behavior of bolted structures under fixed tightening torque. In the course of engineering a structure pre-tightening forces always change, and few studies have been carried out to analyze the vibration performance of bolted joints under varying tightening torque.

In this paper, bolt connection modeling and parameter identification in engineering are investigated. The bolt interface is modeled based on the thin layer element method. The target function and mode updating method are used to identify the thin layer element parameters. The relationships between the identified parameters and the change in pre-tightening torque are obtained. The resulting curves can provide guidance for the accurate modeling of the same type of bolt connections in engineering.

2. Basic Theory

2.1. Thin-Layer Element

A common interface element based on the node element was proposed by Goodman, Taylor and Brekke [21]. The element formula is derived from the relative node displacement of the solid element around the interface element, as shown in Figure 1.

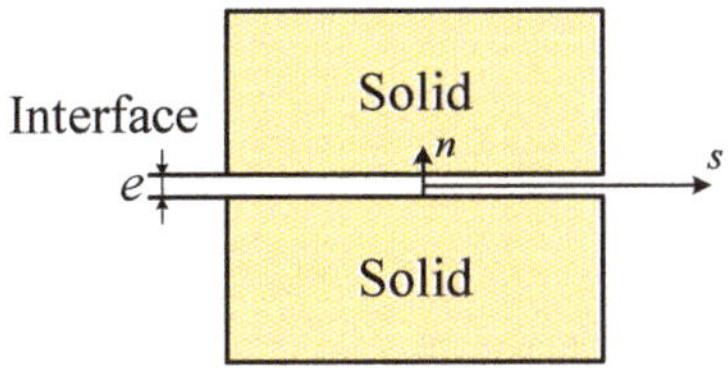

Figure 1. Interface element with relative nodal displacement.

In two-dimensional analysis, the constitutive relation or stress-relative displacement relation of the interface behavior is expressed as

$$\left\{ \begin{array}{c} \sigma_n \\ \tau \end{array} \right\} = \left[\begin{array}{cc} k_n & 0 \\ 0 & k_s \end{array} \right] \left\{ \begin{array}{c} v_r \\ u_r \end{array} \right\} = [C]_i \left\{ \begin{array}{c} v_r \\ u_r \end{array} \right\} \quad (1)$$

where σ_n is the normal stress, τ is the shear stress, k_n denotes the normal stiffness, k_s = shear stiffness, v_r and u_r are the relative normal and the relative shear displacements, respectively, and $[C]_i$ is the constitutive matrix of the interface or joint element. In soil-structure interaction problems, it is usually assumed that the thickness e of the element is zero. In most problems, this formulation can provide satisfactory solutions for stick and slip modes for which the normal stress remains compressive. For some other modes such as debonding, the solutions are often unreliable. An analysis shows that the planar approximation of the element treated as a solid element can provide a satisfactory simulation of the finite-sized interface zone, and at the limit its results approach those of the zero-thickness element [22].

By defining the elements between adjacent contact bodies, simulation of the contact equivalent mechanical properties of the interface using thin layer elements can be implemented [23]. The proposed element essentially represents a solid element of small finite thickness. It is assumed that the thin layer element is generated by the solid element with eight nodes, as shown in Figure 2, and the displacement of any point of the element $p(x, y, z)$ can be expressed as the following by the knowledge of the FE.

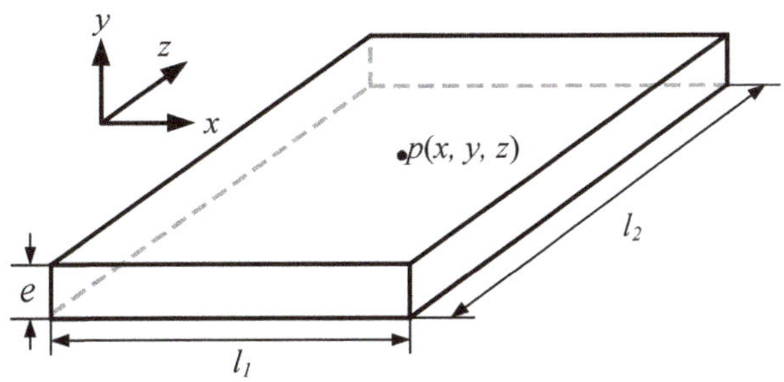

Figure 2. Thin-layer element.

$$\left\{ \begin{array}{l} x = \sum\limits_{i=1}^{8} N_i.x_i \\ y = \sum\limits_{i=1}^{8} N_i.y_i \\ z = \sum\limits_{i=1}^{8} N_i.z_i \end{array} \right. \quad (2)$$

In the formula, x_i, y_i, z_i, is the coordinate of the nodes, while N_i is the shape function.

The relationships between the element stress, the strain, and the node p are shown as follows:

$$\varepsilon = \boldsymbol{B} \cdot \boldsymbol{p} \quad (3)$$

$$\sigma = \boldsymbol{D} \cdot \boldsymbol{B} \cdot \boldsymbol{p} \tag{4}$$

$\boldsymbol{B}$ and $\boldsymbol{D}$ are the geometric matrix and elastic matrix, respectively.

l_1 and l_2 are the length and the width of the thin-layer elements, and e is the thickness in z-direction, which is shown in Figure 2. The thickness e is much smaller than the sizes l_1 and l_2 of the other two directions of the element. The in-plane strain components (ε_x, ε_y, γ_{xy}) are ignored under the circumstances, as well as the stress components (σ_x, σ_y, τ_{xy}) of the element. The values of partial derivatives of the form functions of thin layer elements of different sizes with respect to the local coordinates at Gaussian points are analyzed and compared in Table 1.

Table 1. The value of partial derivatives of the shape functions of thin layer elements with respect to the local coordinates at Gaussian points.

Gaussian Points	$l_1 \times l_2 \times e = 1 \times 1 \times 0.1$			$l_1 \times l_2 \times e = 5 \times 5 \times 0.1$		
	$\partial N_i/\partial x$	$\partial N_i/\partial y$	$\partial N_i/\partial z$	$\partial N_i/\partial x$	$\partial N_i/\partial y$	$\partial N_i/\partial z$
$i = 1$	−0.16667	−0.62201	−1.66667	−0.03333	−0.12440	−1.66667
$i = 2$	0.16667	−0.16667	−0.44658	0.03333	−0.03333	−0.44658
$i = 3$	0.62201	0.16667	−1.16667	0.12440	0.03333	−1.66667
$i = 4$	−0.62201	0.62201	−6.22008	−0.12440	0.12440	−6.22008
$i = 5$	−0.04466	−0.16667	1.66667	−0.00893	−0.03333	1.66667
$i = 6$	0.04466	−0.04466	0.44658	0.00893	−0.00893	0.44658
$i = 7$	0.16667	0.04466	1.66667	0.03333	0.00893	1.66667
$i = 8$	−0.16667	0.16667	6.22008	−0.03333	0.03333	6.22008

As can be seen from the data in Table 1, $\partial N_i/\partial z$ is greater than $\partial N_i/\partial x$ and $\partial N_i/\partial y$. When the size in the z direction is much smaller than the size in the x and y directions, it can be considered that $\partial Ni/\partial x = \partial N_i/\partial y \approx 0$ and the strain component $\varepsilon_x = \varepsilon_y = \gamma_{xy} \approx 0$. In other words, only three strain components are not zero at the Gauss point. The strain component can be simplified to $\varepsilon = [\varepsilon_z \ \gamma_{yz} \ \gamma_{zx}]^T$. Synthesizing the above analysis, it is assumed that the normal contact characteristics and tangential contact characteristics of the interface are independent of each other. The constitutive equation of the thin layer element which characterizes interface contact is:

$$\begin{Bmatrix} \sigma_n \\ \tau_{tx} \\ \tau_{ty} \end{Bmatrix} = \begin{bmatrix} E_n & 0 & 0 \\ 0 & G_t & 0 \\ 0 & 0 & G_t \end{bmatrix} \begin{Bmatrix} \varepsilon_n \\ \gamma_{tx} \\ \gamma_{ty} \end{Bmatrix} \tag{5}$$

E_n and G_t are the normal elastic modulus and tangential shear modulus, respectively. If the tangential and normal contact properties are coupled, the coupling term can be added to Equation (5). If the contact properties are coupled to normals and tangents, the coupling term can be added to the constitutive relation (5). The constitutive equation is used when using isotropic constitutive material to simulate the thin layer element.

$$\begin{Bmatrix} \sigma_{xx} \\ \sigma_{yy} \\ \sigma_{zz} \\ \tau_{xy} \\ \tau_{yz} \\ \tau_{zx} \end{Bmatrix} = \begin{bmatrix} \lambda + 2G & \lambda & \lambda & & & \\ & \lambda + 2G & \lambda & & & \\ & & \lambda + 2G & & & \\ & & & G & & \\ & & & & G & \\ sym & & & & & G \end{bmatrix} \begin{Bmatrix} \varepsilon_{xx} \\ \varepsilon_{yy} \\ \varepsilon_{zz} \\ \gamma_{xy} \\ \gamma_{yz} \\ \gamma_{zx} \end{Bmatrix} \tag{6}$$

$$\lambda = \frac{G(E - 2G)}{E - 3G} = \frac{E\nu}{(1+\nu)(1-2\nu)} \tag{7}$$

λ is the Lamé constant and $G = E/2(1 + \nu)$ is the shear modulus. As we know, the number of independent material parameters of the isotropic material is 2. When the thickness of the cell is much smaller than the feature size of the other two directions, let $\varepsilon_x = \varepsilon_y = \varepsilon_{xy} \approx 0$. Finally, the constitutive equation of the material can be written as follows:

$$\left\{ \begin{array}{c} \sigma_z \\ \tau_{yz} \\ \tau_{zx} \end{array} \right\} = \begin{bmatrix} \lambda + 2G & 0 & 0 \\ 0 & G & 0 \\ 0 & 0 & G \end{bmatrix} \left\{ \begin{array}{c} \varepsilon_z \\ \gamma_{yz} \\ \gamma_{zx} \end{array} \right\} \tag{8}$$

In this equation, the normal elastic constant and the tangent elastic constant are not independent.

Combined with the constitutive relation from the aforementioned content, the element stiffness matrix of $\boldsymbol{K}$ can be obtained according to the principle of virtual work.

$$\delta W = \int_0^{l_1}\int_0^{l_2}\int_0^{d} \sigma^T \delta\varepsilon \mathrm{d}x\mathrm{d}y\mathrm{d}z = \delta \boldsymbol{u}_{nodal}^T \boldsymbol{K} \boldsymbol{u}_{nodal} \tag{9}$$

Isoparametric transformation is used to calculate the stiffness matrix $\boldsymbol{K}$ of the thin layer element.

$$\boldsymbol{K} = \int_{V_e} \boldsymbol{B}^T \boldsymbol{D}\boldsymbol{B} \mathrm{dV} = \int_{-1}^{1}\int_{-1}^{1}\int_{-1}^{1} \boldsymbol{B}^T \boldsymbol{D}\boldsymbol{B} det(\boldsymbol{J}) \mathrm{d}\xi \mathrm{d}\eta \mathrm{d}\zeta \tag{10}$$

Figure 3 shows the isoparametric transformation of the thin layer element. ξ, η, and ζ are global coordinate symbols, and $\boldsymbol{J}$ is the Jacobian matrix. When the global coordinate system is consistent with the local coordinate system, the following expression is given:

$$\boldsymbol{J} = \begin{bmatrix} \partial x/\partial \xi & \partial y/\partial \xi & \partial z/\partial \xi \\ \partial x/\partial \eta & \partial y/\partial \eta & \partial z/\partial \eta \\ \partial x/\partial \zeta & \partial y/\partial \zeta & \partial z/\partial \zeta \end{bmatrix} = \begin{bmatrix} l_1/2 & 0 & 0 \\ 0 & l_2/2 & 0 \\ 0 & 0 & d/2 \end{bmatrix} \tag{11}$$

By using the two-node Gaussian integral method, the stiffness matrix $\boldsymbol{K}$ of the thin layer element is expressed as follows:

$$\boldsymbol{K} = \sum_{i=1}^{2}\sum_{j=1}^{2}\sum_{k=1}^{2} \boldsymbol{B}^T(\xi_i, \eta_j, \zeta_k) \boldsymbol{C}\boldsymbol{B}(\xi_i, \eta_j, \zeta_k) det(\boldsymbol{J}(\xi_i, \eta_j, \zeta_k)) w_{\xi,i} w_{\eta,j} w_{\zeta,k} \tag{12}$$

where w_ξ, w_η, and w_ζ are the Gaussian integral weight function.

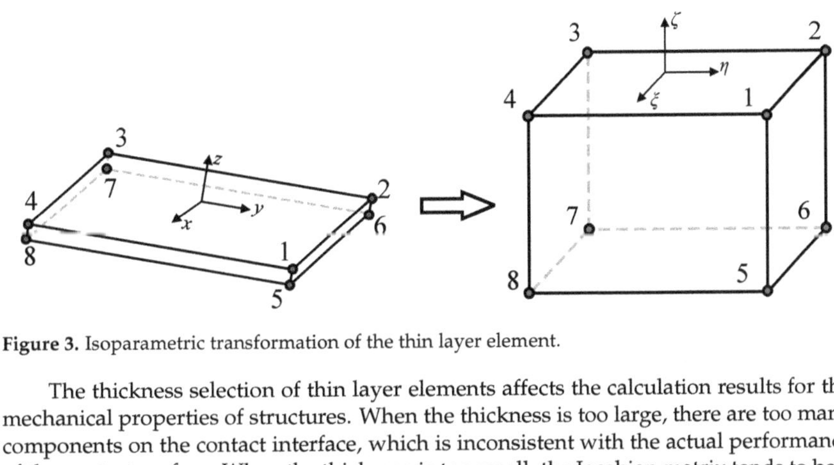

Figure 3. Isoparametric transformation of the thin layer element.

The thickness selection of thin layer elements affects the calculation results for the mechanical properties of structures. When the thickness is too large, there are too many components on the contact interface, which is inconsistent with the actual performance of the contact surface. When the thickness is too small, the Jacobian matrix tends to be a singular matrix, and the error is too large to be used to calculate the displacement strain relationship. The ratio coefficient is used as the selection condition for the thickness of thin layer element modeling, as shown below:

$$R = \frac{max(l_1, l_2)}{e} \tag{13}$$

Desai et al. [24] proposed that when the range of R is 10–100 we can get an exact result, as studied in the context of the static contact problem where the thin layer element is based on a linear constitutive. Sharma and Desai [22] think reasonable results can be obtained when R is 5. In this paper, R in the case study is 10.

The bolted joints fix the interface; thus, the determination of the contact stiffness is the core of the dynamic analysis of the bolted structure. Sharma and Desai obtain the normal contact stiffness and tangential contact stiffness by testing the relationship between the stress and the displacement of the contact interface:

$$\begin{cases} k_\tau = \frac{e\tau}{eu_r} \\ k_n = \frac{e\sigma_n}{ev_r} \end{cases} \tag{14}$$

where k_n is the normal contact stiffness, k_τ is the tangential contact stiffness, e is thickness, u_r and v_r are the normal displacement and tangential displacement of the thin layers, respectively, and σ and τ are normal stress and tangential stress. The interface is generated by the Hexahedron element with linear constitutive $\boldsymbol{C}$, which is as follows:

$$\boldsymbol{C} = \frac{E}{(1+\nu)(1-2\nu)} \begin{bmatrix} 1-\nu & \nu & 0 \\ \nu & 1-\nu & \nu \\ 0 & 0 & (1-2\nu)/2 \end{bmatrix} \tag{15}$$

where E is the elastic Modulus, ν is the Poisson's ratio.

Schmidt and Bograd [25] obtained the relationship between the equivalent shear modulus of the contact surface and the tangential stiffness of the contact surface by testing tangential force of the bolted joints' structure:

$$G = \frac{ke}{A} \tag{16}$$

where A is the real contact area and k is the tangential stiffness of the connection which correlates with the surface pressure and friction properties of the contact interface.

The approximate elastic parameters of the thin layer element can be obtained by the above two methods of testing the contact stiffness. The thin layer element can provide the computation, distribution, and concentration of stresses and strain within the thin finite zones, and hence, can permit evaluation of progressive damage and failure as they occur in many engineering problems. In the case of this paper, the first four order bending modal data are used. The parameter identification method is used to determine the parameters of the thin layer element.

2.2. Parameter Identification

Parameter identification methods are widely used in modeling problems [26–29]. The key to solving parameter identification problems is optimization of the algorithm, the essence of which is an optimization problem to minimize the discrepancies between the measured and predicted parameters. The objective function and the constraint are defined as:

$$\begin{cases} \text{Min } J(\mathbf{p}) = \varepsilon^T \mathbf{W} \varepsilon = \|\mathbf{W}^{1/2}(\mathbf{z}^m - \mathbf{z}^a(\mathbf{p}))\|_2^2 \\ s.t. \quad \mathbf{p}_1 \le \mathbf{p} \le \mathbf{p}_2 \end{cases} \tag{17}$$

$\mathbf{p}$ is a vector for identifying parameters. The expression of the objective function $J(\mathbf{p})$ is the weighted and squared residual difference between the experimental and calculated modal parameters. The domain of the function $J(\mathbf{p})$ is defined in a reasonable range of

$\mathbf{p}_1 \leq \mathbf{p} \leq \mathbf{p}_2$. The minimum of the function is calculated in this domain. The ε is the residual difference of modal parameters, and $\mathbf{z}^m$, and $\mathbf{z}^a(\mathbf{p})$ are the modal parameters of the test and calculation, respectively. The weighted matrix **W** is a diagonal matrix reflecting the relative weight of the residual difference of the modal parameters. Sensitivity analysis is used to solve the structural optimization and model updating problems with the iterative method. The *j*-th iteration can be described as follows:

$$\mathbf{W}^{1/2}\left(\mathbf{z}^m - \mathbf{z}_j^a\right) = \mathbf{S}_j\left(\mathbf{p}_{j+1} - \mathbf{p}_j\right) \tag{18}$$

$\mathbf{S}_j = \mathbf{W}^{1/2}\partial\mathbf{z}_j/\partial\mathbf{p}_j$ represents the sensitivity matrix, which is the weighted Jacobian matrix of modal parameters. Using the numerical calculus, $\mathbf{p}_{j+1}$, can be obtained by Equation (18). The modal parameters are taken as the objective function and the iterative optimization algorithm is used for the calculation so that the loss function is continuously reduced and the parameter p converges. Finally, the precise material parameters of the contact surface can be obtained.

Through optimization algorithm iteration, the parameters of thin layer elements can be converged and identified. The implementation procedure can be illustrated as follows:

(1) Initialize $j = 0$, construct an initial FE model of the bolt connection using the thin-layer element;
(2) Select the elastic parameters **p** through relative sensitivity analysis using the initial FE model;
(3) Pair the experimental and numerical modal shapes using modal correlation analysis, calculating the modal assurance criterion (MAC) to achieve this;
(4) Calculate the residual $\mathbf{z}^m - \mathbf{z}^a{}_j$ between the experimental and numerical modal data, solve the iteration format Equation (18), and obtain the variation $\Delta\mathbf{p} = \mathbf{p}_{j+1} - \mathbf{p}_j$;
(5) If the variables $\Delta\mathbf{p}$ are converged, go to step (6); otherwise set $j = j + 1$ and go to step (2), with the stop criterion $\|\mathbf{p}_{j+1} - \mathbf{p}_j\| \leq 10^{-6}$;
(6) The exact elastic parameters are obtained.

2.3. Pre-Tightening Torque Dependent Parameters

The identification method of pre-tightening torque dependent parameters for empirical modeling of bolted joints in this paper is mainly divided into the following steps

(1) Modelling. The interface of the bolted joint is characterized by the isotropic thin layer element, ignoring the influence of the bolt.
(2) Experiment. The torque wrench is used to get modal data under different pre-tightening torque within a range.
(3) Identification. Under each pre-tightening force, different parameters of the thin layer unit can be obtained through parameter identification, and the precise thin layer element under this pre-tightening force can be obtained.
(4) Prediction. All parameter values of thin-layer elements within a certain range of pre-tightening can be predicted by taking appropriate pre-tightening step and curve fitting.
(5) Verification. New experimental data are used to verify the predictions.
(6) The flowchart of this article is shown in Figure 4.

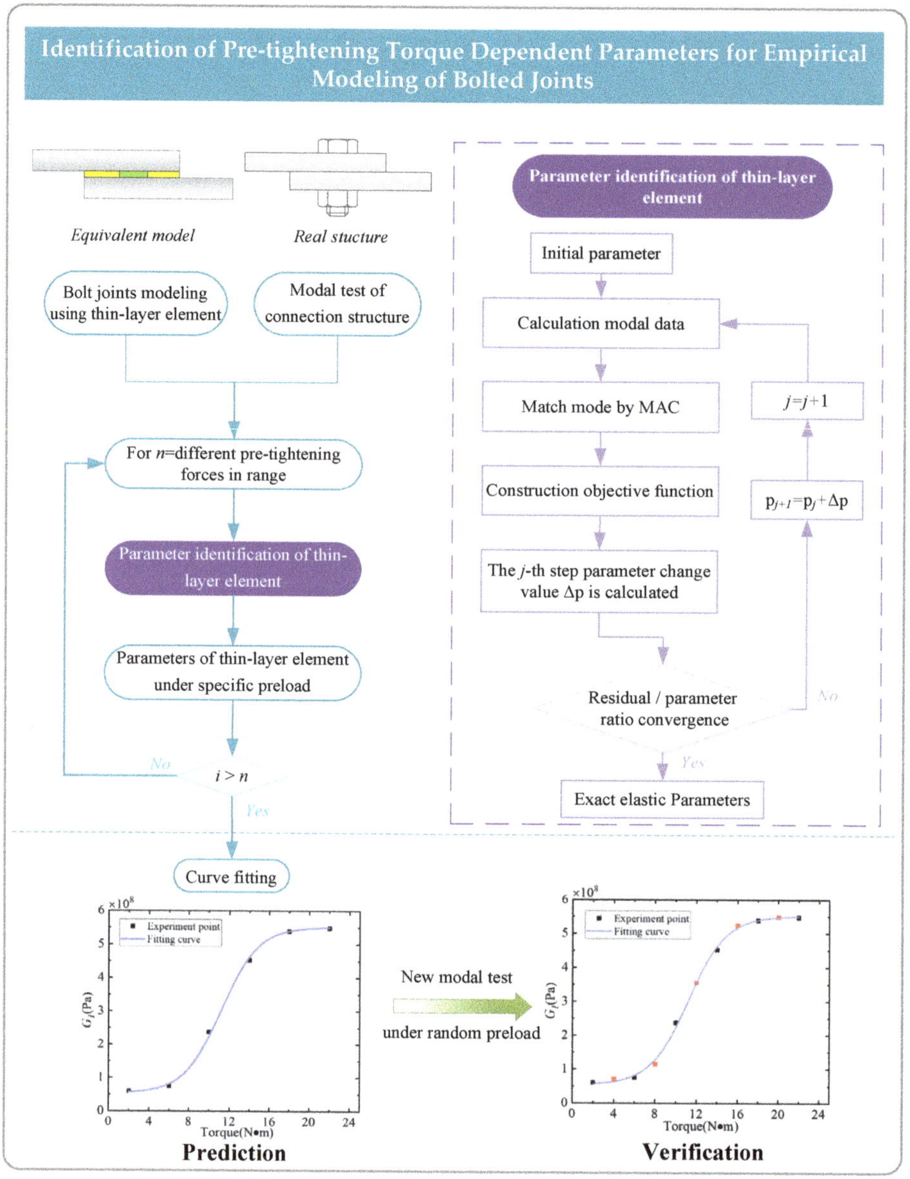

Figure 4. The flowchart for the parameter identification method of bolted joints under different pre-tightening torque.

3. Case Study

In this section, the effectiveness of the proposed method is verified by a bolt pre-tightening test and simulation analysis.

3.1. Modal Test

As shown in Figure 5, the study object is a bolted joint structure which is composed of two lap plates and four bolts. The size of the lap plate is measured at about 200 mm long × 80 mm wide × 8 mm deep and the connecting length is 80 mm. The lap plate is made of aluminum alloy. Bolts and nuts are made from low carbon steel of which the nominal diameter is M10. The material parameters of the above two components are shown in Table 2.

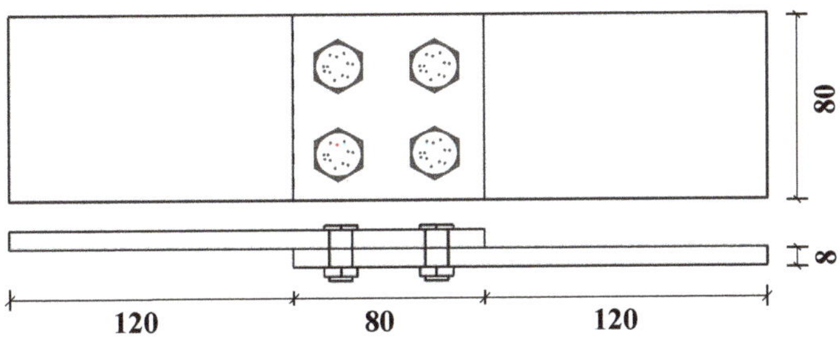

Figure 5. Bolted lap structure (mm).

Table 2. The material parameters of components.

Material	ρ/(kg/m^3)	E/(Pa)	G/(Pa)
Aluminum alloy	2750	6.9×10^{10}	2.69×10^{10}
Low-carbon steel	7900	2.1×10^{11}	8.08×10^{10}

The test sample bolted structure was made by machining operation. A torque wrench was used to control the pre-tightening torque of a single bolt in the group. In this way, the modal test is carried out by hammering under different pre-tightening torque as shown in Figure 6.

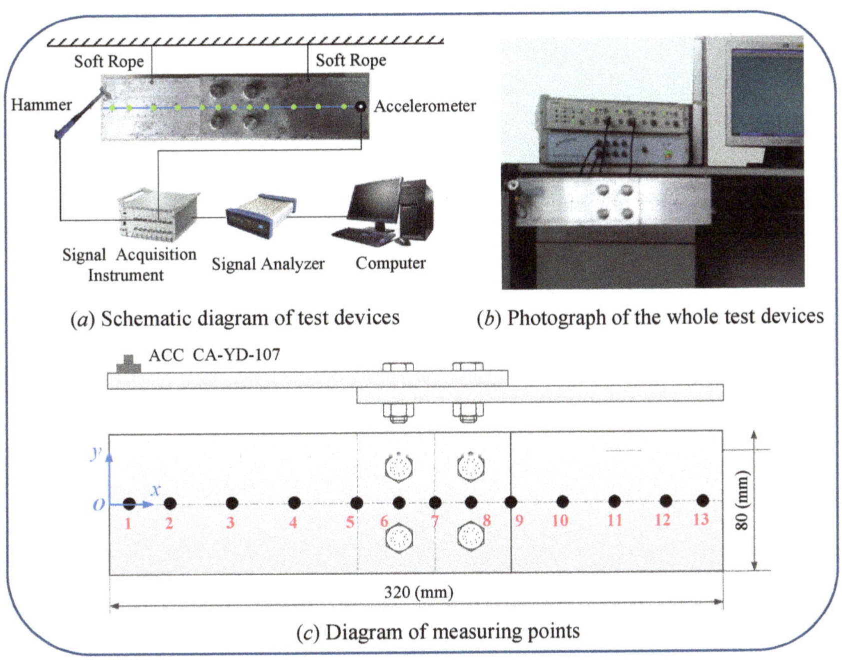

Figure 6. Modal test of the connection structure.

The piezoelectric accelerometers used in this modal test, had a mass of 20 g, charge sensitivity of 6.05 pC/ms^2, and frequency range from 0.5 Hz to 5 kHz. The CL–YD–303 hammer had a reference sensitivity of 3.99 pC/N. CRAS V7.0, a vibration and dynamic signal acquisition and analysis system developed by Nanjing AnZheng Software Engineering Company, was used as the analytical instrument, and MaCras was used as the modal analysis software. To simulate free boundary conditions, a spring rope suspension was applied to counteract gravity; this method is effective in avoiding imported errors which affect the dynamic characteristics of the structure. The suspension plane is orthogonal to the test direction to avoid the effect of suspension conditions on the test results. The accelerometer is arranged at the end of the structure to avoid the vibration type node. Thirteen test points are arranged along the length direction of the lapping structure. The sampling frequency is set to 5000 Hz. The arrangement of measuring points is shown in Table 3.

Table 3. Distance of measuring points from origin (point *O*).

Measuring Point	1	2	3	4	5	6	7	8	9	10	11	12	13
Position (mm)	1	3	6	9	12	14	16	18	20	23	26	29	31

The pre-tightening torque T_N = 2, 4, 6, 8, 10, 12, 14, 16, 18, 20, 22, 25, 28 N·m are applied to the bolt through a torque wrench. The first four order modal frequencies under different pre-tightening forces are obtained through multiple modal tests, which are all shown in Table 4. The experimental modal data are plotted and the relationship between the test frequency and the pre-tightening torque is obtained, as shown in Figure 7.

Table 4. Experimental modal frequency under different pre-tightening torque.

Pre-Tightening Torque (N·m)	Modal Frequency (Hz)			
	Mode 1	Mode 2	Mode 3	Mode 4
2	375.04	962.47	2032.03	2844.67
4	387.64	975.00	2068.01	2910.88
6	387.50	974.89	2118.68	2952.14
8	399.89	986.96	2151.37	2969.46
10	412.56	1000.19	2216.35	3052.07
12	424.63	999.94	2222.10	3105.54
14	424.99	1012.38	2253.68	3118.02
16	425.00	1012.59	2276.85	3128.10
18	424.96	1012.53	2284.80	3137.43
20	424.91	1012.50	2277.04	3153.12
22	424.80	1012.51	2303.74	3147.28

As can be seen from Figure 7, with the increase of pre-tightening torque the first four order modal frequencies also increase at the same time. The modal frequencies of the first, the third and the fourth order gradually increase with the increase of the pre-tightening torque, while the second modal frequencies tend to be stable when the pre-tightening torque comes to 14 N·m.

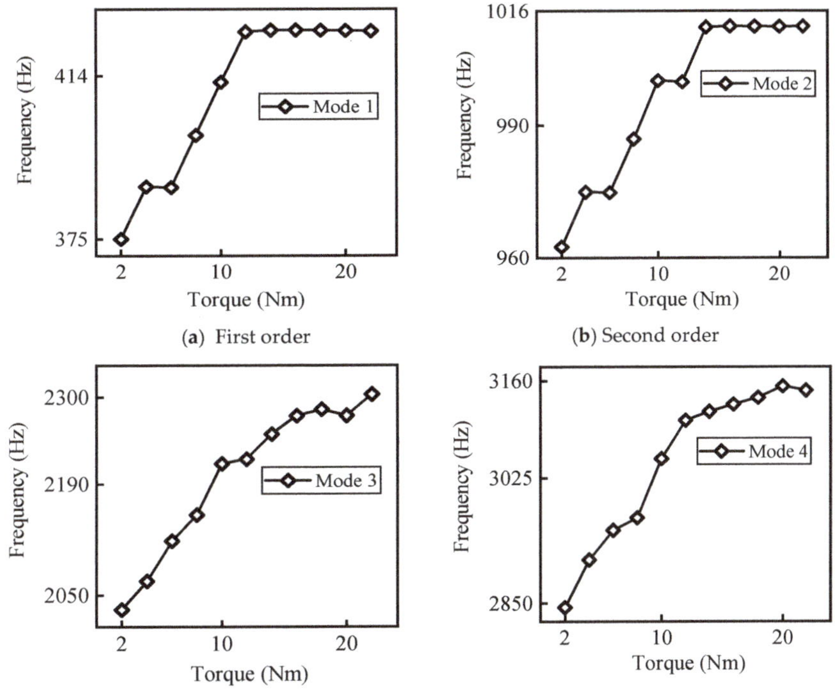

Figure 7. The relationship between the experimental frequency and the pre-tightening torque.

The damping ratio of the bolted connection structure is related to many factors, such as material performance, medium condition, surface roughness, processing method, pre-tightening force state and loading frequency, the influence of which is mostly reflected in nonlinear relations. In addition, the measurement errors will also have an impact on the damping ratio. When measuring the frequency response functions, the measured result of the damping ratio is larger than the actual result due to the effect of accelerometer weighting functions. The first four order damping ratios under different pre-tightening forces were obtained, and are all shown in Table 5. Furthermore, the relationship between the experimental damping ratio and the pre-tightening torque was obtained as is shown in Figure 8.

Table 5. Experimental damping ratio under different pre-tightening torque.

Pre-Tightening Torque (N·m)	Damping Ratio (%)			
	Mode 1	**Mode 2**	**Mode 3**	**Mode 4**
2	0.44	0.07	0.99	0.72
4	0.49	0.03	0.80	0.98
6	0.20	0.11	1.46	0.53
8	0.01	0.50	1.78	0.48
10	0.48	0.24	1.23	0.41
12	1.80	0.14	0.84	1.11
14	0.09	0.78	1.32	0.54
16	0.06	0.18	1.77	0.44
18	0.18	0.12	1.77	0.42
20	0.17	0.04	1.48	0.93
22	0.93	0.11	1.05	0.37

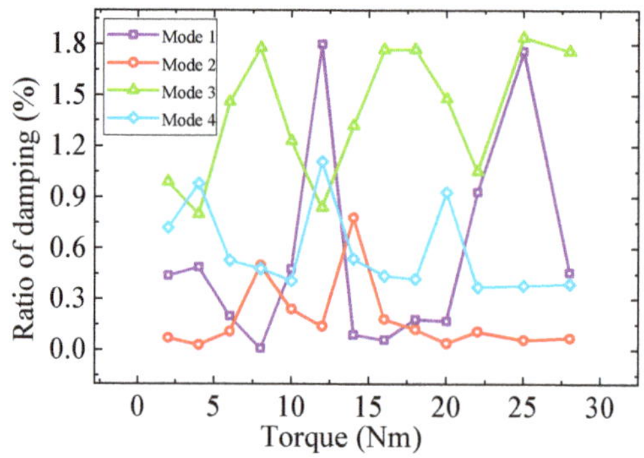

Figure 8. Experimental damping ratio depends on the pre-tightening torque.

As shown in Figure 8, the relationship between the damping ratio of the connection structure and the pre-tightening torque is generally non-linear, and there is no rule to be found. Meanwhile, the relationship between the first four order experimental modal shapes and the pre-tightening torque is shown in Figure 9.

As can be seen from Figure 9, the modal shapes are generally smooth. On the one hand, the connection stiffness is enough to provide good linearity to the structure; on the other hand, it is impossible to measure the detailed characteristics of the modal shapes in the connecting part due to the limited test precision. At the same time, it can be found that the mechanical properties of the bolt connection tend to be stable with the increase of the pre-tightening force.

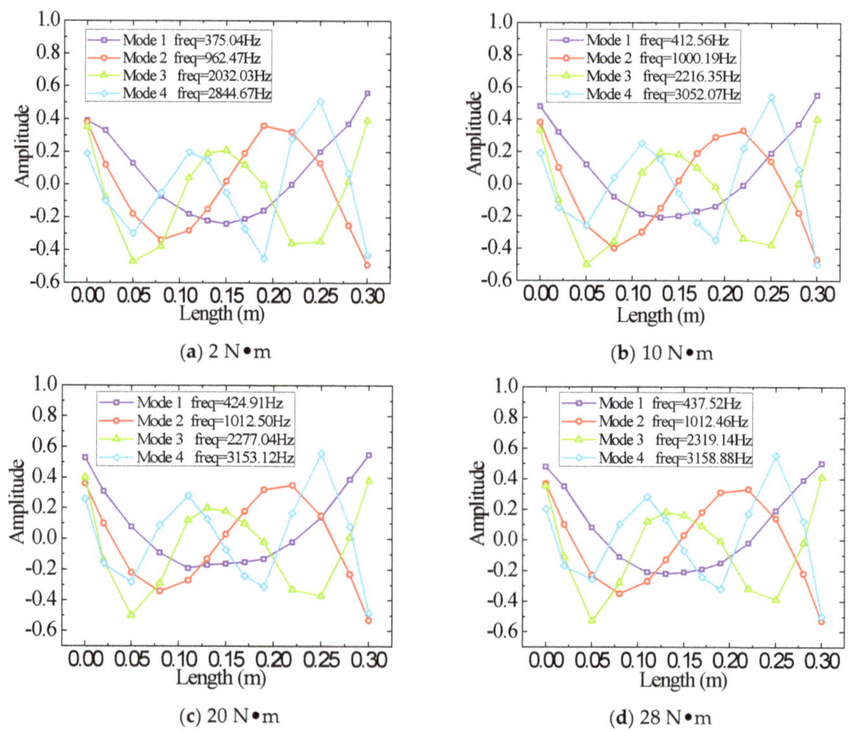

Figure 9. The first four order experimental modal shapes under different pre-tightening torque.

3.2. Modelling and Parameter Identification

When thin layer elements are used to model bolted structures, the role of bolts on the structure is replaced by thin layer elements established on the contact surface. A schematic diagram of the method is shown in Figure 10.

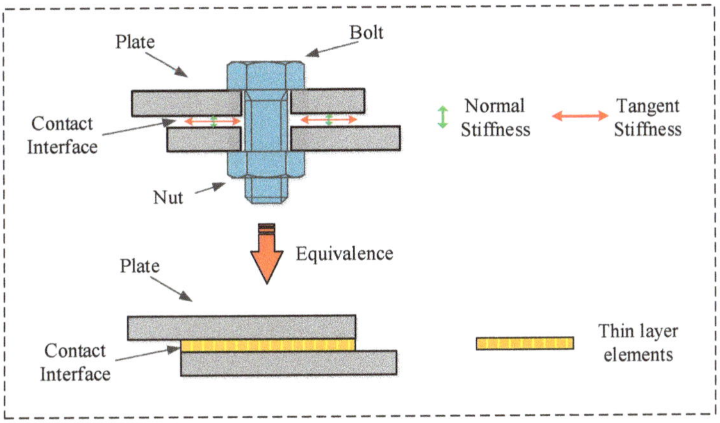

Figure 10. Parameterization of contact interface.

However, the pressure distribution on the bolt joint surface is not uniform. The pressure gradually decreases along the radial direction, as shown in Figure 11. The pressure distribution of the joint surface is shown in Equations (19)–(21).

$$P_{\max} = \frac{3F_b}{\pi(r_m^2 + r_m r_i + r_i^2)} \tag{19}$$

$$P(r) = -\frac{P_{\max}}{r_m - r_i}(r - r_m) \tag{20}$$

$$r_m = r_b + h \cdot \tan\alpha \tag{21}$$

where F_b is the bolt pre-tightening force, r_i is the radius of the bolt hole, r_m is the maximum contact radius, r_b is the radius of the bolt head, and h is the thickness of the connected piece.

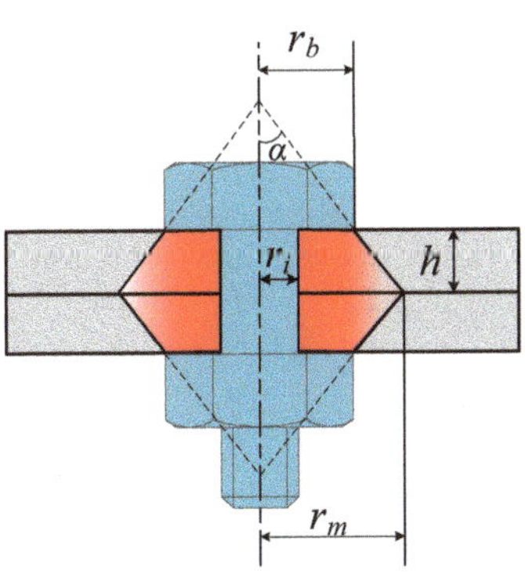

Figure 11. Schematic diagram of pressure distribution in bolted joint.

We assuming that the normal dynamic stiffness K_n and tangential dynamic stiffness K_τ of the joint surface per unit area under uniform pressure are shown in Equations (22) and (23)

$$K_n = \alpha_n \cdot P(r)^{\beta_n} \tag{22}$$

$$K_\tau = \alpha_\tau \cdot P(r)^{\beta_\tau} \tag{23}$$

where $P(r)$ is the pressure of the joint surface, α_n and β_n are normal characteristic parameters of the joint, α_τ and β_τ are tangential characteristic parameters of the joint.

The relationship between the E and G of the gradient connecting layer and K is shown in Equations (24) and (25)

$$E(r) = \frac{K_n h}{A} = \alpha_n P(r)^{\beta_n} h \tag{24}$$

$$G(r) = \frac{K_\tau h}{A} = \alpha_\tau P(r)^{\beta_\tau} h \tag{25}$$

where A is the area of the connecting layer and h is the thickness of the connecting layer. The influence area of the bolt connection is divided into three areas, as shown in Figure 12.

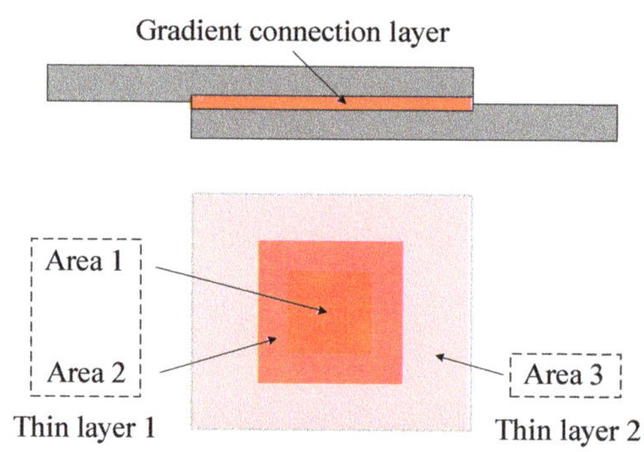

Figure 12. The influence area of the bolt connection, which is divided into three areas.

In order to facilitate calculation and ignore the influence of the screw hole, the gradient connection layer is divided into two areas and changed into a square in this paper. The contact pressure of each area is the average pressure of each layer.

The FEM of the connection structure as shown in Figure 13 is established by ignoring the influence of the bolt weight and screw hole. The lap plate is simulated with a solid element, and the interface is simulated with an isotropic thin-layer element. The contact stiffness close to the bolt is higher than that away from the bolt, and two different isotropic constitutive relations are used to simulate the contact surface and then used to identify the parameters. The parameters to be identified near the bolt area (the red area in Figure 13) are elastic modulus E_1 and shear modulus G_1. For another area in the contact area, the parameters to be recognized away from the bolt are the elastic modulus E_2 and shear modulus G_2. The ratio coefficient of the thin-layer element is Ratio = 10, and the density of the material is 0 t/mm^3.

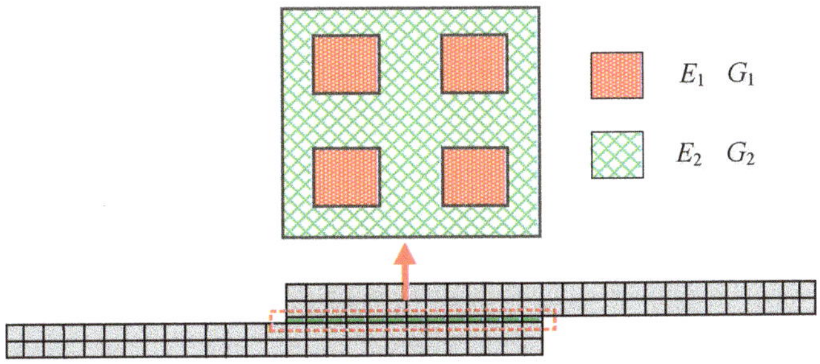

Figure 13. FEM of the connection structure.

It is important to note the setting of the initial value of the identification parameter. The parameters (E_1, G_1) of the thin layer element are of the same order of magnitude as the material parameters of the bolt where close to the bolt. However, the parameter selection (E_2, G_2) is an order of magnitude lower than the material parameters where away from the bolt. The initial value as well as the identification results are listed in Table 6. The iterative convergence curves of the identification parameters are as shown in Figure 14, when the pre-tightening torque is T_N = 20 N·m.

Table 6. Initial value of the identified parameters.

Identified Parameters (N/m^2)	E_1	G_1	E_2	G_2
Initial value	3.00×10^9	1.15×10^9	3.00×10^6	1.15×10^6
Identified value	8.63×10^8	3.42×10^8	9.33×10^5	4.46×10^5

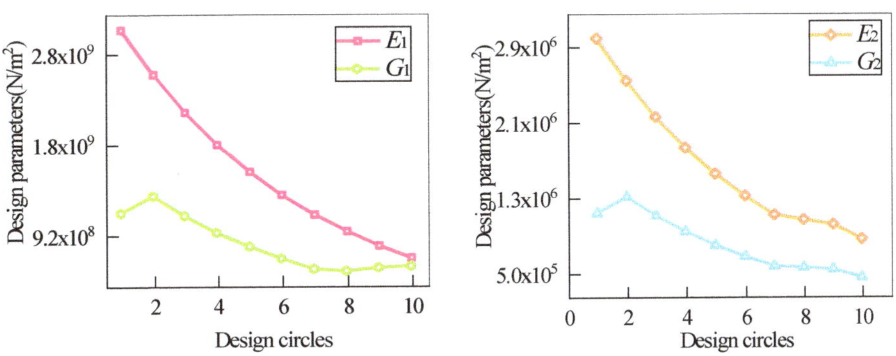

(**a**) Iteration curve of E_1, G_1 (**b**) Iteration curve of E_2, G_2

Figure 14. Iterative convergence curves of parameter selection.

Table 7 lists the modal frequency parameters before and after identification as well as the error between the above parameters and the experimental frequency. It can be seen from the identification results that the maximum updated error is no more than 2.5%. The results have comparatively higher identification accuracy.

Table 7. Comparison of modal frequency errors.

Order	Experiment (Hz)	Initial (Hz)	Error %	Updated (Hz)	Error %
1	424.91	437.55	2.97	427.01	0.49
2	1012.50	1044.80	3.19	1037.05	2.42
3	2277.04	2314.00	1.62	2261.19	−0.69
4	3153.12	3209.10	1.78	3136.38	−0.53

The elastic modulus E is greater than the shear modulus G, which reflects the characteristics of normal contact stiffness more than the tangential stiffness. The identification results (E_1, G_1) near the bolt area are far greater than that of the bolt area (E_2, G_2), which better reflects the actual situation. Therefore, the connection performance of the overlapping structure with multiple bolts can be well simulated by the of isotropic thin layer element.

Similarly, the material parameters of the thin layer element under different pre-tightening torque are obtained by the iterative solution method. Finally, the relationship between the identification parameters and the different pre-tightening torque is achieved, as shown in Figure 15.

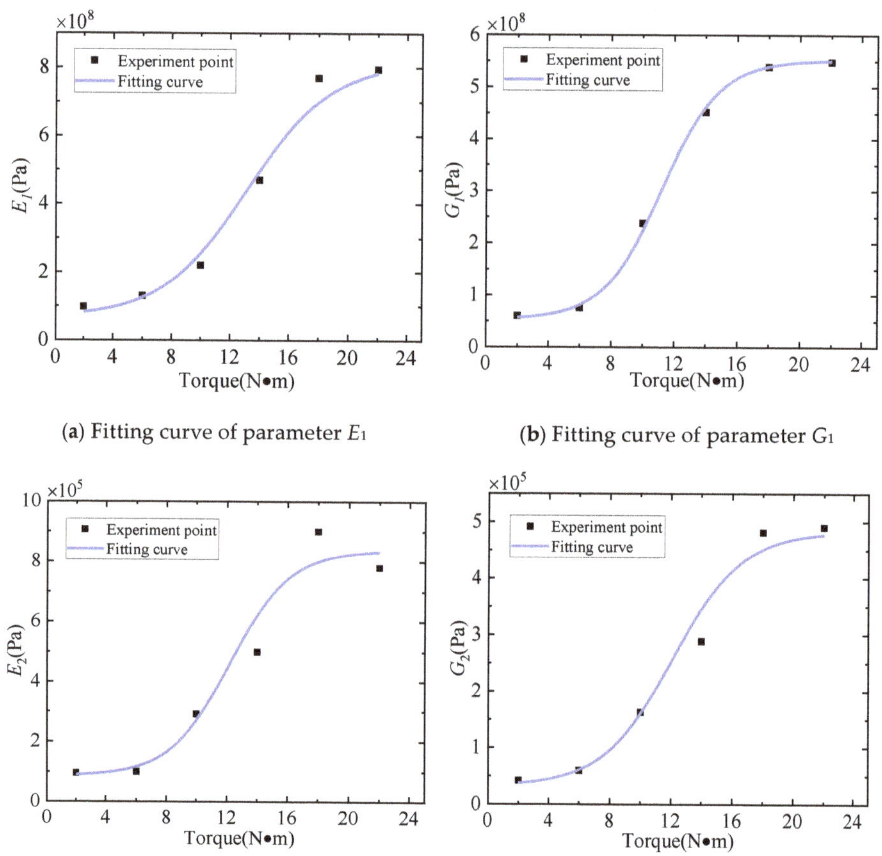

(**a**) Fitting curve of parameter E_1 (**b**) Fitting curve of parameter G_1

(**c**) Fitting curve of parameter E_2 (**d**) Fitting curve of parameter G_2

Figure 15. The relation curves of identification parameters change with pre-tightening torque.

In order to verify the correctness of the method, we selected five points evenly in the interval of the fitting curve, obtained E_1, G_1, E_2 and G_2 through the parameter identification method, and put these values into the curve for verification. In Figure 16, the red dots are the verification points. It can be seen that the error between the red dots and the fitting curve is small, which proves the effectiveness of the method.

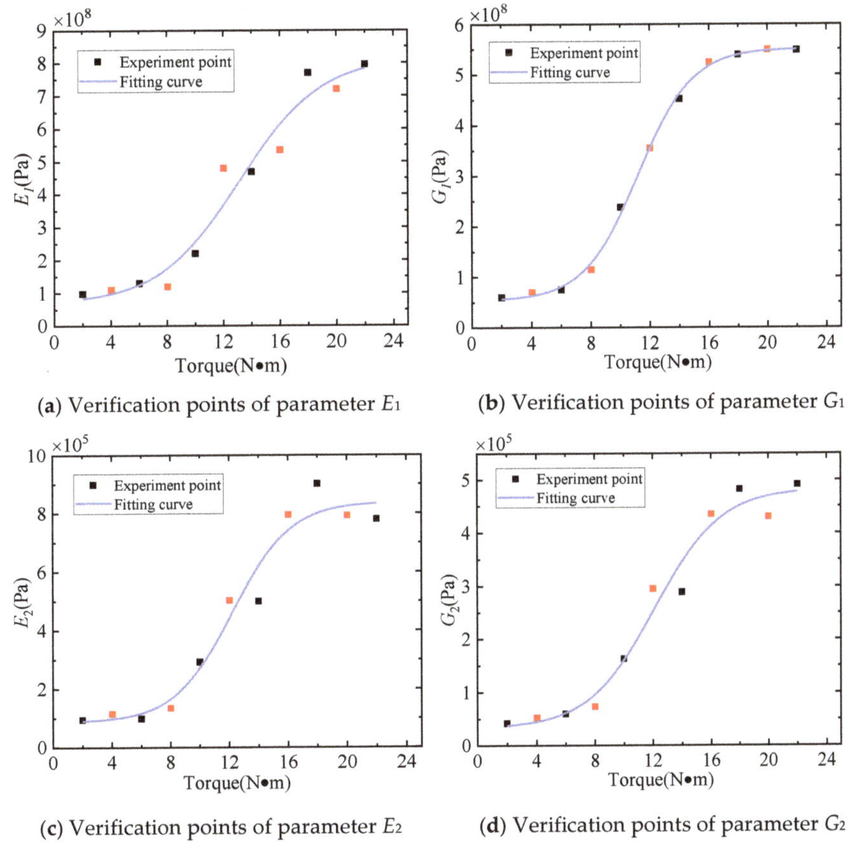

Figure 16. The relation curves of identification parameters change with pre-tightening torque and verification points.

The curve obtained by this method is of high precision when a certain number of points and intervals are reasonably selected for fitting. The more points used, the more accurate the curve.

Through curve fitting, the relationship between the elastic modulus E_1 and E_2 and the shear modulus G_1 and G_2 under different pre-tightening torque was obtained. The resulting curves can guide the modeling of the same type of bolted joints. The specific operation is as follows: (*i*) The torque wrench can be used to determine the pre-tightening torque of the bolt. (*ii*) The parameter values of the thin layer element material E_1, G_1, E_2 and G_2 are drawn from the relation curves. Thus, an accurate dynamic model of the bolted joint structure could be established.

4. Conclusions

On the basis of the thin layer element with isotropic constitutive relation, this paper presents a method to simulate the stiffness of the joint surface of the bolt connection considering the variability of pre-tightening torque. The proposed method in this paper is verified by the pre-tightening test, and the following conclusions can be drawn:

(1) Through curve fitting after the parameter identification, the relationship between the pre-tightening torque and the identification parameters can be obtained, which can guide accurate modeling of the same type of bolt connection.
(2) The thin element method with linear constitutive relation makes modeling and parameter identification very efficient. It can also accurately reflect the contact performance of the joint structure of the bolt lap joint. The identification results for the contact surface material near the bolt area are far greater than for the distance away from the bolt area, which better reflects the inhomogeneity of the contact interface stiffness distribution.
(3) This method can accurately describe the mechanical properties of the contact surface of the bolt lap structure and guide numerical analysis of the bolt connection structure, which has great engineering significance.

Author Contributions: Conceptualization, D.J.; Funding acquisition, D.J.; Methodology, H.Q. and Z.C.; Resources, H.Q.; Software, Y.T.; Validation, Y.T. and Z.C.; Writing—original draft, Y.T.; Writing—review and editing, H.Q., Z.C., D.Z. and D.J. All authors have read and agreed to the published version of the manuscript.

Funding: This research was funded by National Natural Science Foundation of China, grant number No. 11602112.This research was funded by Natural Science Research Project of Higher Education in Jiangsu Province, grant number 20KJB460003. This research was funded by Qing Lan Project.

Institutional Review Board Statement: Not applicable.

Informed Consent Statement: Not applicable.

Data Availability Statement: Not applicable.

Acknowledgments: The authors are grateful for the support from the National Natural Science Foundation of China (No. 11602112), The Natural Science Foundation of the Jiangsu Higher Education Institutions of China (20KJB460003) and the Qing Lan Project of China.

Conflicts of Interest: The authors declare that there are no conflicts of interest regarding the publication of this paper.

References

1. Farhad, M.; Majid, M.; Aref, A. Nonlinear behavior of single bolted flange joints: A novel analytical model. *Eng. Struct.* **2018**, *173*, 908–917. [CrossRef]
2. Bograd, S.; Reuss, P.; Schmidt, A.; Gaul, L.; Mayer, M. Modeling the dynamics of mechanical joints. *Mech. Syst. Signal Process.* **2011**, *25*, 2801–2826. [CrossRef]
3. Ibrahim, R.A.; Pettit, C.L. Uncertainties and dynamic problems of bolted joints and other fasteners. *J. Sound Vib.* **2005**, *279*, 857–936. [CrossRef]
4. Luan, Y.; Guan, Z.Q.; Cheng, G.D.; Liu, S. A simplified nonlinear dynamic model for the analysis of pipe structures with bolted flange joints. *J. Sound Vib.* **2012**, *331*, 325–344. [CrossRef]
5. Wang, Z.C.; Xin, Y.; Ren, W.X. Nonlinear structural joint model updating based on instantaneous characteristics of dynamic responses. *Mech. Syst. Signal Process.* **2016**, *76–77*, 476–496. [CrossRef]
6. Li, Z.; Soga, K.; Wang, F.; Wright, P.; Tsuno, K. Behaviour of cast-iron tunnel segmental joint from the 3D FE analyses and development of a new bolt-spring model. *Tunn. Undergr. Space Technol.* **2014**, *41*, 176–192. [CrossRef]
7. Vilela, P.M.L.; Carvalho, H.; Grilo, L.F.; Montenegro, P.A.; Calçada, R.B. Unitary model for the analysis of bolted connections using the finite element method. *Eng. Fail. Anal.* **2019**, *104*, 308–320. [CrossRef]
8. Fukuoka, T. Finite Element Analysis of the Mechanical Behavior of Multibolted Joints Subjected to Shear Loads. *J. Press. Vessel. Technol.* **2018**, *140*, 051201. [CrossRef]
9. Yan, A.M.; Wang, X.F.; Yang, H.; Huang, F.L.; Pi, A.G. Research on the Shock Response Characteristics of a Threaded Connection Using the Thin-Layer Element Method. *Appl. Sci.* **2021**, *11*, 5611. [CrossRef]
10. Ahmadian, H.; Jalali, H. Generic element formulation for modelling bolted lap joints. *Mech. Syst. Signal Process.* **2007**, *21*, 2318–2334. [CrossRef]
11. Yao, X.Y.; Wang, J.J.; Zhai, X. Research and application of improved thin-layer element method of aero-engine bolted joints. *Proc. Inst. Mech. Eng. Part G J. Aerosp. Eng.* **2017**, *231*, 823–839. [CrossRef]

12. Cooper, C.; Turvey, G. Effects of joint geometry and bolt torque on the structural performance of single bolt tension joints in pultruded GRP sheet material. *Compos. Struct.* **1995**, *32*, 217–226. [CrossRef]
13. Zhao, Y.S.; Yang, C.; Cai, L.G.; Shi, W.M.; Hong, Y. Stiffness and Damping Model of Bolted Joints with Uneven Surface Contact Pressure Distribution. *Stroj. Vestn. J. Mech. Eng.* **2016**, *62*, 665–677. [CrossRef]
14. Zhao, G.; Xiong, Z.; Xin, J.; Hou, L.; Gao, W. Prediction of contact stiffness in bolted interface with natural frequency experiment and FE analysis. *Tribol. Int.* **2018**, *127*, 157–164. [CrossRef]
15. Yang, K.T.; Park, Y.S. Joint structural parameter identification using a subset of frequency response function measurements. *Mech. Syst. Signal Process.* **1993**, *7*, 509–530. [CrossRef]
16. Jiang, D.; Wu, S.Q.; Shi, Q.F.; Fei, Q.G. Contact interface parameter identification of bolted joint structure with uncertainty using thin layer element method. *Eng. Mech.* **2015**, *32*, 220–227. [CrossRef]
17. Ren, Y.; Beards, C. Identification of joint properties of a structure using FRF data. *J. Sound Vib.* **1995**, *186*, 567–587. [CrossRef]
18. Tsai, J.S.; Chou, Y.F. Modeling of the dynamic characteristics of two-bolt-joints. *J. Chin. Inst. Eng.* **1988**, *11*, 235–245. [CrossRef]
19. Yang, T.; Fan, S.; Lin, C.S. Joint stiffness identification using FRF measurements. *Comput. Struct.* **2003**, *81*, 2549–2556. [CrossRef]
20. Xu, Y.; Cao, Z.; Jiang, D. Boundary condition identification of a clamped honeycomb sandwich panel based on thin-layer element. *J. Vibroeng.* **2019**, *21*, 2286–2295. [CrossRef]
21. Goodman, R.E. A model for the mechanics of jointed rock. *J. Soil Mech. Found. Div.* **1968**, *94*, 637–659. [CrossRef]
22. Sharma, K.G.; Desai, C.S. Analysis and Implementation of Thin-Layer Element for Interfaces and Joints. *J. Eng. Mech.* **1992**, *118*, 2442–2462. [CrossRef]
23. Givoli, D. Finite element modeling of thin layers. *CMES Comput. Modeling Eng. Sci.* **2004**, *5*, 497–514. [CrossRef]
24. Desai, C.; Zaman, M.; Lightner, J.; Siriwardane, H. Thin-layer element for interfaces and joints. *Int. J. Numer. Anal. Methods Geomech.* **1984**, *8*, 19–43. [CrossRef]
25. Schmidt, A.; Bograd, S.; Gaul, L. Measurement of join patch properties and their integration into finite-element calculations of assembled structures. *Shock. Vib.* **2012**, *19*, 1125–1133. [CrossRef]
26. Jiang, D.; Xu, Y.; Zhu, D.; Cao, Z. Temperature-dependent thermo-elastic parameter identification for composites using thermal modal data. *Adv. Mech. Eng.* **2019**, *11*, 1687814019884165. [CrossRef]
27. Wu, Y.; Zhu, R.; Cao, Z.; Liu, Y.; Jiang, D. Model Updating Using Frequency Response Functions Based on Sherman–Morrison Formula. *Appl. Sci.* **2020**, *10*, 4985. [CrossRef]
28. Jiang, D.; Zhang, D.; Fei, Q.; Wu, S. An approach on identification of equivalent properties of honeycomb core using experimental modal data. *Finite Elem. Anal. Des.* **2014**, *90*, 84–92. [CrossRef]
29. Adel, F.; Shokrollahi, S.; Jamal-Omidi, M.; Ahmadian, H. A model updating method for hybrid composite/aluminum bolted joints using modal test data. *J. Sound Vib.* **2017**, *396*, 172–185. [CrossRef]

applied sciences

MDPI

Article

Mesh-Free Surrogate Models for Structural Mechanic FEM Simulation: A Comparative Study of Approaches

Johannes G. Hoffer [1,*], Bernhard C. Geiger [1], Patrick Ofner [1] and Roman Kern [2,*]

[1] Know-Center GmbH, Research Center for Data-Driven Business & Big Data Analytics, Inffeldgasse 13, 8010 Graz, Austria; bgeiger@know-center.at (B.C.G.); patrick@ofner.science (P.O.)
[2] Institute of Interactive Systems and Data Science, Graz University of Technology, Inffeldgasse 16c, 8010 Graz, Austria
* Correspondence: jhoffer@know-center.at (J.G.H.); rkern@tugraz.at (R.K.)

Abstract: The technical world of today fundamentally relies on structural analysis in the form of design and structural mechanic simulations. A traditional and robust simulation method is the physics-based finite element method (FEM) simulation. FEM simulations in structural mechanics are known to be very accurate; however, the higher the desired resolution, the more computational effort is required. Surrogate modeling provides a robust approach to address this drawback. Nonetheless, finding the right surrogate model and its hyperparameters for a specific use case is not a straightforward process. In this paper, we discuss and compare several classes of mesh-free surrogate models based on traditional and thriving machine learning (ML) and deep learning (DL) methods. We show that relatively simple algorithms (such as k-nearest neighbor regression) can be competitive in applications with low geometrical complexity and extrapolation requirements. With respect to tasks exhibiting higher geometric complexity, our results show that recent DL methods at the forefront of literature (such as physics-informed neural networks) are complicated to train and to parameterize and thus, require further research before they can be put to practical use. In contrast, we show that already well-researched DL methods, such as the multi-layer perceptron, are superior with respect to interpolation use cases and can be easily trained with available tools. With our work, we thus present a basis for the selection and practical implementation of surrogate models.

Keywords: FEM; surrogate modeling; mesh-free; machine learning; deep learning

Citation: Hoffer, J.G.; Geiger, B.C.; Ofner, P.; Kern, R. Mesh-Free Surrogate Models for Structural Mechanic FEM Simulation: A Comparative Study of Approaches. *Appl. Sci.* **2021**, *11*, 9411. https://doi.org/10.3390/app11209411

Academic Editors: Jin-Gyun Kim, Jae-Hyuk Lim and Peter Persson

Received: 15 September 2021
Accepted: 3 October 2021
Published: 11 October 2021

1. Introduction

Assessing the properties of mechanical structures with real physical experiments is expensive, as it costs both time and resources. To reduce these costs of knowledge enrichment in the field of structural analysis, computer simulations of structural mechanics have become crucial. An essential simulation method is the finite element method (FEM) in which the simulation domain space is represented by a finite number of connected elements. Space- and time-dependent behavior between connected elements and within the elements themselves is governed by physical equations. Observation of real physical experiments provides the coefficients for these governing equations. Since most geometries and use cases cannot be solved analytically, an approximation of the proposed physical equations is obtained by numerical methods [1]. However, solving complex problems with FEM is time-consuming and computationally expensive. In order to reduce the computational effort, surrogate modeling offers a promising solution [2].

Surrogate models are trained in a supervised manner and are designed to learn the function mapping between inputs and outputs from a given FEM simulation use case. With a sufficient amount of training data with respect to the use case, an according model is able to substitute for the FEM simulation use case up to a certain accuracy.

There is already a considerable number of related work concerning surrogate modeling of structural mechanics simulations with machine learning (ML) or deep learning (DL)

approaches. In the following, we want to present the most important works for this paper. Artificial neural networks (ANN) are used in the work of Roberts et al. [3] to predict damage development in forged brake discs reinforced with Al-SiC particles, using damage maps. The ANN is a multilayer perceptron (MLP), and training data are obtained from FEM simulations using the commercial DEFORM simulation software. For rapid estimation of forming and cutting forces for given process parameters, Hans Raj et al. [4] investigate a method using MLP models. The researchers focus on two processes: hot upsetting and extrusion. Each process, represented by a MLP, is trained with FEM simulation results from the FORGE2 commercial FEM simulation software. García-Crespo et al. [5] predict the projectile response after impact with steel armor using a MLP; their surrogate model studied is trained with data from FEM simulations of the use case. Nourbakhsh et al. [6] explore generalizable surrogate models for 3D trusses, using MLP and FEM training data. Chan et al. [7] estimate the performance of hot-forged product designs, using a MLP trained on FEM results obtained with the commercial software DEFORM. D'Addona and Antonelli [8] use single-layer feedforward ANNs instead of FEM as a metamodel in a sequential approximate optimization (SAO) algorithm. In a case study on hot forging of a steel disk, they compare their results with an ANN trained on FEM simulation results and the FEM simulation software QForm3D. Gudur and Dixit [9] predict the velocity field and location of neutral point of cold flat rolling with a MLP trained with rigid-plastic FEM simulation results. Pellicer-Valero et al. [10] predict the mechanical behavior of different livers with MLPs trained from FEM simulations.

Abueidda et al. [11] estimate the mechanical properties of a two-dimensional checkerboard composite using a convolutional neural network (CNN) trained with FEM results. Regarding mesh-based approaches, Pfaff et al. [12] present a framework to train graph neural networks (GNN) on mesh-based simulations and show the applicability in aerodynamics, structural mechanics, and fabric.

Surrogate models were also obtained using classical, i.e., non-neural ML, approaches. For example, the authors of [3] apply Gaussian process regression (GPR) besides ANN in their approach. Loghin and Ismonov [13] predict the stress intensity factors, using GPR trained with FEM results of a classical bolt-nut assembly. Ming et al. [14] model the electrical discharge machining process with GPR trained from data generated with numerical FEM simulation.

Using support vector regression (SVR), Pan et al. [15] construct a metamodel in an optimization approach for lightweight vehicle design. Training data are generated, using design of experiment approaches with FEM simulations. To predict the stress at the implant–bone interface, Li et al. [16] utilize SVR in order to replace FEM simulation. Hu and Li [17] estimate cutting coefficients in a mechanistic milling force model with SVR trained with FEM simulation data.

Employing tree-based models, Martínez-Martínez et al. [18] estimate the biomechanical behavior of breast tissue under compression, using three different tree-based models trained from FEM simulations. The models are trained with FEM data in terms of nodal coordinates and nodal tissue membership. Zhang et al. [19] estimate the base failure stability for braced excavations in anisotropic clay using extreme gradient boosting, random forest regression (RFR) and data obtained from FEM simulation results. Qi et al. [20] utilize a decision tree regressor to predict the mechanical properties of carbon fiber reinforced plastics with data obtained from FEM simulations. Besides MLPs Pellicer-Valero et al. [10] utilize RFRs to predict the biomechanics of livers.

A recent neural network–based approach are physics informed neural networks (PINNs). PINNs are trained simultaneously on data and governing differential equations and can be used for the solution and inversion of equations governing physical systems. Utilizing PINNs, Haghighat and Juanes [21] substitute a particular FEM simulation of a perforated strip under uniaxial extension. In [22], Haghighat et al. present a surrogate modeling approach with PINNs and a specific use case. Focusing on consistency, Shin [23] evaluates findings regarding PINNs with Poisson's equation and the heat equation. Yin et al. [24] use PINNs to predict permeability and viscoelastic modulus from thrombus deformation data, described by the fourth-order Cahn–Hilliard and Navier–Stokes equations. In addition to the application of PINNs in structural mechanics problems, there is also a considerable number of papers, especially in computational fluid dynamics [25–29].

Related work shows capabilities of surrogate modeling, thus demonstrating the feasibility of supervised learning models trained with FEM simulations. From our analysis of the existing literature, we identify the following drawbacks:

- In most cases, the surrogate model only substitutes for a subset of the considered computational domain. Thus, such an approach focuses only on a region of interest and cannot be used to evaluate the entire computational domain (notable exceptions are [12,22]).
- Surrogate models representing the complete discretized computational domain (mesh) are solely fitted and evaluated on one use case—generalization to unseen data is only achieved with respect to the discretization of the computational domain, but not with respect to other use case specific parameters (notable exception concerning material parameters [22]).
- Due to differences in FEM use cases and data, the comparison of related work is useful only in some cases.
- Replication of published experiments is often not achievable because important parameters are not reported, e.g., number of finite elements, type of finite elements (bilinear, biquadratic, reduced integration etc.), method of discretization (meshing), as well as hyperparameters of the ML models, such as learning or activation functions.

To address these drawbacks, we present the following contributions of our paper:

1. We present the main DL and ML methods together with a compact description and mathematical notation to equip practitioners with a reference to surrogate FEM simulation mesh-free and assess the feasibility and maturity of the novel PINNs method.
2. We utilize three classic use cases in structural mechanics and evaluate these models in terms of performance on unseen configurations (inter- and extrapolation) in order to assess their ability to generalize across different use case specific parameters.
3. We discuss the characteristics of all DL and ML models, and their practical implications, in the context of the use cases.

With our work, we pave the way of mesh-free surrogate modeling for practical use: we provide a basis for efficient model and hyperparameters selection regarding use case and performance metrics. These insights shall not only assist the domain expert during model selection, but will also help in consolidating the current research in mesh-free surrogate modeling for structural mechanics applications.

We report all information to make our experiments reproducible. If certain model settings are not mentioned, they are left at default values. Moreover, our FEM simulations are performed with Abaqus Student Edition 2019 (Dassault Systèmes, Velizy-Villacoublay, France), and thus, the process of data generation is not limited to commercial software, which makes it possible for everyone to connect to our research.

The remainder of this paper is organized as follows. In Section 2, we present the materials and methods of our experiments, first providing insights into the process of data generation, using the FEM simulations in Section 2.1, then describing the datasets obtained from the FEM simulations in Section 2.2, followed by the ML and DL models used in Section 2.3. Section 3 shows the results, which are discussed in Section 4. In Section 5, we present the conclusion of our work and an outlook for the future.

2. Materials and Methods

In this section, we present all relevant information about the methodology of our experiments. First, Section 2.1 provides an overview of the data generation process, using three classic FEM simulation use cases. Then, Section 2.2 describes the datasets used from the FEM simulations, and Section 2.3 presents the ML and DL models used. A more detailed overview of the mathematical background and assumptions of the ML and DL models can be found in the Appendix. When predicting a particular use case with a surrogate model, the individual nodes discretizing the particular geometry of the use case (i.e., mesh) are sequentially input into the surrogate model with the appropriate generalization variable. The surrogate model then predicts the output of each node in sequence; see Figure 1.

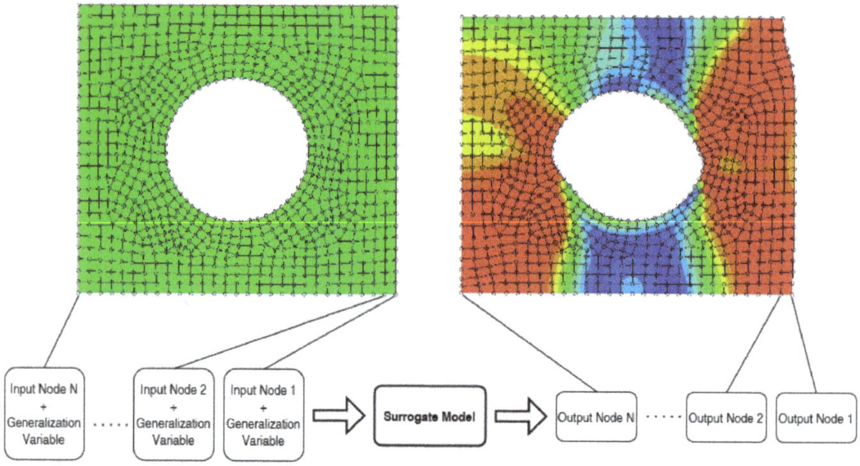

Figure 1. Principle of our surrogate model approach: all N nodes (i.e., their coordinates), together with the respective generalization variable, are sequentially entered into a surrogate model, which then sequentially predicts the outcome of the respective coordinates (i.e., the displacements, strains, and stresses of the respective node).

It should be noted that there are no constraints on the discretization (mesh), i.e., the node coordinates can be freely chosen within the simulation domain and nodes are not connected to each other. Therefore, we refer to our approach as mesh-free, but we want to clearly distinguish ourselves from other mesh-free methods, such as smoothed particle hydrodynamics, the diffuse element method, the moving particle finite element method, etc. The predictions of the individual nodes together constitute the prediction for the simulation domain of the particular use case. By adding the nodal displacement outputs of the surrogate model to the initial node coordinates, we obtain the new deformed geometry. Further surrogate model outputs (e.g., stresses, strains) describe the queried nodes and thus the complete simulation domain in more detail.

2.1. FEM Use Cases

For illustration, we base our evaluation on three classic use cases from structural mechanics. We consider the (1) tensile load, (2) bending load and (3) compressive load:

1. Elongation of a plate with a perforation;
2. Bending of a beam;
3. Compression of a block with four perforations.

See Table 1 and Figure 2. We utilize an isotropic elasto-plastic rate-independent material model (i.e., a perfectly plastic material). The kinematic relations for our 2D plane strain use cases are defined by the total strain components $\varepsilon_{xx} = \frac{\partial u_x}{\partial x}$, $\varepsilon_{yy} = \frac{\partial u_y}{\partial y}$, $\varepsilon_{xy} = \frac{1}{2}(\frac{\partial u_x}{\partial y} + \frac{\partial u_y}{\partial x})$, $\varepsilon_{zz} = 0$ with displacements u_x and u_y and deviatoric strain components $e_{xx} = \varepsilon_{xx} - \frac{\varepsilon_{vol}}{3}$, $e_{yy} = \varepsilon_{yy} - \frac{\varepsilon_{vol}}{3}$, $e_{xy} = \varepsilon_{xy}$ and $e_{zz} = -\frac{\varepsilon_{vol}}{3}$. Since there is no volumetric plastic strain in the von Mises yield function, the volumetric strain can be expressed as

$\varepsilon_{vol} = \text{trace}(\varepsilon)$ s.t. $\varepsilon_{vol} = \varepsilon_{xx} + \varepsilon_{yy}$. The deviatoric stress components are defined by $s_{xx} = \sigma_{xx} - (\frac{\sigma_{xx}+\sigma_{yy}+\sigma_{zz}}{3})$, $s_{yy} = \sigma_{yy} - (\frac{\sigma_{xx}+\sigma_{yy}+\sigma_{zz}}{3})$, $s_{xy} = \sigma_{xy}$ and $s_{zz} = \sigma_{zz} - (\frac{\sigma_{xx}+\sigma_{yy}+\sigma_{zz}}{3})$, where σ_{ij} $(i,j \in \{x,y\})$ are the components of the Cauchy stress tensor. The plastic strain components are defined by $\varepsilon^{pl}_{xx} = \bar{\varepsilon}^{pl}\frac{3}{2}\frac{s_{xx}}{q}$, $\varepsilon^{pl}_{yy} = \bar{\varepsilon}^{pl}\frac{3}{2}\frac{s_{yy}}{q}$, $\varepsilon^{pl}_{xy} = \bar{\varepsilon}^{pl}\frac{3}{2}\frac{s_{xy}}{q}$ and $\varepsilon^{pl}_{zz} = \bar{\varepsilon}^{pl}\frac{3}{2}\frac{s_{zz}}{q}$ with equivalent plastic strain of the von Mises model as $\bar{\varepsilon}^{pl} = \bar{\varepsilon} - \frac{\sigma_Y}{3\mu} \geq 0$, where σ_Y is the yield stress and μ the second Lamé parameter. The total equivalent strain is defined by $\bar{\varepsilon} = \sqrt{\frac{2}{3}\sum_{i,j\in\{x,y\}} e_{ij}e_{ij}}$ with deviatoric strain components e_{ij}. The decomposition of the strain is $\varepsilon_{ij} = \varepsilon^{el}_{ij} + \varepsilon^{pl}_{ij}$ with elastic component ε^{el}_{ij} and plastic component ε^{pl}_{ij} of the respective strain matrices. The equivalent stress is defined by $q = \sqrt{\frac{3}{2}s_{ij}s_{ij}}$. In our PINN approach, we utilize the definitions of the total strain components, deviatoric strain and stress components and plastic strain components in the respective regularization term.

We use quarter symmetry in use cases 1 and 3 to make efficient use of computational resources. Additional information regarding the variation of parameters in the simulations is presented in Table 2, where simulations marked in bold are used for the test and evaluation of the surrogate models and are not in the training dataset. Conversely, simulations not marked in bold represent the training dataset and are not in the test dataset. In use cases exhibiting varied geometry parameters (i.e., elongation of a plate and compression of a block use cases), the mesh is also different in each simulation. Thus, we train and evaluate the surrogate models on use cases with different meshes (i.e., in each simulation, the node coordinates differ).

Table 1. Classic FEM use cases. Overview of the three use cases and their main change and types of deformations. In the first two use cases, only a single change is conduced, while in the last use case, a combination of changes is studied.

Use Case	Change	Deformation
Plate	Geometry	Elongation
Beam	Material Properties	Bending
Block	Geometry, Material Properties	Compression

The first use case, a perforated steel strip under tensile load, is similar to the nonlinear solid mechanics use case of [21,22]. However, in our approach, we evaluate the generalization over the perforation diameter and use material properties for steel and a top edge displacement of 5 mm in positive y-axis to consider a more challenging use case.

We execute different simulation settings, where the generalization variable (diameter of perforation) is changed in each simulation; see Figure 2a and Table 2. In our second use case, we simulate a bending beam that end is displaced about 5 mm in the positive x-direction; see Figure 2b. We vary the yield stress generalization variable in each simulation setting; see Table 2. In our third use case, we simulate a quarter-symmetric block with four perforations under compressive load, which is compressed about 5 mm in the negative y-axis; see Figure 2c. In this use case, we vary two generalization variables (yield stress and width of the block) in each simulation; see Table 2.

We evaluate our models on interpolation (i.e., that the generalization variables for testing are within the range of the generalization variables observed during training) and extrapolation (i.e., that the generalization variables for testing are outside the range of the generalization variables observed during training) tasks. In Table 2, we mark interpolation tasks with superscript (i) and extrapolation tasks with superscript (e).

In Figure 3, we present the perfect nonlinear elastoplastic material behavior of our use cases. The Young's modulus is 210 GPa, Poisson's ratio 0.3 and the yield stress 900 MPa. In our first use case, the perforated plate, we use this setting in each simulation. In the other two use cases, the yield stress varies, while the remaining material parameters stay the same.

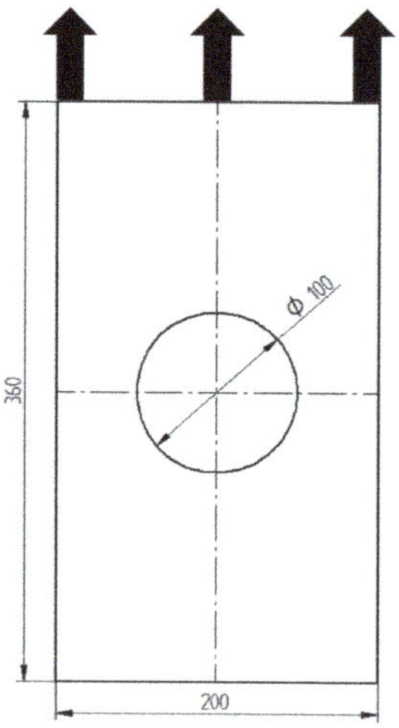

(**a**) Dimensions of the plate with one perforation

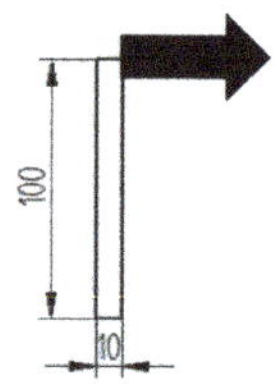

(**b**) Dimensions of the beam

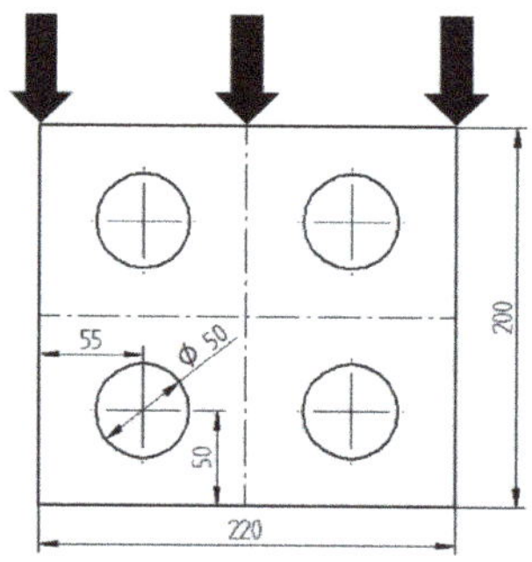

(**c**) Dimensions of the block with four perforations

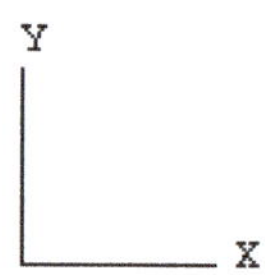

(**d**) Coordinate system

Figure 2. The three use cases: (**a**) elongation of a plate (diameter = 100 mm) about 5 mm at the top end in positive y-direction, (**b**) bending of a beam by a displacement at the top end about 5 mm in positive x-direction, (**c**) compression of a block with four perforations in the center of the quarter-symmetric parts (width = 220 mm) about 5 mm in negative y-direction and (**d**) the considered coordinate system.

All parts are meshed, using plane strain 4-node bilinear quadrilateral elements with reduced integration and hourglass control. Please note that although [22] recommends the use of larger order elements for the approximation of body forces, we use bilinear elements since we do not use body forces in our surrogate modeling approaches. We create a finer mesh near additional geometric details (i.e., perforations in the plate and block use cases) and seed the perforation edge of the plate with an approximate size of 3.8 mm and the remaining edges with an approximate size of 5 mm. The perforation edges of the block are seeded with an approximate size of 3 mm and the remaining edges with an approximate size of 4 mm. The beam exhibits no comparable geometric details; thus, we seed all edges with an approximate size of 1.5 mm.

Table 2. Dataset generation by executing several different simulations with varying generalization variables (Plate: perforation *Diameter*, Beam: *Yield Stress* and Block: *Yield Stress* and *Width*), bold marked simulations are not in the training dataset and only used for test and evaluation. Interpolation tasks are marked with superscript (i) and extrapolation tasks with superscript (e).

						Plate							
Simulation ID	**1** $^{(e)}$	2	3	**4** $^{(i)}$	5	**6** $^{(i)}$	7	8	**9** $^{(e)}$				
Diameter [mm]	**60**	70	80	**90**	100	**110**	120	130	**140**				
						Beam							
Simulation ID	**1** $^{(e)}$	2	3	**4** $^{(i)}$	5	**6** $^{(i)}$	7	8	**9** $^{(e)}$				
Yield Stress [MPa]	**850**	900	950	**1000**	1050	**1100**	1150	1200	**1250**				
						Block							
Simulation ID	**1** $^{(e)}$	**2** $^{(e)}$	3	4	5	6	**7** $^{(i)}$	8	9	10	11	**12** $^{(e)}$	**13** $^{(e)}$
Yield Stress [MPa]	**750**	**750**	900	900	900	1050	**1050**	1050	1200	1200	1200	**1350**	**1350**
Width [mm]	**180**	**260**	200	220	240	200	**220**	240	200	220	240	**180**	**260**

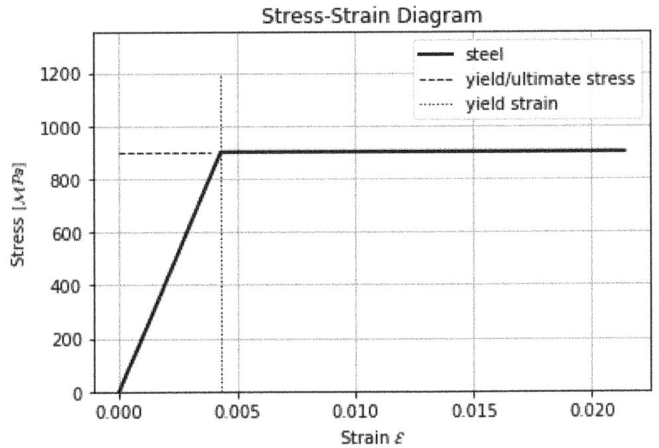

Figure 3. Perfect nonlinear elastoplastic material properties for a Young's modulus of 210 GPa, Poisson's ratio of 0.3 and yield stress of 900 MPa. The yield stress varies in simulations regarding the beam and block use cases.

We obtain our FEM simulation results in the context of general static simulations. Details of the simulation steps are shown in Table 3. Simulation control parameters that are not listed are left at default values.

Table 3. Abaqus FEM simulation control parameters.

Abaqus FEM Simulation Settings	
Simulation type	Static, General
Time period	1
Nlgeom	On
Max number of increments	100
Initial increment size	1
Min increment size	1×10^{-5}
Max increment size	1
Equation solver method	Direct
Solution technique	Full Newton

2.2. Dataset

The nodal data from our Abaqus FEM simulations constitute the datasets. For each use case, the nodal data are split into training and test dataset, respectively. The training dataset $D = \{X_1, \ldots, X_n\}$ with number of training instances n and the test dataset $T = \{X_{n+1}, \ldots, X_{n+m}\}$ with number of test instances m are generated from several FEM simulations; see Tables 2 and 6, where bold marked simulations belong to T and the remaining to D. Thus, we split our data due to different generalization variables and not randomly. We denote each instance with index i, $i \in \{1, 2, \ldots, n+m\}$. An instance $X_i = (x_i, y_i)$ is generated of an input vector $x_i \in \mathbb{R}^p$ and output vector $y_i \in \mathbb{R}^q$. Each input vector x_i is composed of the initial x- and y-coordinates of a FEM node and the respective generalization variable (i.e., perforated plate: *Diameter*, beam: *Yield Stress*, block with four perforations: *Width* and *Yield Stress*) of the FEM simulation; see Table 4. Thus, we have $p = 3$ in the plate and beam use case, and $p = 4$ in the block use case.

Table 4. Surrogate model input variables. Data obtained from FEM simulations are transformed so that each FEM node (represented by its x- and y-coordinates) with the respective generalization variable is an instance.

Simulation	Plate	Beam	Block
Input variables	x-coordinate y-coordinate Diameter	x-coordinate y-coordinate Yield Stress	x-coordinate y-coordinate Yield Stress Width

In our setting, each output vector y_i contains 13 ($q = 13$) output variables obtained from FEM simulation with input x_i, namely the ε^t_{xx}, ε^t_{xy} and ε^t_{yy} total strain components, the ε^p_{xx}, ε^p_{xy}, ε^p_{yy} and ε^p_{zz} plastic strain components, the σ_{xx}, σ_{xy}, σ_{yy} and σ_{zz} principal and shear stress components and the displacement in x- and y-directions u and v of each node; see Table 5 and Figure 4. We split the data in a training and test dataset (see Table 6) and standardized the data by removing the mean and scaling to unit variance.

Table 5. Surrogate model output variables. For each input FEM node, a surrogate model predicts its respective strains, stresses and displacements.

Output Variables			
ε^t_{xx}	ε^p_{xx}	σ_{xx}	u
ε^t_{xy}	ε^p_{xy}	σ_{xy}	v
ε^t_{yy}	ε^p_{yy}	σ_{yy}	
	ε^p_{zz}	σ_{zz}	

In Figure 4, we present graphical results with visible mesh obtained from Abaqus FEM simulation of the output variables used for a block use case.

Table 6. Dataset splits: number of training instances n and test instances m due to the data generation from Table 2.

	Plate	Beam	Block
Training dataset D	4447	2720	6722
Test dataset T	3534	2176	4107

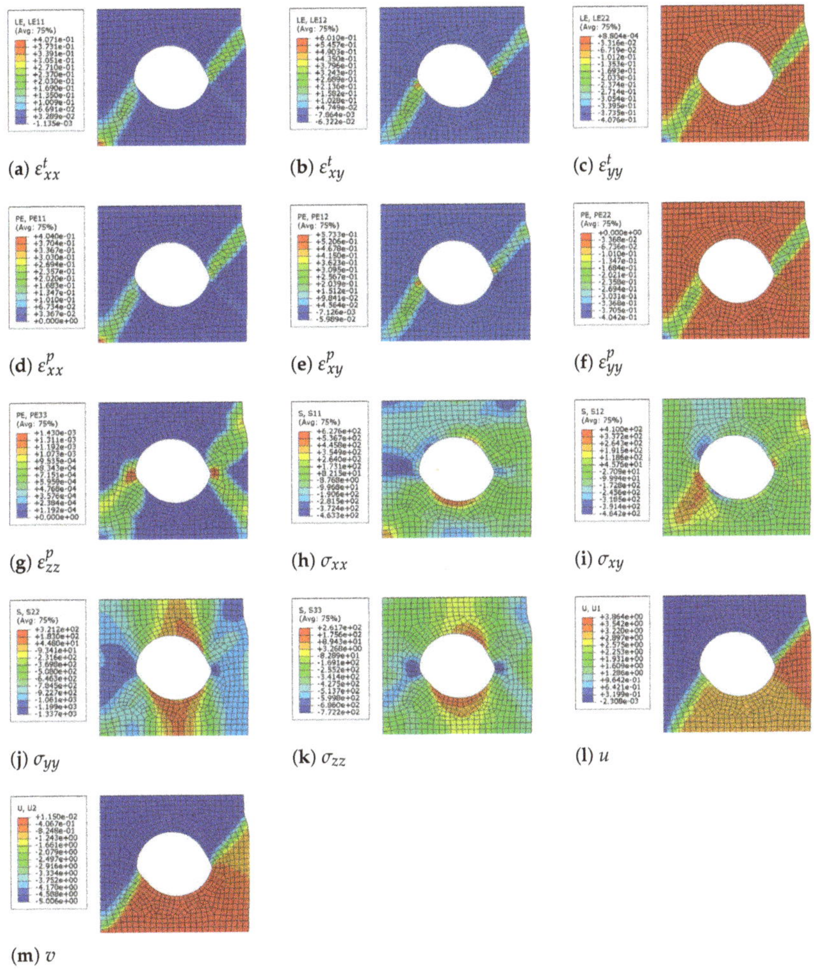

(a) ε^t_{xx} (b) ε^t_{xy} (c) ε^t_{yy} (d) ε^p_{xx} (e) ε^p_{xy} (f) ε^p_{yy} (g) ε^p_{zz} (h) σ_{xx} (i) σ_{xy} (j) σ_{yy} (k) σ_{zz} (l) u (m) v

Figure 4. Block use case: Abaqus FEM results that our surrogate models should predict.

2.3. Surrogate Models

In this section, we give an overview of the surrogate models used and their general assumptions; to highlight the differences as well as the advantages and disadvantages between them, we present a detailed mathematical background in Appendix A. We have selected models from different learning paradigms:

1. Gradient boosting decision tree regressor (GBDTR): piecewise constant model.
2. K-nearest neighbor regressor (KNNR): distance-based model.
3. Gaussian process regressor (GPR): Bayesian model.
4. Support vector regressor (SVR): hyperplane-based model.
5. Multi layer perceptron (MLP): classic feedforward neural network model.
6. Physics informed neural network (PINN): neural network model with physics-based regularization.

3. Results

For evaluation, we split the data into a training and test dataset to fit and test our surrogate models; see Table 6 for the dataset sizes and Table 2 for more details regarding the data split.

As a next step, we need to define hyperparameters for each model and each use case. We performed hyperparameter optimization using only training data; no test data were used. In our PINN approaches, the adaptation of hyperparameters was based on the work of [21,22]. Our MLPs were designed to be similar to our PINNs to allow for fair comparisons. We varied hyperparameters in our neural network approaches (MLP and PINN) following best practices and guidelines, where we optimized the number of hidden layers, number of neurons per hidden layer, activation function, validation split, earlystopping patience and the size of the batch per training epoch. Regarding the rest of our models, we applied a grid-search with a five fold cross-validation, utilizing the training data to obtain the best hyperparameters. The hyperparameters for each use case are in Appendix B and Tables A1–A6.

Our evaluation is based on R2-scores with respect to the FEM results and inference time. For models that contain inherent randomness, such as MLPs, GBDTR and PINNs, a five-fold cross-validation was conducted. For these models, we report the mean values and standard deviation of the R2-score. For the sake of brevity, we report only the average R2-scores across all 13 targets in this section; see Tables 7–9. The R2-scores for individual targets are provided in Appendix C. The inference times are based on the mean value of three measurements. Inferences were run on a machine with 16 GB RAM, 8 CPUs and Intel(R) i7-8565 2.0GHz processor. To compare the inference time of our surrogate models with the computation time required to run FEM simulations, we have included the latter also in Tables 7–9.

Table 7. Plate: averaged results, bold values indicate the best performing surrogate models. Values in parentheses are the corresponding standard deviations of the average R2-scores due to repeated experiments of stochastic process models. For further information concerning simulations, see Table 2.

Model	MLP	PINN	SVR	GBDTR	KNNR	GPR	FEM
			Simulation 1				
R2	**0.9900** (6.155×10^{-9})	0.7797 (8.709×10^{-2})	0.6188	0.6606 (3.959×10^{-8})	0.8164	0.6131	-
Inference time [s]	0.0523	0.0746	1.15	0.311	0.00722	0.151	9.01
			Simulation 4				
R2	**0.9978** (1.970×10^{-4})	0.9089 (2.598×10^{-2})	0.7174	0.9014 (6.310×10^{-2})	0.9298	0.8761	-
Inference time [s]	0.0781	0.0638	1.20	0.271	0.00734	0.139	9.08
			Simulation 6				
R2	**0.9920** (1.889×10^{-3})	0.8470 (6.309×10^{-2})	0.7251	0.8503 (1.005×10^{-1})	0.9219	0.8676	-
Inference time [s]	0.0595	0.0641	1.10	0.251	0.00797	0.131	9.88
			Simulation 9				
R2	**0.9786** (2.970×10^{-5})	0.7562 (1.046×10^{-1})	0.6568	0.7263 (9.780×10^{-9})	0.8045	0.5651	-
Inference time [s]	0.0665	0.0715	1.02	0.251	0.00734	0.139	10.03

For graphical results, we chose simulations that cover the error situation quite well in order to make statements about the performance of each model. In addition to the absolute errors (Figures 5a–f–10a–f), the corresponding FEM simulations of the basis are shown in Figures 5g–10g.

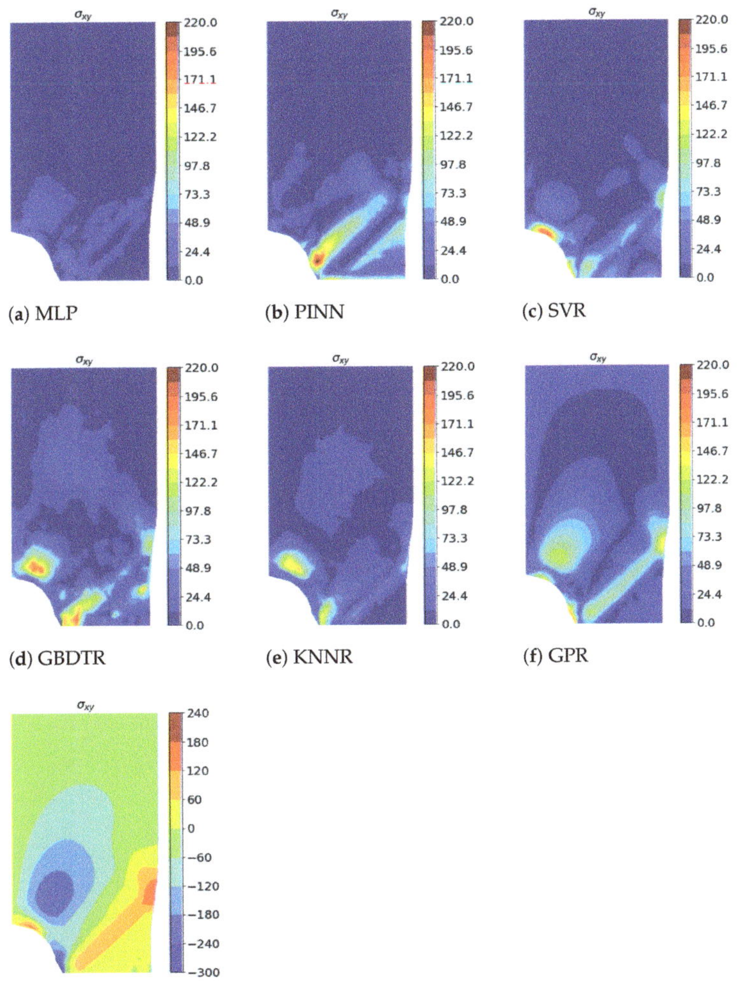

(**a**) MLP (**b**) PINN (**c**) SVR

(**d**) GBDTR (**e**) KNNR (**f**) GPR

(**g**) FEM result

Figure 5. Elongation of a perforated plate, Simulation 1 (extrapolation): absolute errors of different surrogate models (**a**–**f**) and ground truth Abaqus FEM simulation (**g**) of σ_{xy}.

GBDTR, KNNR, GPR and SVR algorithms were implemented with the scikit-learn library version 0.24.0 in Python. The SVR and GBDTR algorithms were constructed with MultiOutputRegressor scikit-learn API to fit one regressor per target. Regarding our DL algorithms, the utilized MLPs were implemented with the keras API version 2.4.3 and our PINNs were implemented with the sciann API version 0.5.5.0 in Python 3.8.5. We used the PDEs from [21,22], but instead of the inversion part, we trained our PINNs additionally with plastic strain data, same as for the rest of the surrogate models.

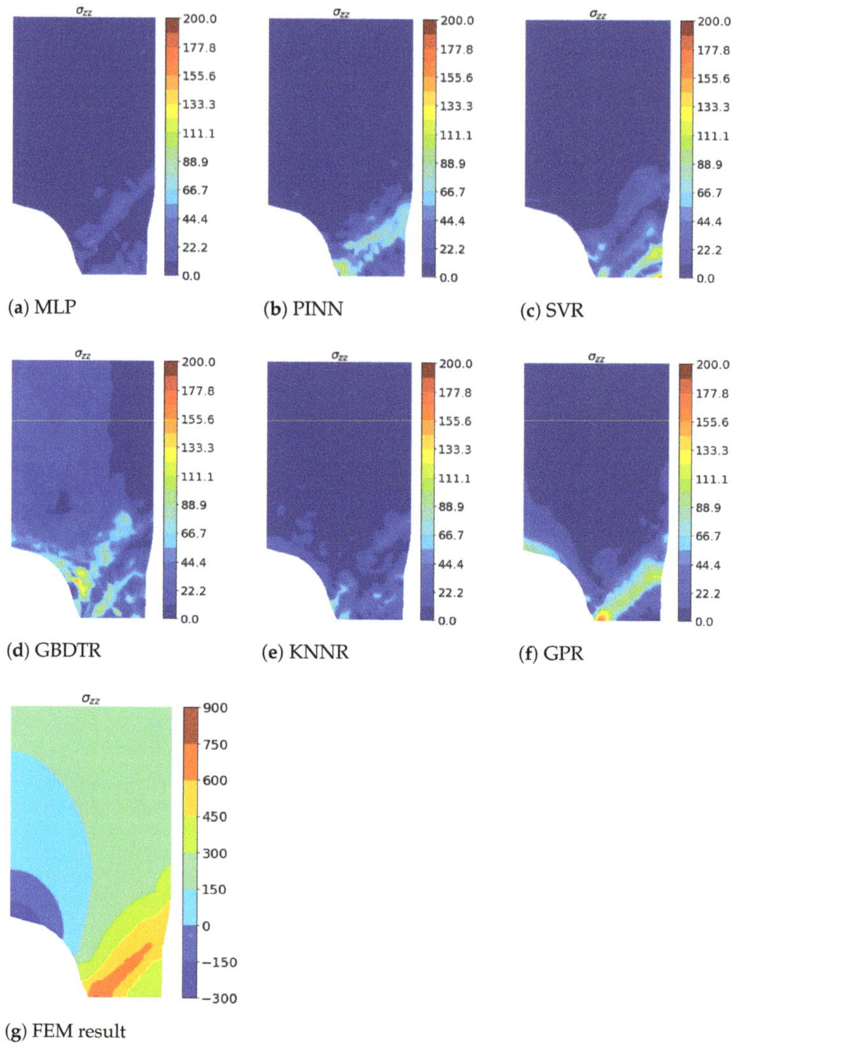

Figure 6. Elongation of a perforated plate, Simulation 4 (interpolation): absolute errors of different surrogate models (**a**–**f**) and ground truth Abaqus FEM simulation (**g**) of σ_{zz}.

In the elongation of a perforated plate use case, our approach is based on a total of nine FEM simulations. We used five simulations for training and four simulations to evaluate the fitted models; see Table 2. We report the average of R2-scores across all outputs in Table 7 with the corresponding inference times.

Regarding extrapolation, the absolute errors of each surrogate model with respect to σ_{xy} of Simulation 1 are shown in Figure 5. We plot the absolute errors of each surrogate model of σ_{zz} of Simulation 4 in Figure 6 as an example of interpolation. In addition, we show in both figures the ground truth of the corresponding output variable obtained from the FEM simulation. For both interpolation and extrapolation, the errors are large near the shear band. As far as extrapolation is concerned, in addition to the errors near the shear band, most models have significant errors near the maximum negative xy shear stresses; see blue areas in Figure 5g. GBDTR performs well overall, though the error increases in various locations; while PINNs have a similar average performance, they perform better outside the shear band regarding absolute errors. MLP overall shows the best results followed by KNNR.

In the bending beam use case, similar to the perforated plate use case, we trained our models on five simulations and tested them using the remaining four, see Table 2. We present the average R2-scores across all outputs and inference times in Table 8 for the test simulations 1, 4, 6 and 9.

Table 8. Beam: averaged results, bold values indicate the best performing surrogate models. Values in parentheses are the corresponding standard deviations of the average R2-scores due to repeated experiments of stochastic process models. For further information concerning simulations, see Table 2.

Model	MLP	PINN	SVR	GBDTR	KNNR	GPR	FEM
				Simulation 1			
R2	0.6682 (7.345×10^{-6})	0.6165 (1.648×10^{-2})	0.5122	**0.7120** (1.088×10^{-8})	0.6288	0.5377	-
Inference time [s]	0.0781	0.0638	1.20	0.271	0.00734	0.139	9.08
				Simulation 4			
R2	**0.9979** (1.319×10^{-3})	0.9379 (1.042×10^{-3})	0.7558	0.9640 (6.751×10^{-4})	0.9621	0.8243	-
Inference time [s]	0.0457	0.110	0.418	0.0541	0.0790	0.221	6.81
				Simulation 6			
R2	**0.9981** (8.315×10^{-4})	0.9314 (8.396×10^{-4})	0.7406	0.9516 (1.368×10^{-4})	0.9617	0.8059	-
Inference time [s]	0.0442	0.0668	0.402	0.0569	0.0705	0.212	6.61
				Simulation 9			
R2	−1196.2920 (6.166×10^{3})	−1305.9226 (3.269×10^{1})	**−107.7646**	−830.8926 (4.2984)	−322.4940	−420.9029	-
Inference time [s]	0.0781	0.0638	1.20	0.271	0.00734	0.139	9.08

We provide a graphical representation of the absolute error of the surrogate models regarding ε^t_{yy} in Figure 7a–f with the FEM simulation result in (g) as one instance of interpolation. Absolute errors of the surrogate models regarding ε^p_{xx} and extrapolation are shown in Figure 8. Overall higher errors can be observed near the encastred boundary condition of the beam for some models for that output. While the PINN shows a competitive average R2-score regarding interpolation, on this single target, its performance shows significant weaknesses.

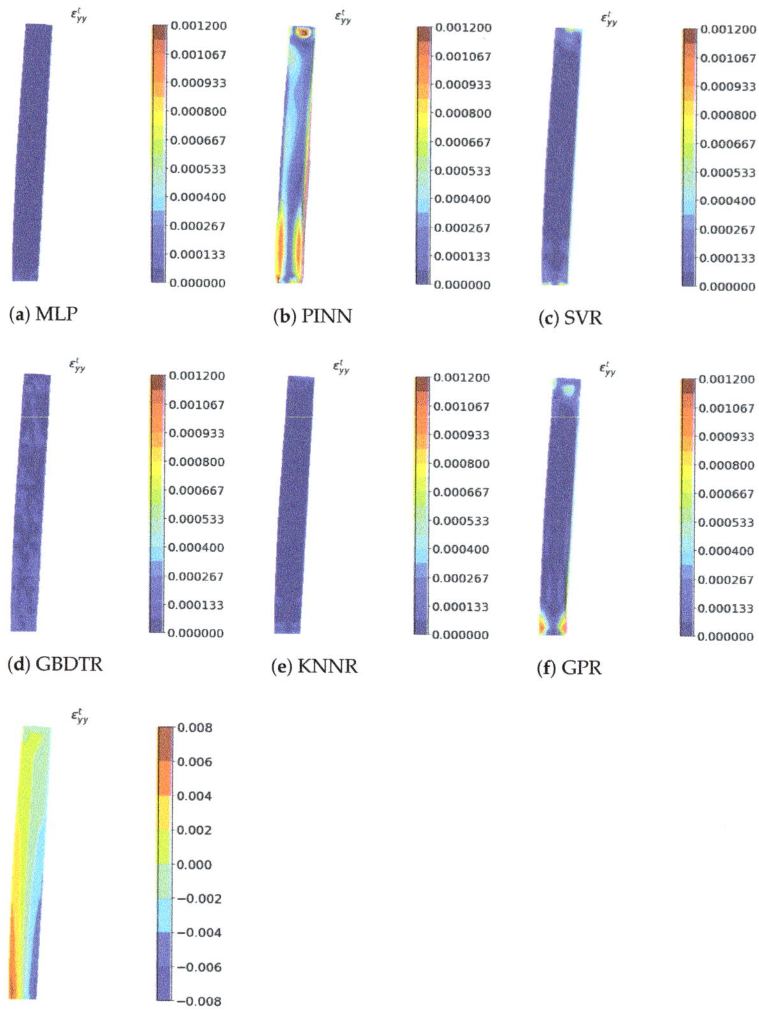

(**a**) MLP (**b**) PINN (**c**) SVR

(**d**) GBDTR (**e**) KNNR (**f**) GPR

(**g**) FEM result

Figure 7. Bending of a beam, Simulation 6 (interpolation): absolute errors of different surrogate models (**a**–**f**) and ground truth Abaqus FEM simulation (**g**) of ε^t_{yy}.

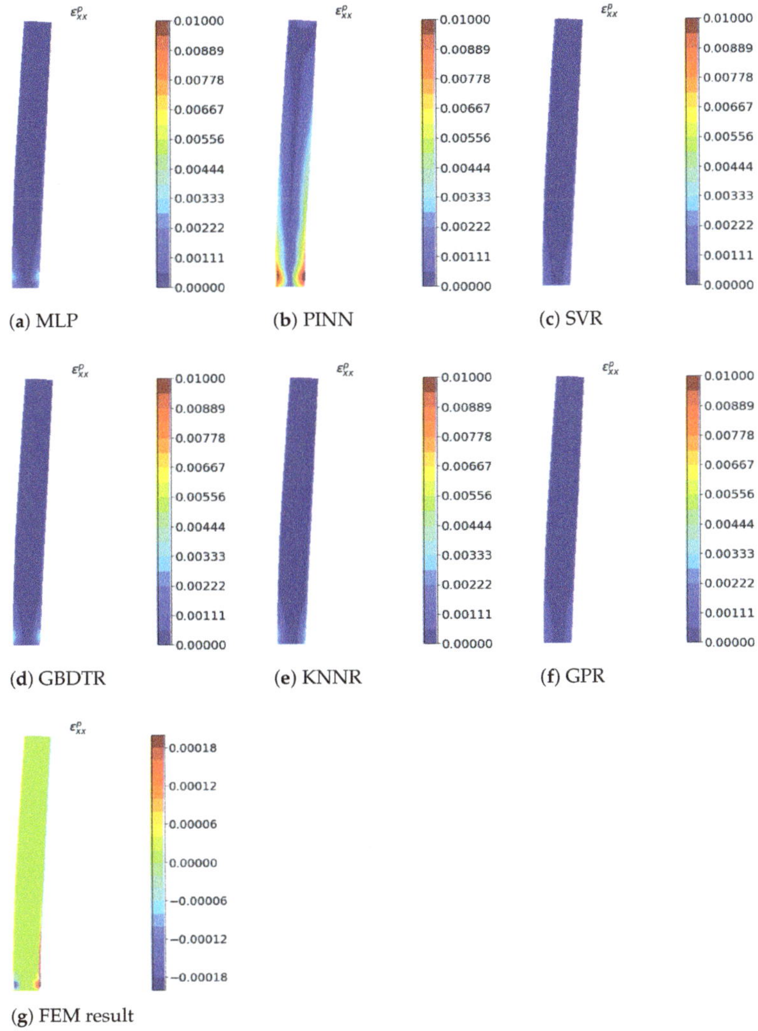

Figure 8. Bending of a beam, Simulation 9 (extrapolation): absolute errors of different surrogate models (**a–f**) and ground truth Abaqus FEM simulation (**g**) of ε^p_{xx}.

The compression of a block with four perforations use case presents a more complex setting because we generalize by two generalization variables (yield stress and block width). Therefore, we utilize more training data for this use case; see Table 2. We report the average results of R2-scores with corresponding standard deviations, if applicable, in Table 9.

Table 9. Block: averaged results, bold values indicate the best performing surrogate models. Values in parentheses are the corresponding standard deviations of the average R2-scores, due to repeated experiments of stochastic process models. For further information concerning simulations, see Table 2.

Model	MLP	PINN	SVR	GBDTR	KNNR	GPR	FEM
				Simulation 1			
R2	0.5562 (1.952×10^{-1})	−0.4441 (6.066×10^{-1})	−0.2463	0.5695 (1.665×10^{-3})	**0.7808**	0.1059	-
Inference time [s]	0.0781	0.0638	1.20	0.271	0.00734	0.139	9.08
				Simulation 2			
R2	0.3768 (3.803×10^{-1})	0.1850 (5.531×10^{-2})	−0.1800	0.5149 (3.320×10^{-4})	**0.7366**	0.1409	-
Inference time [s]	0.0457	0.110	0.418	0.0541	0.0790	0.221	6.81
				Simulation 7			
R2	**0.9976** (8.258×10^{-5})	0.9410 (4.310×10^{-3})	0.6415	0.9702 (1.884×10^{-2})	0.9767	0.5200	-
Inference time [s]	0.0442	0.0668	0.402	0.0569	0.0705	0.212	6.61
				Simulation 12			
R2	0.6303 (3.204×10^{-1})	−0.4480 (6.652×10^{-1})	−0.2230	0.5702 (6.266×10^{-4})	**0.7797**	0.1553	-
Inference time [s]	0.0442	0.0668	0.402	0.0569	0.0705	0.212	6.61
				Simulation 13			
R2	0.6475 (3.326×10^{-1})	−0.1122 (3.265×10^{-1})	−0.0745	0.5894 (1.579×10^{-4})	**0.7687**	0.1771	-
Inference time [s]	0.0781	0.0638	1.20	0.271	0.00734	0.139	9.08

As an instance for interpolation, the absolute errors regarding ε^p_{xx} can be seen in Figure 9a–f with Abaqus FEM simulation result (g). Respectively, an instance for extrapolation is shown in Figure 10 with absolute errors (a–f) and FEM ground truth (g). Some models show higher prediction errors near shear bands (high ε^p_{xx} regions) regarding the interpolation task. However, SVR and GPR cannot extract meaningful information from the training data, especially in the space free of plastic deformation. This is indicated by the low average R2-scores, compared to the other models. Considering absolute errors of σ_{xy} and extrapolation the MLP, which is otherwise performing well, shows weaknesses and is in general outperformed by the KNNR.

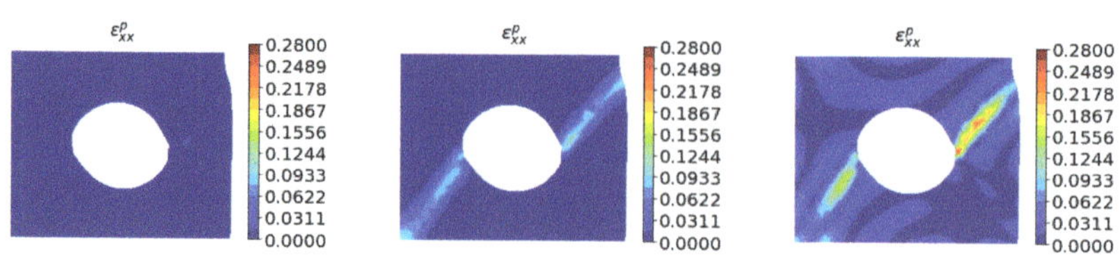

(**a**) MLP (**b**) PINN (**c**) SVR

Figure 9. *Cont.*

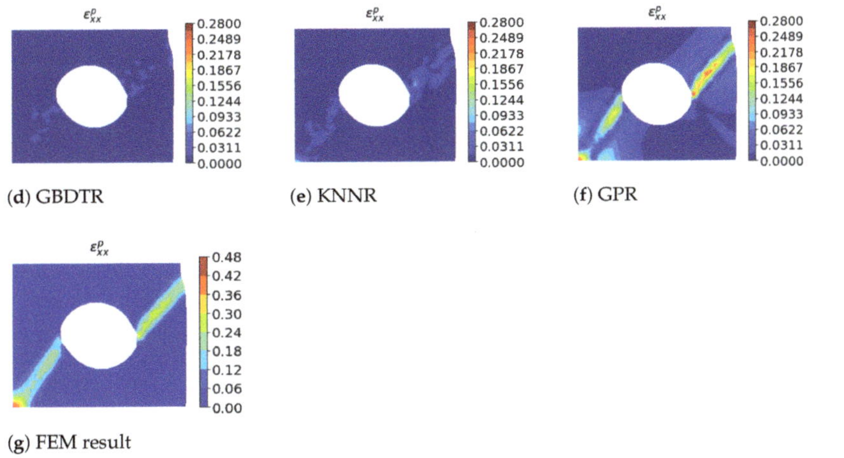

Figure 9. Compression of a block, Simulation 7 (interpolation): absolute errors of different surrogate models (**a–f**) and ground truth Abaqus FEM simulation (**g**) of ε^p_{xx}.

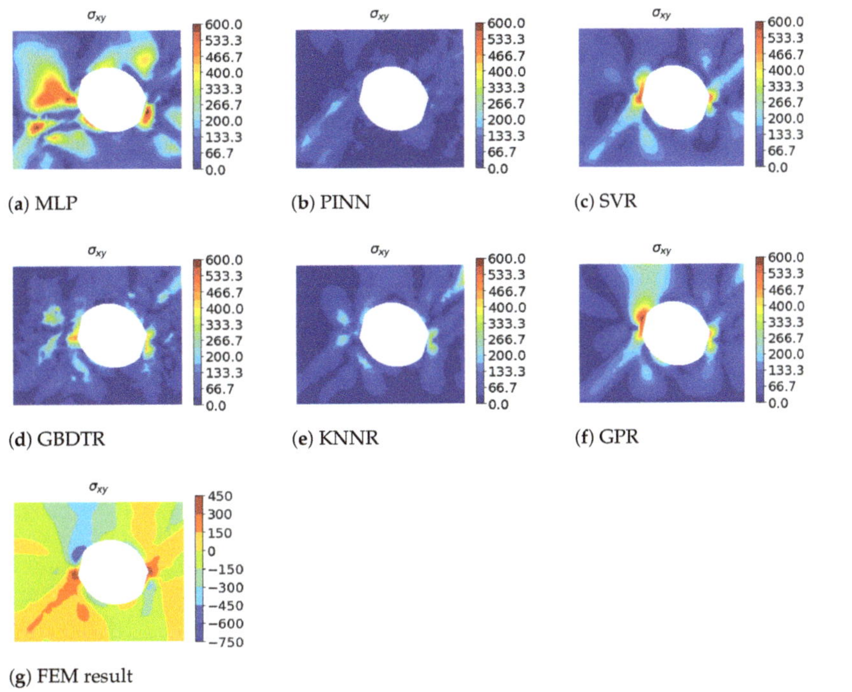

Figure 10. Compression of a block, Simulation 13 (extrapolation): absolute errors of different surrogate models (**a–f**) and ground truth Abaqus FEM simulation (**g**) of σ_{xy}.

4. Discussion

All classes of surrogate models that we considered in this work share several key characteristics: (1) they are mesh-free and thus, can deliver results with infinite resolution; (2) the computation time required to obtain the target values at predefined positions is orders of magnitude lower than for FEM simulations; (3) since for each simulation setup, where the geometry changes, a different mesh is created during FEM simulations, our results indicate that all classes of surrogate models generalize (interpolate) reasonably well across training data positions; (4) furthermore, all surrogate model classes generalize at least to some extent across use case parameters, such as changes in geometry or material parameters. Finally, all surrogate model classes must be used with care, as they do not extrapolate well to data positions and/or use case parameters unseen during training. Our findings show this in the extrapolation result of the beam use case, Simulation 9: due to the greater yield stress, almost no plastic deformation occurs; thus, the surrogate models are not able to learn such material behavior. Similar findings can be seen from the extrapolation results of the block use case, Simulation 1, 2, 12 and 13: approaches utilizing PINNs and SVRs are not able to predict acceptable strain components, leading to overall worse averaged R2-scores. In general, it can be stated that the surrogate models used show similar behavior with respect to inter- and extrapolation, but differ with respect to individual components, i.e., some models are better at predicting individual components (e.g., strains) for unknown generalization variables (e.g., yield strength) than others. Another example would be the symmetric nature of the use case, making it redundant to evaluate, e.g., stresses at negative x-positions, the proposed surrogate models will certainly respond with such stress values, which consequently, cannot be considered meaningful. Similarly, while the surrogate models may well be evaluated at physically meaningless use case parameters, e.g., negative radii, the thus obtained results must be considered meaningless as well. Therefore, all surrogate models must be treated with this in mind, which is a fundamental difference to FEM simulations that do not offer such modes of failure. With these considerations in mind, we now turn to discuss specific characteristics of each surrogate model class.

Our KNNR approach, which can be considered simple compared to the other algorithms, gave competitive results; moreover, this approach showed the best results regarding extrapolation (i.e., Simulations 1, 2, 12 and 13) in the block use case.

Algorithms we constructed with MultiOutputRegressor (SVR and GBDTR) could give better results if the hyperparameters are tuned to each target separately. However, we did not do this for fairness reasons since our other algorithms are also fitted to the overall use case and not to each target individually. We intend to monitor this in the future.

In our setting, the GPR algorithm did not deliver good results. Tuning the kernel function could deliver better results; however, we do not believe that it would be practical to modify for each new simulation use case. Thus, we not intend to head in this direction. However, we plan to investigate whether other Bayesian methods (e.g., Bayesian neural network [30] or neural processes [31]) could be beneficial.

Our MLPs approaches delivered the overall best results in our comparison, especially regarding interpolation (i.e., in the plate and beam use cases Simulations 4 and 6 and in the block use case, Simulation 7). They achieved high accuracies (R2-score > 0.992), while reducing the inference time by a factor of over 100 in comparison to FEM simulations. As mentioned before, designing the architecture is not a straightforward process; however, if the network is deep enough and suitable optimization methods are available (e.g., Adam optimizer) the network can be also efficiently trained utilizing early stopping.

As already reported in literature [32–35], we experienced in our setting that PINNs are not straightforward to design and train. Due to several plateaus in the loss function, early stopping did not prove to be effective. Therefore, we set a fixed number of training epochs. One reason for our observation could be the existence of a non-convex Pareto frontier [36]. In the multi-objective optimization problem, the optimizer might attempt to adjust the model parameters while situated between the different losses, leading it to favor one loss at the expense of the other [37]. Possible approaches to overcome this problem are adaptive optimizers [38], adaptive loss [39], and adaptive activation functions [40]. Moreover, PINNs are objects of current research and will gain more and more attention in the future. Besides other fundamental methods, we additionally plan to aim in that direction for improved surrogate modeling.

5. Conclusions

In this work, we deliver a comprehensive evaluation of generalizable and mesh-free ML and DL surrogate models based on FEM simulation and show that surrogate modeling leads to fast predictions with infinite resolution for practical use. In the context of our evaluation, we show which ML and DL models are target oriented at which level of complexity with respect to prediction accuracy and inference time, which can serve as a basis for the practical implementation of surrogate models (in, for example, production for real-time prediction, cyber–physical systems, and process design).

In future work, we plan to conduct more complex experiments, e.g., generalizing across more input variables regarding geometry (e.g., consideration of all component dimensions) and material parameters (e.g., non-perfect nonlinear material behavior, time-dependent material properties, grain growth, and phase transformation). We will moreover explore extended surrogate models with more complex output variables (e.g., grain size, grain structure, and phase transformation).

Author Contributions: Conceptualization, J.G.H. and P.O.; methodology, J.G.H. and B.C.G.; software, J.G.H.; validation, J.G.H., B.C.G. and P.O.; formal analysis, J.G.H.; investigation, J.G.H.; resources, J.G.H., B.C.G., P.O. and R.K.; data curation, J.G.H.; writing—original draft preparation, J.G.H.; writing—review and editing, J.G.H., B.C.G., P.O. and R.K.; visualization, J.G.H.; supervision, R.K.; project administration, J.G.H. and P.O.; funding acquisition, J.G.H., P.O. and R.K. All authors have read and agreed to the published version of the manuscript.

Funding: This research was funded by Österreichische Forschungsförderungsgesellschaft (FFG) Grant No. 881039.

Institutional Review Board Statement: Not applicable.

Informed Consent Statement: Not applicable.

Data Availability Statement: Not applicable.

Acknowledgments: The project BrAIN—Brownfield Artificial Intelligence Network for Forging of High Quality Aerospace Components (FFG Grant No. 881039) is funded in the framework of the program 'TAKE OFF', which is a research and technology program of the Austrian Federal Ministry of Transport, Innovation and Technology. The Know-Center is funded within the Austrian COMET Program—Competence Centers for Excellent Technologies—under the auspices of the Austrian Federal Ministry of Transport, Innovation and Technology, the Austrian Federal Ministry of Economy, Family and Youth and by the State of Styria. COMET is managed by the Austrian Research Promotion Agency FFG. The authors would also like to thank the developers of the sciann API for making their work available and for responding promptly to our questions.

Conflicts of Interest: The authors declare no conflict of interest.

Appendix A. Surrogate Models

We follow the notation introduced in Section 2.2 with data instance $X_i = (x_i, y_i)$ containing input vector x_i and output vector y_i, the number of training instances is n and the number of test instances is m. Notations regarding individual models are introduced when needed.

Appendix A.1. GBDTR

Boosting methods are powerful techniques in which the final "strong" regressor model is based on an iteratively formed ensemble of "weak" base regressor models [41]. The main idea behind boosting is to sequentially add new models to the ensemble, iteratively refining the output. In GBDTR models, boosting is applied to arbitrary differentiable loss functions. In general, GBDTR models are additive models, where the samples are modified so that the labels are set to the negative gradient, while the distribution is held constant [42].

The additive method of GBDTR is the following:

$$\hat{y}_i = F_G(x_i) = \sum_{g=1}^{G} h_g(x_i) \tag{A1}$$

where $\hat{y}_i$ is the prediction for a given input x_i, and h_g are the fitted base tree regressors. The constant G is the number of base tree regressors. The GBDTR algorithm is greedy, where a newly added tree regressor h_g is fitted to minimize the loss L_g of the resulting ensemble $F_g = F_{g-1} + h_g$, i.e.,

$$h_g = arg \min_h L_g = arg \min_h \sum_{i=1}^{n} l(y_i, F_{g-1}(x_i) + h(x_i)) \tag{A2}$$

Here, $l(y_i, F(x_i))$ is defined by the loss parameters, and $h(x_i)$ is the candidate base regressor. With the utilization of a first-order Taylor approximation:

$$l(z) \approx l(a) + (z-a)\frac{\partial l(a)}{\partial a} \tag{A3}$$

where z corresponds to $F_{g-1}(x_i) + h_g(x_i)$ and a corresponds to $F_{g-1}(x_i)$, we can approximate the value of l with the following:

$$l(y_i, F_{g-1}(x_i) + h_g(x_i)) \approx l(y_i, F_{g-1}(x_i)) + h_g(x_i)\left[\frac{\partial l(y_i, F(x_i))}{\partial F(x_i)}\right]_{F=F_{g-1}} \tag{A4}$$

We denote the derivative of the loss with g_i and remove constant terms:

$$h_m \approx arg \min_h \sum_{i=1}^{n} h(x_i) g_i \tag{A5}$$

h_m is minimized if $h(x_i)$ is fitted to predict a value proportional to the negative gradient.

Appendix A.2. KNNR

The KNNR algorithm is a relatively simple method mathematically, compared to other algorithms presented here. Here, the model stores all available use cases from the training dataset D and predicts the numerical target $\hat{y}_j$ of a test query instance x_j with $n < j \leq (n+m)$ based on a similarity measure (e.g., distance functions). The algorithm computes the distance-weighted average of the numerical targets of the K nearest neighbors of x_j in D [43].

Specifically, we introduce a distance metric d that measures the distance between all training instances x_i with $i \leq n$ and a test instance x_j. Next, the training instances are sorted w.r.t. their respective distance in ascending order to the test instance, i.e., there is a permutation π_j of the training indices i such that $d(x_{\pi_j(1)}, x_j) \leq d(x_{\pi_j(2)}, x_j) \leq \cdots \leq d(x_{\pi_j(n)}, x_j)$. Then, the estimate $\hat{y}_j(x_j)$ is given as the following:

$$\hat{y}_j(x_j) = \frac{1}{K}\sum_{i=1}^{K} y_{\pi_j(i)} \tag{A6}$$

where K must be specified as a hyperparameter.

Appendix A.3. GPR

Gaussian process regression modeling is a non-parametric Bayesian approach [44]. In general, a Gaussian process is a generalization of the Gaussian distribution. The Gaussian distribution describes random variables or random vectors, while a Gaussian process describes a function $f(x)$ [45].

In general, a Gaussian process is completely specified by its mean function $\mu(x)$ and covariance function $K(x, x')$ (also called kernel).

If the function $f(x)$ under consideration is modeled by a Gaussian process, i.e., if $f(x) \sim \mathcal{GP}(\mu(x), K(x, x'))$, then we have the following

$$\mathbb{E}[f(x)] = \mu(x) \tag{A7}$$

$$\mathbb{E}[(f(x) - \mu(x))(f(x') - \mu(x'))] = K(x, x') \tag{A8}$$

for all x and x'. Thus, we can define the Gaussian process as the following:

$$f(x) \sim \mathcal{N}(\mu(x), K(x, x)) \tag{A9}$$

We use the notation that matrix $D = (X_D, Y_D)$ contains the training data with input data matrix $X_D = (x_1, \ldots, x_n)$ and output data matrix $Y_D = (y_1, \ldots, y_n)$, and test data matrix $T = (X_T, Y_T)$ contains the test data with $X_T = (x_{n+1}, \ldots, x_{n+m})$ as input and $Y_T = (y_{n+1}, \ldots, y_{n+m})$ as output. We can define that they are jointly Gaussian and zero mean with consideration of the prior distribution:

$$\begin{bmatrix} Y_D \\ Y_T \end{bmatrix} \sim \mathcal{N}(0, \begin{bmatrix} K(X_D, X_D)) & K(X_D, X_T)) \\ K(X_T, X_D)) & K(X_T, X_T)) \end{bmatrix}) \tag{A10}$$

The Gaussian process makes a prediction Y_T for X_T in a probabilistic way, where, as stated before, the posterior distribution can be fully described by the mean and the covariance.

$$\begin{aligned} Y_T | X_T, X_D, Y_D \sim \mathcal{N}(&K(X_T, X_D)K(X_D, X_D)^{-1}Y_D, \\ &K(X_T, X_T) - K(X_T, X_D)K(X_D, X_D)^{-1}K(X_D, X_T)) \end{aligned} \tag{A11}$$

Appendix A.4. SVR

The SVR approach is a generalization of the SVM classification problem by introducing an ϵ-sensitive region around the approximated function, also called an ϵ-tube. The optimization task in SVR contains two steps: first, finding a convex ϵ-insensitive loss function that need to be minimized, and second, finding the smallest ϵ-tube that contains the most training instances.

The convex optimization has a unique solution and is solved using numerical optimization algorithms. One of the main advantages of SVR is that the computational complexity does not depend on the dimensionality of the input space [46]. To deal with otherwise intractable constraints of the optimization problem, we introduce slack variables ξ_i and ξ_i^* [47]. The positive constant C determines the trade-off between the flatness of the function and the magnitude up to which deviations greater than ϵ are allowed. The primal quadratic optimization problem of SVR is defined as the following:

$$\underset{\omega, b}{\text{minimize}} \ \frac{1}{2}||\omega||^2 + C\sum_{i=1}^{n}(\xi_i + \xi_i^*) \tag{A12}$$

$$subject\ to\ the\ following: \begin{cases} y_i - \omega^T x_i - b & \leq & \epsilon + \xi_i \\ \omega^T x_i + b - y_i & \leq & \epsilon + \xi_i^* \\ \xi_i, \xi_i^* & \geq & 0 \end{cases} \tag{A13}$$

Here, ω is the weight and b the bias to be adjusted. The constrained quadratic optimization problem can be solved by minimizing the Lagrangian with non-negative Lagrange multipliers $\lambda_i, \lambda_i^*, \alpha_i, \alpha_i^*, i \in \{1, \ldots, n\}$:

$$\begin{aligned} \mathcal{L}(\omega, \xi^*, \xi, \lambda, \lambda^*, \alpha, \alpha^*) = &\frac{1}{2}||\omega||^2 + C\sum_{i=1}^{n}\xi_i + \xi_i^* + \sum_{i=1}^{n}\alpha_i^*(y_i - \omega^T x_i - \varepsilon - \xi_i^*) \\ &+ \sum_{i=1}^{n}\alpha_i(-y_i + \omega^T x_i - \varepsilon - \xi_i) - \sum_{i=1}^{n}\lambda_i\xi_i + \lambda_i^*\xi_i^* \end{aligned} \tag{A14}$$

The minimum of $\mathcal{L}$ can be found by taking the partial derivatives with respect to the variables and making them equal to zero (*Karush-Kuhn-Tucker* (KKT) conditions). With the final KKT condition, we can state the following:

$$\begin{aligned} \alpha_i(-y_i + \omega^T x_i - \varepsilon - \xi_i) &= 0 \\ \alpha_i^*(-y_i + \omega^T x_i - \varepsilon - \xi_i^*) &= 0 \\ \lambda_i \xi_i &= 0 \\ \lambda_i^* \xi_i^* &= 0 \end{aligned} \tag{A15}$$

The Lagrange multipliers that are zero correspond to the inside of the ε-tube, while the support vectors have non-zero Lagrange multipliers. The function estimate depends only on the support vectors, hence this representation is sparse. More specifically, we can derive the following function approximation to predict $\hat{y}_j(x_j)$:

$$\hat{y}_j(x_j) = \sum_{i=1}^{n_{SV}} (\alpha_i^* - \alpha_i) x_i^T x_j \tag{A16}$$

with $\alpha_i, \alpha_i^* \in [0, C]$ and the number of support vectors n_{SV}. For nonlinear SVR we replace $\omega^T x_i$ in (12)–(15) by $\omega^T \phi(x_i)$ and the inner product in (16) by the kernel $K(x_i, x_j)$.

Appendix A.5. MLP

A neural network is a network of simple processing elements, also called neurons. The neurons are arranged in layers. In a fully-connected multi-layer network, a neuron in one layer is connected to every neuron in the layer before and after it. The number of neurons in the input layer is the number of input features p and the number of neurons in the output layer is the number of targets q [48]. MLPs have several theoretical advantages, compared to other ML algorithms. Due to the universal approximation theorem, an MLP can approximate any function if the activation functions of the network are appropriate [49–51]. The MLP makes no prior assumptions about the data distribution, and in many cases, can be trained to generalize to new data not yet seen [52]. However, finding the right architecture and finding the setting of training parameters is not straightforward and usually done by trial and error influenced by the literature and guidelines.

A neural network output $\hat{y}$ corresponding to an input x can be represented as a composition of functions, where the output of layer $L-1$ acts as input to the following layer L. For example, for non-linear activation function σ_L, weight matrix W_L, and bias vector b_L of the respective layer L, we obtain the following:

$$\hat{y}(x) = t_L(x) = \sigma_L(W_L^T t_{L-1}(x) + b_L) \tag{A17}$$

With the neural network estimate $\hat{y}(x)$ and the respective target y of an input x, we can denote a loss function $\mathcal{L}$. A very common loss function for MLPs for regression tasks is the mean-squared error:

$$\mathcal{L}(W, b) = \frac{1}{n} \sum_{i=1}^{n} (\hat{y}(x_i) - y_i)^2 \tag{A18}$$

where W and b are the collections of all weight matrices and bias terms, respectively. Optimal weight W^* and bias b^* terms for each layer are identified with minimizing the loss function $\mathcal{L}$ via back-propagation [53].

$$W^* / b^* = \underset{W,b}{\operatorname{argmin}}\ \mathcal{L}(W, b) \tag{A19}$$

Appendix A.6. PINN

In PINNs, the network is trained simultaneously on data and governing differential equations. PINNs are regularized such that their function approximation $\hat{y}(x)$ obeys known laws of physics that apply to the observed data. This type of network is well suited for solving and inverting equations that control physical systems and find application in fluid and solid mechanics as well as in dynamical systems [21,35].

PINNs share similarities with common ANNs, but the loss function has an additional part that describes the physics behind the use case setting. More specifically, the loss $\mathcal{L}$ is composed of the data-driven loss $\mathcal{L}_{data}$ and the physics-informed loss $\mathcal{L}_{physics}$:

$$\mathcal{L} = \mathcal{L}_{data} + \mathcal{L}_{physics} \tag{A20}$$

While the data-driven loss is often a standard mean-squared error, the physics-informed loss accounts for the degree to which the function approximation solves a given system of governing differential equations. For further details, we refer the reader to [23,35,54] in general and to the Python package of [21,22] in particular for simple implementation of structural mechanics use cases.

Appendix B. Hyperparameters

Table A1. Best performing hyperparameters GBDTR.

	Plate	Beam	Block
loss	ls	ls	ls
criterion	friedman_mse	mse	friedman_mse
max_features	auto	log2	auto
n_estimators	400	1000	2000

Table A2. Best performing hyperparameters KNNR.

	Plate	Beam	Block
n_neighbors (K)	7	5	10
weights	distance	distance	distance
algorithm	brute	ball_tree	auto
leaf_size	1	5	1
p_value	1	2	5

Table A3. Best performing hyperparameters GPR.

	Plate	Beam	Block
kernel	Matern()**2	RationalQuadratic()**2	RationalQuadratic()**2
alpha	10^{-13}	10^{-13}	10^{-14}

Table A4. Best performing hyperparameters SVR.

	Plate	Beam	Block
kernel	rbf	rbf	rbf
gamma	scale	scale	scale
epsilon	0.005	0.005	0.4
C	95	5	105

Table A5. Best performing hyperparameters MLP.

	Plate	Beam	Block
hidden layers	3	2	4
neurons	100-100-100	100-100	100-100-100-100
activation function	relu	relu	relu
batch size	32	32	64
validation split	0.1	0.1	0.1
early stopping patience	5000	5000	7500
max epochs	100,000	100,000	100,000
stopped at	27,693	26,383	43,272

Table A6. Best hyperparameters PINN.

	Plate	Beam	Block
hidden layers	4	4	4
neurons	100-100-100-100	100-100-100-100	100-100-100-100
activation function	tanh	tanh	tanh
batch size	64	64	64
epochs	50,000	50,000	50,000

Appendix C. Detailed Results

Table A7. Detailed results for the plate elongation use case Simulation 1.

SIMULATION 1										
		MLP		PINN		SVR	GBDTR		KNNR	GPR
		mean	std	mean	std		mean	std		
R2	ε^t_{xx}	0.9923	3.121×10^{-6}	0.8331	5.206×10^{-2}	4.117×10^{-1}	5.296×10^{-1}	1.788×10^{-6}	7.973×10^{-1}	5.936×10^{-1}
	ε^t_{xy}	0.9900	3.681×10^{-7}	0.4748	1.814×10^{-1}	5.390×10^{-2}	3.846×10^{-1}	2.065×10^{-5}	5.121×10^{-1}	5.039×10^{-1}
	ε^t_{yy}	0.9924	3.055×10^{-6}	0.8749	9.156×10^{-2}	4.169×10^{-1}	5.281×10^{-1}	1.243×10^{-7}	7.992×10^{-1}	5.950×10^{-1}
	ε^p_{xx}	0.9923	3.269×10^{-6}	0.7385	3.795×10^{-2}	4.079×10^{-1}	5.120×10^{-1}	2.215×10^{-7}	7.964×10^{-1}	5.933×10^{-1}
	ε^p_{xy}	0.9901	2.085×10^{-6}	0.6195	3.218×10^{-1}	4.889×10^{-2}	3.509×10^{-1}	1.003×10^{-5}	5.014×10^{-1}	5.011×10^{-1}
	ε^p_{yy}	0.9923	3.243×10^{-6}	0.7463	3.195×10^{-2}	4.127×10^{-1}	5.069×10^{-1}	5.126×10^{-8}	7.976×10^{-1}	5.941×10^{-1}
	ε^p_{zz}	0.9865	5.307×10^{-6}	0.3235	4.347×10^{-1}	8.349×10^{-1}	7.773×10^{-1}	7.044×10^{-10}	9.194×10^{-1}	6.734×10^{-1}
	σ_{xx}	0.9798	9.128×10^{-6}	0.9496	1.676×10^{-2}	8.886×10^{-1}	7.682×10^{-1}	1.685×10^{-10}	8.854×10^{-1}	7.017×10^{-1}
	σ_{xy}	0.9760	2.915×10^{-7}	0.8405	9.858×10^{-2}	8.373×10^{-1}	6.984×10^{-1}	3.676×10^{-9}	8.605×10^{-1}	6.706×10^{-1}
	σ_{yy}	0.9908	2.639×10^{-6}	0.8574	6.011×10^{-2}	9.822×10^{-1}	8.925×10^{-1}	2.420×10^{-9}	9.120×10^{-1}	5.432×10^{-1}
	σ_{zz}	0.9914	2.684×10^{-6}	0.9484	1.407×10^{-2}	9.208×10^{-1}	8.774×10^{-1}	2.448×10^{-8}	9.326×10^{-1}	6.558×10^{-1}
	u	0.9981	3.358×10^{-8}	0.9690	1.919×10^{-2}	9.095×10^{-1}	8.721×10^{-1}	2.230×10^{-7}	9.443×10^{-1}	6.629×10^{-1}
	v	0.9976	5.954×10^{-9}	0.9610	3.203×10^{-2}	9.195×10^{-1}	8.903×10^{-1}	1.258×10^{-11}	9.549×10^{-1}	6.810×10^{-1}
mean		0.9900	6.155×10^{-9}	0.7797	8.709×10^{-2}	0.6188	0.6606	3.959×10^{-8}	0.8164	0.6131
MSE	ε^t_{xx}	2.916×10^{-5}	4.483×10^{-11}	6.325×10^{-4}	1.973×10^{-4}	2.230×10^{-3}	1.783×10^{-3}	2.569×10^{-11}	7.683×10^{-4}	1.540×10^{-3}
	ε^t_{xy}	3.237×10^{-5}	3.865×10^{-12}	1.702×10^{-3}	5.879×10^{-4}	3.066×10^{-3}	1.994×10^{-3}	2.168×10^{-10}	1.581×10^{-3}	1.608×10^{-3}
	ε^t_{yy}	2.927×10^{-5}	4.523×10^{-11}	4.815×10^{-4}	3.523×10^{-4}	2.244×10^{-3}	1.816×10^{-3}	1.841×10^{-12}	7.727×10^{-4}	1.558×10^{-3}
	ε^p_{xx}	2.945×10^{-5}	4.752×10^{-11}	9.969×10^{-4}	1.447×10^{-4}	2.257×10^{-3}	1.861×10^{-3}	3.220×10^{-12}	7.764×10^{-4}	1.551×10^{-3}
	ε^p_{xy}	2.974×10^{-5}	1.886×10^{-11}	1.144×10^{-3}	9.679×10^{-4}	2.861×10^{-3}	1.952×10^{-3}	9.069×10^{-11}	1.499×10^{-3}	1.500×10^{-3}
	ε^p_{yy}	2.954×10^{-5}	4.803×10^{-11}	9.764×10^{-4}	1.230×10^{-4}	2.260×10^{-3}	1.898×10^{-3}	7.593×10^{-13}	7.789×10^{-4}	1.562×10^{-3}
	ε^p_{zz}	1.921×10^{-9}	1.071×10^{-19}	2.604×10^{-3}	1.673×10^{-3}	2.345×10^{-8}	3.163×10^{-8}	1.421×10^{-23}	1.145×10^{-8}	4.638×10^{-8}
	σ_{xx}	1.599×10^{2}	5.736×10^{2}	3.997×10^{2}	1.329×10^{2}	8.831×10^{2}	1.837×10^{3}	1.059×10^{-2}	9.082×10^{2}	2.364×10^{3}
	σ_{xy}	1.196×10^{2}	7.221	7.941×10^{2}	4.906×10^{2}	8.099×10^{2}	1.501×10^{3}	9.107×10^{-2}	6.941×10^{2}	1.640×10^{3}
	σ_{yy}	5.013×10^{2}	7.911×10^{3}	7.809×10^{3}	3.291×10^{3}	9.766×10^{2}	5.883×10^{3}	7.253	4.818×10^{3}	2.501×10^{4}
	σ_{zz}	1.896×10^{2}	1.297×10^{3}	1.135×10^{3}	3.094×10^{2}	1.741×10^{3}	2.696×10^{3}	1.183×10^{1}	1.481×10^{3}	7.567×10^{3}
	u	4.736×10^{-3}	2.001×10^{-7}	7.558×10^{-2}	4.684×10^{-2}	2.210×10^{-1}	3.123×10^{-1}	1.329×10^{-6}	1.361×10^{-1}	8.230×10^{-1}
	v	0.0055	3.012×10^{-8}	0.0877	7.203×10^{-2}	1.810×10^{-1}	2.467×10^{-1}	6.361×10^{-11}	1.015×10^{-1}	7.174×10^{-1}

Table A8. Detailed results for the plate elongation use case Simulation 4.

		SIMULATION 4								
		MLP		PINN		SVR	GBDTR		KNNR	GPR
		mean	std	mean	std		mean	std		
R2	ε^t_{xx}	0.9994	1.890×10^{-5}	0.9615	1.029×10^{-2}	0.6110	0.9061	8.599×10^{-2}	0.9354	0.8478
	ε^t_{xy}	0.9984	1.387×10^{-4}	0.6064	2.507×10^{-1}	0.1553	0.8252	1.696×10^{-1}	0.7558	0.6991
	ε^t_{yy}	0.9994	1.450×10^{-5}	0.9817	8.052×10^{-3}	0.6169	0.9067	9.160×10^{-2}	0.9361	0.8495
	ε^p_{xx}	0.9994	1.634×10^{-5}	0.8183	5.486×10^{-2}	0.6087	0.8379	3.180×10^{-2}	0.9346	0.8457
	ε^p_{xy}	0.9984	1.217×10^{-5}	0.9410	7.109×10^{-3}	0.1468	0.7309	1.302×10^{-1}	0.7502	0.6967
	ε^p_{yy}	0.9994	1.572×10^{-5}	0.8881	8.563×10^{-4}	0.6117	0.8487	3.739×10^{-2}	0.9351	0.8467
	ε^p_{zz}	0.9934	1.400×10^{-3}	0.7888	2.487×10^{-3}	0.9349	0.9317	1.525×10^{-2}	0.9790	0.9440
	σ_{xx}	0.9957	1.192×10^{-4}	0.9903	3.579×10^{-4}	0.9326	0.9572	4.072×10^{-2}	0.9742	0.9404
	σ_{xy}	0.9930	6.390×10^{-4}	0.9753	1.977×10^{-4}	0.8643	0.8846	1.103×10^{-1}	0.9417	0.8974
	σ_{yy}	0.9985	9.487×10^{-5}	0.8972	3.716×10^{-3}	0.9932	0.9660	3.214×10^{-2}	0.9909	0.9736
	σ_{zz}	0.9972	1.784×10^{-4}	0.9813	2.062×10^{-4}	0.9813	0.9726	2.625×10^{-2}	0.9901	0.9667
	u	0.9995	4.181×10^{-5}	0.9934	3.368×10^{-4}	0.9297	0.9710	2.879×10^{-2}	0.9792	0.9370
	v	0.9997	6.148×10^{-5}	0.9927	1.157×10^{-3}	0.9392	0.9792	2.026×10^{-2}	0.9857	0.9454
mean		0.9978	1.970×10^{-4}	0.9089	2.598×10^{-2}	0.7174	0.9014	6.310×10^{-2}	0.9298	0.8761
MSE	ε^t_{xx}	2.150×10^{-6}	6.723×10^{-8}	1.370×10^{-4}	3.662×10^{-5}	1.384×10^{-3}	3.200×10^{-4}	3.200×10^{-4}	2.298×10^{-4}	5.412×10^{-4}
	ε^t_{xy}	4.496×10^{-6}	3.991×10^{-7}	1.132×10^{-3}	7.213×10^{-4}	2.430×10^{-3}	4.955×10^{-4}	4.955×10^{-4}	7.027×10^{-4}	8.656×10^{-4}
	ε^t_{yy}	2.186×10^{-6}	5.247×10^{-8}	6.610×10^{-5}	2.914×10^{-5}	1.386×10^{-3}	3.345×10^{-4}	3.345×10^{-4}	2.312×10^{-4}	5.447×10^{-4}
	ε^p_{xx}	2.173×10^{-6}	5.875×10^{-8}	6.531×10^{-4}	1.972×10^{-4}	1.407×10^{-3}	3.485×10^{-4}	3.484×10^{-4}	2.352×10^{-4}	5.547×10^{-4}
	ε^p_{xy}	4.228×10^{-6}	3.139×10^{-8}	1.522×10^{-4}	1.834×10^{-5}	2.202×10^{-3}	5.152×10^{-4}	5.152×10^{-4}	6.446×10^{-4}	7.827×10^{-4}
	ε^p_{yy}	2.194×10^{-6}	5.701×10^{-8}	4.060×10^{-4}	3.106×10^{-6}	1.409×10^{-3}	3.422×10^{-4}	3.422×10^{-4}	2.356×10^{-4}	5.562×10^{-4}
	ε^p_{zz}	7.663×10^{-10}	1.627×10^{-10}	7.659×10^{-4}	9.020×10^{-6}	7.573×10^{-9}	4.990×10^{-9}	4.726×10^{-9}	2.438×10^{-9}	6.512×10^{-9}
	σ_{xx}	5.480×10^{1}	1.515	1.226×10^{2}	4.547	8.561×10^{2}	5.339×10^{2}	5.272×10^{2}	3.283×10^{2}	7.575×10^{2}
	σ_{xy}	3.438×10^{1}	3.155	1.218×10^{2}	9.760×10^{-1}	6.699×10^{2}	5.597×10^{2}	5.547×10^{2}	2.880×10^{2}	5.066×10^{2}
	σ_{yy}	1.267×10^{2}	8.027	8.700×10^{3}	3.144×10^{2}	5.743×10^{2}	3.081×10^{3}	2.514×10^{3}	7.740×10^{2}	2.236×10^{3}
	σ_{zz}	6.863×10^{1}	4.437	4.654×10^{2}	5.129	4.640×10^{2}	6.858×10^{2}	6.491×10^{2}	2.456×10^{2}	8.285×10^{2}
	u	8.003×10^{-4}	6.664×10^{-5}	1.046×10^{-2}	5.368×10^{-4}	1.120×10^{-1}	4.637×10^{-2}	4.578×10^{-2}	3.320×10^{-2}	1.005×10^{-1}
	v	4.276×10^{-4}	8.849×10^{-5}	1.057×10^{-2}	1.666×10^{-3}	8.756×10^{-2}	2.954×10^{-2}	2.952×10^{-2}	2.062×10^{-2}	7.857×10^{-2}

Table A9. Detailed results for the plate elongation use case Simulation 6.

		SIMULATION 6								
		MLP		PINN		SVR	GBDTR		KNNR	GPR
		mean	std	mean	std		mean	std		
R2	ε^t_{xx}	0.9958	2.566×10^{-3}	0.9336	1.934×10^{-2}	0.6182	0.8405	1.562×10^{-1}	0.9228	0.8366
	ε^t_{xy}	0.9930	4.453×10^{-3}	0.2877	3.328×10^{-1}	0.1915	0.6924	3.030×10^{-1}	0.7280	0.6696
	ε^t_{yy}	0.9959	2.522×10^{-3}	0.9211	5.104×10^{-2}	0.6238	0.8344	1.648×10^{-1}	0.9238	0.8385
	ε^p_{xx}	0.9958	2.606×10^{-3}	0.5414	3.173×10^{-1}	0.6149	0.7529	8.924×10^{-2}	0.9221	0.8352
	ε^p_{xy}	0.9932	4.276×10^{-3}	0.9131	2.635×10^{-2}	0.1831	0.5966	1.726×10^{-1}	0.7218	0.6668
	ε^p_{yy}	0.9958	2.590×10^{-3}	0.7744	8.943×10^{-2}	0.6178	0.7799	9.094×10^{-2}	0.9228	0.8362
	ε^p_{zz}	0.9901	2.688×10^{-3}	0.8264	2.288×10^{-2}	0.9590	0.8938	2.641×10^{-3}	0.9860	0.9487
	σ_{xx}	0.9837	1.326×10^{-2}	0.9839	3.098×10^{-4}	0.9527	0.9600	3.838×10^{-2}	0.9799	0.9407
	σ_{xy}	0.9768	8.968×10^{-3}	0.9655	2.023×10^{-3}	0.8687	0.8571	1.381×10^{-1}	0.9431	0.9030
	σ_{yy}	0.9890	1.196×10^{-2}	0.9103	1.363×10^{-3}	0.9908	0.9729	2.571×10^{-2}	0.9938	0.9709
	σ_{zz}	0.9900	8.749×10^{-3}	0.9851	1.518×10^{-3}	0.9818	0.9682	3.102×10^{-2}	0.9923	0.9636
	u	0.9980	1.261×10^{-3}	0.9835	1.344×10^{-3}	0.9077	0.9366	6.320×10^{-2}	0.9717	0.9336
	v	0.9988	6.926×10^{-4}	0.9849	6.032×10^{-3}	0.9161	0.9685	3.118×10^{-2}	0.9760	0.9359
mean		0.9920	1.889×10^{-3}	0.8470	6.309×10^{-2}	0.7251	0.8503	1.005×10^{-1}	0.9219	0.8676
MSE	ε^t_{xx}	1.563×10^{-5}	9.386×10^{-6}	2.475×10^{-4}	7.210×10^{-5}	1.423×10^{-3}	5.884×10^{-4}	5.884×10^{-4}	2.878×10^{-4}	6.090×10^{-4}
	ε^t_{xy}	2.402×10^{-5}	1.372×10^{-5}	2.360×10^{-3}	1.103×10^{-3}	2.680×10^{-3}	1.012×10^{-3}	1.012×10^{-3}	9.014×10^{-4}	1.095×10^{-3}
	ε^t_{yy}	1.579×10^{-5}	9.392×10^{-6}	2.994×10^{-4}	1.936×10^{-4}	1.427×10^{-3}	6.268×10^{-4}	6.268×10^{-4}	2.890×10^{-4}	6.128×10^{-4}
	ε^p_{xx}	1.596×10^{-5}	9.635×10^{-6}	1.725×10^{-3}	1.194×10^{-3}	1.449×10^{-3}	6.327×10^{-4}	6.327×10^{-4}	2.929×10^{-4}	6.201×10^{-4}
	ε^p_{xy}	2.135×10^{-5}	1.199×10^{-5}	2.627×10^{-4}	7.965×10^{-5}	2.469×10^{-3}	8.705×10^{-4}	8.705×10^{-4}	8.407×10^{-4}	1.007×10^{-3}
	ε^p_{yy}	1.599×10^{-5}	9.654×10^{-6}	8.559×10^{-4}	3.392×10^{-4}	1.450×10^{-3}	5.899×10^{-4}	5.899×10^{-4}	2.930×10^{-4}	6.214×10^{-4}
	ε^p_{zz}	1.341×10^{-9}	6.361×10^{-10}	6.584×10^{-4}	8.677×10^{-5}	4.377×10^{-9}	5.918×10^{-9}	5.708×10^{-9}	1.492×10^{-9}	5.484×10^{-9}
	σ_{xx}	4.784×10^{2}	5.454×10^{2}	2.162×10^{2}	4.170	6.368×10^{2}	5.300×10^{2}	5.246×10^{2}	2.709×10^{2}	7.981×10^{2}
	σ_{xy}	2.621×10^{2}	2.590×10^{2}	1.620×10^{2}	9.491	6.161×10^{2}	6.614×10^{2}	6.566×10^{2}	2.669×10^{2}	4.552×10^{2}
	σ_{yy}	2.230×10^{3}	2.811×10^{3}	8.488×10^{3}	1.291×10^{2}	8.701×10^{2}	2.741×10^{3}	2.264×10^{3}	5.900×10^{2}	2.755×10^{3}
	σ_{zz}	4.078×10^{2}	4.385×10^{2}	3.842×10^{2}	3.926×10^{1}	4.705×10^{2}	8.263×10^{2}	7.996×10^{2}	1.981×10^{2}	9.411×10^{2}
	u	2.471×10^{-3}	1.295×10^{-3}	1.953×10^{-2}	1.591×10^{-3}	1.093×10^{-1}	7.515×10^{-2}	7.473×10^{-2}	3.349×10^{-2}	7.862×10^{-2}
	v	1.594×10^{-3}	3.341×10^{-4}	1.621×10^{-2}	6.487×10^{-3}	9.022×10^{-2}	3.373×10^{-2}	3.371×10^{-2}	2.578×10^{-2}	6.895×10^{-2}

Table A10. Detailed results for the plate elongation use case Simulation 9.

		SIMULATION 9								
		MLP		PINN		SVR	GBDTR		KNNR	GPR
		mean	std	mean	std		mean	std		
R2	ε^t_{xx}	0.9902	4.611×10^{-7}	0.8149	1.175×10^{-1}	5.305×10^{-1}	7.066×10^{-1}	4.844×10^{-7}	8.255×10^{-1}	5.599×10^{-1}
	ε^t_{xy}	0.9699	1.932×10^{-5}	0.3356	9.434×10^{-2}	5.673×10^{-2}	3.932×10^{-1}	1.518×10^{-8}	4.667×10^{-1}	3.628×10^{-1}
	ε^t_{yy}	0.9903	4.339×10^{-7}	0.8484	1.445×10^{-1}	5.344×10^{-1}	7.197×10^{-1}	3.018×10^{-8}	8.272×10^{-1}	5.611×10^{-1}
	ε^p_{xx}	0.9904	3.884×10^{-7}	0.6097	7.907×10^{-2}	5.180×10^{-1}	7.195×10^{-1}	4.415×10^{-11}	8.235×10^{-1}	5.571×10^{-1}
	ε^p_{xy}	0.9704	1.921×10^{-5}	0.5937	3.410×10^{-1}	4.617×10^{-2}	4.245×10^{-1}	3.808×10^{-8}	4.609×10^{-1}	3.586×10^{-1}
	ε^p_{yy}	0.9904	3.972×10^{-7}	0.7357	4.504×10^{-2}	5.204×10^{-1}	7.128×10^{-1}	7.477×10^{-7}	8.242×10^{-1}	5.579×10^{-1}
	ε^p_{zz}	0.9628	7.728×10^{-6}	0.3747	4.181×10^{-1}	9.292×10^{-1}	8.309×10^{-1}	6.767×10^{-9}	9.016×10^{-1}	6.839×10^{-1}
	σ_{xx}	0.9633	3.899×10^{-4}	0.9577	1.764×10^{-2}	9.258×10^{-1}	9.264×10^{-1}	1.662×10^{-9}	9.516×10^{-1}	6.847×10^{-1}
	σ_{xy}	0.9438	1.369×10^{-3}	0.8168	1.079×10^{-1}	8.161×10^{-1}	6.292×10^{-1}	1.792×10^{-7}	7.731×10^{-1}	5.730×10^{-1}
	σ_{yy}	0.9876	5.783×10^{-5}	0.9443	4.324×10^{-3}	9.779×10^{-1}	9.280×10^{-1}	1.478×10^{-11}	9.367×10^{-1}	5.502×10^{-1}
	σ_{zz}	0.9865	1.062×10^{-5}	0.9526	2.420×10^{-2}	9.704×10^{-1}	9.328×10^{-1}	3.200×10^{-10}	9.407×10^{-1}	6.108×10^{-1}
	u	0.9898	3.302×10^{-6}	0.9168	6.792×10^{-2}	8.492×10^{-1}	6.756×10^{-1}	2.928×10^{-7}	8.302×10^{-1}	6.322×10^{-1}
	v	0.9859	2.747×10^{-6}	0.9300	6.483×10^{-2}	8.630×10^{-1}	8.433×10^{-1}	2.501×10^{-8}	8.963×10^{-1}	6.544×10^{-1}
mean		0.9786	2.970×10^{-5}	0.7562	1.046×10^{-1}	0.6568	0.7263	9.780×10^{-9}	0.8045	0.5651
MSE	ε^t_{xx}	3.323×10^{-5}	5.335×10^{-12}	6.296×10^{-4}	3.996×10^{-4}	1.597×10^{-3}	9.979×10^{-4}	5.605×10^{-12}	5.937×10^{-4}	1.497×10^{-3}
	ε^t_{xy}	1.069×10^{-4}	2.439×10^{-10}	2.360×10^{-3}	3.352×10^{-4}	3.351×10^{-3}	2.156×10^{-3}	1.916×10^{-13}	1.895×10^{-3}	2.264×10^{-3}
	ε^t_{yy}	3.349×10^{-5}	5.193×10^{-12}	5.245×10^{-4}	5.000×10^{-4}	1.611×10^{-3}	9.698×10^{-4}	3.612×10^{-13}	5.977×10^{-4}	1.518×10^{-3}
	ε^p_{xx}	3.322×10^{-5}	4.646×10^{-12}	1.350×10^{-3}	2.735×10^{-4}	1.667×10^{-3}	9.703×10^{-4}	5.282×10^{-16}	6.106×10^{-4}	1.532×10^{-3}
	ε^p_{xy}	9.449×10^{-5}	1.962×10^{-10}	1.298×10^{-3}	1.090×10^{-3}	3.048×10^{-3}	1.839×10^{-3}	3.889×10^{-13}	1.723×10^{-3}	2.050×10^{-3}
	ε^p_{yy}	3.332×10^{-5}	4.818×10^{-12}	9.206×10^{-4}	1.569×10^{-4}	1.670×10^{-3}	1.000×10^{-3}	9.069×10^{-12}	6.123×10^{-4}	1.540×10^{-3}
	ε^p_{zz}	2.279×10^{-9}	2.901×10^{-20}	2.178×10^{-3}	1.456×10^{-3}	4.339×10^{-9}	1.036×10^{-8}	2.540×10^{-23}	6.026×10^{-9}	1.937×10^{-8}
	σ_{xx}	3.733×10^{2}	4.030×10^{4}	4.303×10^{2}	1.794×10^{2}	7.548×10^{2}	7.480×10^{2}	1.718×10^{-1}	4.916×10^{2}	3.206×10^{3}
	σ_{xy}	1.946×10^{2}	1.639×10^{4}	6.338×10^{2}	3.734×10^{2}	6.364×10^{2}	1.283×10^{3}	2.146	7.852×10^{2}	1.478×10^{3}
	σ_{yy}	1.092×10^{3}	4.456×10^{5}	4.890×10^{3}	3.796×10^{2}	1.943×10^{3}	6.320×10^{3}	1.139×10^{-1}	5.556×10^{3}	3.948×10^{4}
	σ_{zz}	2.613×10^{2}	3.972×10^{3}	9.161×10^{2}	4.681×10^{2}	5.727×10^{2}	1.300×10^{3}	1.197×10^{-1}	1.147×10^{3}	7.528×10^{3}
	u	5.525×10^{-3}	9.688×10^{-7}	4.508×10^{-2}	3.679×10^{-2}	8.165×10^{-2}	1.757×10^{-1}	8.590×10^{-8}	9.196×10^{-2}	1.992×10^{-1}
	v	0.0072	7.221×10^{-7}	0.0359	3.324×10^{-2}	7.021×10^{-2}	8.034×10^{-2}	6.573×10^{-9}	5.317×10^{-2}	1.772×10^{-1}

Table A11. Detailed results for the bending beam use case Simulation 1.

		SIMULATION 1								
		MLP		PINN		SVR	GBDTR		KNNR	GPR
		mean	std	mean	std		mean	std		
R2	ε^t_{xx}	0.8367	8.066×10^{-5}	0.7682	1.742×10^{-2}	0.8036	0.8566	2.006×10^{-6}	0.8377	0.6790
	ε^t_{xy}	0.8570	6.906×10^{-5}	0.6632	1.499×10^{-1}	0.4932	0.9030	2.784×10^{-7}	0.6409	0.5727
	ε^t_{yy}	0.9594	1.142×10^{-5}	0.8487	3.318×10^{-4}	0.9520	0.9651	2.776×10^{-9}	0.9607	0.8805
	ε^p_{xx}	0.0647	7.879×10^{-6}	0.0304	1.866×10^{-4}	0.0083	0.1599	1.015×10^{-5}	0.1426	0.0633
	ε^p_{xy}	−0.0091	4.606×10^{-4}	−0.0106	3.336×10^{-4}	−0.0051	0.0906	3.080×10^{-6}	0.0098	0.0652
	ε^p_{yy}	0.0723	2.093×10^{-5}	0.0335	4.720×10^{-4}	0.0051	0.1746	9.704×10^{-7}	0.1568	0.0720
	ε^p_{zz}	0.1157	3.506×10^{-4}	−0.0078	8.842×10^{-4}	0.0152	0.2458	1.254×10^{-6}	0.2373	0.1278
	σ_{xx}	0.9643	3.732×10^{-4}	0.9822	2.748×10^{-3}	0.0291	0.9837	3.203×10^{-7}	0.6293	0.1681
	σ_{xy}	0.9157	1.814×10^{-4}	0.8902	1.515×10^{-2}	0.4227	0.9451	4.972×10^{-6}	0.6324	0.5024
	σ_{yy}	0.9482	8.435×10^{-5}	0.9546	1.249×10^{-3}	0.9618	0.9547	1.283×10^{-7}	0.9540	0.9815
	σ_{zz}	0.9789	1.711×10^{-5}	0.9742	1.819×10^{-3}	0.9766	0.9818	2.750×10^{-7}	0.9781	0.9521
	u	0.9948	6.208×10^{-6}	0.9974	5.933×10^{-4}	0.9978	0.9974	1.963×10^{-10}	0.9972	0.9678
	v	0.9875	2.782×10^{-5}	0.8897	6.238×10^{-2}	0.9976	0.9973	6.646×10^{-8}	0.9976	0.9580
mean		0.6682	7.345×10^{-6}	0.6165	1.648×10^{-2}	0.5122	0.7120	1.088×10^{-8}	0.6288	0.5377
MSE	ε^t_{xx}	3.452×10^{-7}	3.602×10^{-16}	4.899×10^{-7}	3.682×10^{-8}	4.151×10^{-7}	3.077×10^{-7}	9.363×10^{-17}	3.430×10^{-7}	6.783×10^{-7}
	ε^t_{xy}	3.723×10^{-8}	4.679×10^{-18}	8.765×10^{-8}	3.903×10^{-8}	1.319×10^{-7}	2.554×10^{-8}	7.242×10^{-20}	9.346×10^{-8}	1.112×10^{-7}
	ε^t_{yy}	2.876×10^{-7}	5.739×10^{-16}	1.073×10^{-6}	2.352×10^{-9}	3.402×10^{-7}	2.484×10^{-7}	3.517×10^{-18}	2.785×10^{-7}	8.473×10^{-7}
	ε^p_{xx}	4.778×10^{-7}	2.056×10^{-18}	4.953×10^{-7}	9.531×10^{-11}	5.066×10^{-7}	4.288×10^{-7}	4.320×10^{-18}	4.380×10^{-7}	4.785×10^{-7}
	ε^p_{xy}	1.779×10^{-8}	1.431×10^{-19}	1.781×10^{-8}	5.879×10^{-12}	1.772×10^{-8}	1.604×10^{-8}	2.456×10^{-21}	1.745×10^{-8}	1.648×10^{-8}
	ε^p_{yy}	6.251×10^{-7}	9.503×10^{-18}	6.512×10^{-7}	3.180×10^{-10}	6.704×10^{-7}	5.577×10^{-7}	2.451×10^{-18}	5.681×10^{-7}	6.253×10^{-7}
	ε^p_{zz}	1.029×10^{-8}	4.752×10^{-20}	6.790×10^{-7}	5.958×10^{-10}	1.146×10^{-8}	8.783×10^{-9}	8.663×10^{-23}	8.878×10^{-9}	1.015×10^{-8}
	σ_{xx}	9.974×10^{1}	2.906×10^{3}	4.964×10^{1}	7.668	2.709×10^{3}	4.422×10^{1}	5.128×10^{-2}	1.035×10^{3}	2.322×10^{3}
	σ_{xy}	7.615×10^{1}	1.479×10^{2}	9.920×10^{1}	1.368×10^{1}	5.214×10^{2}	5.007×10^{1}	7.083	3.320×10^{2}	4.494×10^{2}
	σ_{yy}	1.295×10^{4}	5.275×10^{6}	1.135×10^{4}	3.123×10^{2}	9.543×10^{3}	1.141×10^{4}	4.827×10^{4}	1.150×10^{4}	4.631×10^{3}
	σ_{zz}	5.921×10^{2}	1.342×10^{4}	7.215×10^{2}	5.094×10^{1}	6.548×10^{2}	5.054×10^{2}	6.585×10^{1}	6.130×10^{2}	1.342×10^{3}
	u	1.251×10^{-2}	3.633×10^{-5}	6.323×10^{-3}	1.435×10^{-3}	5.220×10^{-3}	6.289×10^{-3}	1.284×10^{-8}	6.723×10^{-3}	7.797×10^{-2}
	v	4.059×10^{-4}	2.945×10^{-8}	3.589×10^{-3}	2.030×10^{-3}	7.902×10^{-5}	8.486×10^{-5}	2.337×10^{-11}	7.796×10^{-5}	1.367×10^{-3}

Table A12. Detailed results for the bending beam use case Simulation 4.

		SIMULATION 4								
		MLP		PINN		SVR	GBDTR		KNNR	GPR
		mean	std	mean	std		mean	std		
R2	ε^t_{xx}	0.9992	4.882×10^{-4}	0.9839	6.920×10^{-4}	0.9601	0.9928	1.008×10^{-3}	0.9940	0.9590
	ε^t_{xy}	0.9977	1.512×10^{-3}	0.9643	8.745×10^{-3}	0.5794	0.9941	1.496×10^{-4}	0.9321	0.7219
	ε^t_{yy}	0.9996	2.146×10^{-4}	0.9793	1.865×10^{-4}	0.9970	0.9981	4.564×10^{-4}	0.9995	0.9922
	ε^p_{xx}	0.9965	1.939×10^{-3}	0.8471	6.012×10^{-3}	0.8796	0.8709	2.564×10^{-3}	0.9819	0.8397
	ε^p_{xy}	0.9894	8.291×10^{-3}	0.9911	4.488×10^{-4}	0.0827	0.8513	3.840×10^{-3}	0.7997	0.3642
	ε^p_{yy}	0.9972	1.521×10^{-3}	0.8546	9.650×10^{-4}	0.8968	0.8881	2.980×10^{-3}	0.9855	0.8495
	ε^p_{zz}	0.9985	6.214×10^{-4}	0.7411	1.351×10^{-2}	0.9352	0.9479	2.379×10^{-3}	0.9944	0.8838
	σ_{xx}	0.9981	8.971×10^{-4}	0.9987	5.263×10^{-4}	0.0418	0.9979	7.556×10^{-5}	0.9046	0.4887
	σ_{xy}	0.9974	1.516×10^{-3}	0.9799	1.242×10^{-2}	0.4600	0.9946	4.227×10^{-4}	0.9167	0.6331
	σ_{yy}	0.9997	1.157×10^{-4}	0.9169	2.384×10^{-4}	0.9992	0.9983	1.213×10^{-5}	0.9998	0.9965
	σ_{zz}	0.9997	1.080×10^{-4}	0.9852	2.265×10^{-4}	0.9939	0.9990	1.026×10^{-4}	0.9989	0.9916
	u	0.9998	2.885×10^{-5}	0.9989	3.940×10^{-4}	1.0000	0.9996	7.767×10^{-5}	0.9998	0.9989
	v	0.9997	4.696×10^{-5}	0.9517	1.110×10^{-3}	1.0000	0.9995	4.749×10^{-5}	0.9999	0.9974
mean		0.9979	1.319×10^{-3}	0.9379	1.042×10^{-3}	0.7558	0.9640	6.751×10^{-4}	0.9621	0.8243
MSE	ε^t_{xx}	1.158×10^{-9}	7.461×10^{-10}	2.467×10^{-8}	1.057×10^{-9}	6.097×10^{-8}	1.099×10^{-8}	1.541×10^{-9}	9.198×10^{-9}	6.258×10^{-8}
	ε^t_{xy}	5.680×10^{-10}	3.784×10^{-10}	8.924×10^{-9}	2.188×10^{-9}	1.052×10^{-7}	1.471×10^{-9}	3.743×10^{-11}	1.700×10^{-8}	6.960×10^{-8}
	ε^t_{yy}	2.708×10^{-9}	1.457×10^{-9}	1.404×10^{-7}	1.267×10^{-9}	2.058×10^{-8}	1.267×10^{-8}	3.100×10^{-9}	3.589×10^{-9}	5.309×10^{-8}
	ε^p_{xx}	3.854×10^{-10}	2.155×10^{-10}	1.699×10^{-8}	6.681×10^{-10}	1.337×10^{-8}	1.435×10^{-8}	2.849×10^{-10}	2.012×10^{-9}	1.782×10^{-8}
	ε^p_{xy}	4.304×10^{-11}	3.372×10^{-11}	3.608×10^{-11}	1.825×10^{-12}	3.730×10^{-9}	6.049×10^{-10}	1.562×10^{-11}	8.146×10^{-10}	2.586×10^{-9}
	ε^p_{yy}	4.490×10^{-10}	2.477×10^{-10}	2.367×10^{-8}	1.571×10^{-10}	1.679×10^{-8}	1.822×10^{-8}	4.851×10^{-10}	2.358×10^{-9}	2.449×10^{-8}
	ε^p_{zz}	7.588×10^{-12}	3.101×10^{-12}	4.214×10^{-8}	2.199×10^{-9}	3.233×10^{-10}	2.600×10^{-10}	1.187×10^{-11}	2.771×10^{-11}	5.800×10^{-10}
	σ_{xx}	6.301	2.904	4.242	1.704	3.102×10^{3}	6.781	2.446×10^{-1}	3.087×10^{2}	1.655×10^{3}
	σ_{xy}	2.546	1.489	1.974×10^{1}	1.220×10^{1}	5.306×10^{2}	5.268	4.153×10^{-1}	8.179×10^{1}	3.604×10^{2}
	σ_{yy}	9.601×10^{1}	3.555×10^{1}	2.554×10^{4}	7.326×10^{1}	2.483×10^{2}	5.289×10^{2}	3.727	5.885×10^{1}	1.065×10^{3}
	σ_{zz}	9.587	3.424	4.705×10^{2}	7.179	1.947×10^{2}	3.277×10^{1}	3.251	3.488×10^{1}	2.663×10^{2}
	u	5.949×10^{-4}	7.015×10^{-5}	2.758×10^{-3}	9.579×10^{-4}	1.676×10^{-5}	8.592×10^{-4}	1.889×10^{-4}	3.838×10^{-4}	2.781×10^{-3}
	v	9.229×10^{-6}	1.558×10^{-6}	1.601×10^{-3}	3.683×10^{-5}	5.846×10^{-7}	1.598×10^{-5}	1.576×10^{-6}	2.105×10^{-6}	8.758×10^{-5}

Table A13. Detailed results for the bending beam use case Simulation 6.

		SIMULATION 6								
		MLP		PINN		SVR	GBDTR		KNNR	GPR
		mean	std	mean	std		mean	std		
R2	ε^t_{xx}	0.9997	1.585×10^{-4}	0.9827	1.679×10^{-3}	0.9606	0.9970	3.028×10^{-4}	0.9946	0.9615
	ε^t_{xy}	0.9988	5.440×10^{-4}	0.9636	1.211×10^{-2}	0.6360	0.9956	2.013×10^{-4}	0.9418	0.7492
	ε^t_{yy}	0.9997	1.581×10^{-4}	0.9683	3.948×10^{-4}	0.9979	0.9990	1.438×10^{-4}	0.9997	0.9935
	ε^p_{xx}	0.9973	1.546×10^{-3}	0.8567	3.014×10^{-3}	0.7713	0.8529	1.868×10^{-3}	0.9825	0.7532
	ε^p_{xy}	0.9887	5.094×10^{-3}	0.9788	7.437×10^{-3}	0.1049	0.7773	4.855×10^{-3}	0.7837	0.3351
	ε^p_{yy}	0.9979	1.115×10^{-3}	0.8769	1.114×10^{-2}	0.7980	0.8627	1.675×10^{-3}	0.9851	0.7650
	ε^p_{zz}	0.9980	7.159×10^{-4}	0.6162	1.176×10^{-2}	0.8392	0.8955	3.934×10^{-4}	0.9910	0.8054
	σ_{xx}	0.9985	5.385×10^{-4}	0.9993	1.296×10^{-4}	0.0374	0.9987	1.248×10^{-4}	0.9051	0.4865
	σ_{xy}	0.9984	7.070×10^{-4}	0.9819	1.416×10^{-2}	0.4890	0.9953	2.116×10^{-4}	0.9198	0.6422
	σ_{yy}	0.9997	1.586×10^{-4}	0.9465	1.587×10^{-3}	0.9993	0.9988	1.469×10^{-4}	0.9999	0.9969
	σ_{zz}	0.9997	1.588×10^{-4}	0.9890	3.975×10^{-5}	0.9937	0.9991	1.271×10^{-4}	0.9990	0.9919
	u	0.9997	1.787×10^{-5}	0.9984	2.302×10^{-4}	1.0000	0.9997	6.919×10^{-5}	0.9998	0.9989
	v	0.9997	6.709×10^{-5}	0.9503	1.090×10^{-3}	1.0000	0.9995	1.477×10^{-4}	0.9999	0.9974
mean		0.9981	8.315×10^{-4}	0.9314	8.396×10^{-4}	0.7406	0.9516	1.368×10^{-4}	0.9617	0.8059
MSE	ε^t_{xx}	4.762×10^{-10}	2.159×10^{-10}	2.352×10^{-8}	2.287×10^{-9}	5.373×10^{-8}	4.027×10^{-9}	4.125×10^{-10}	7.327×10^{-9}	5.246×10^{-8}
	ε^t_{xy}	2.949×10^{-10}	1.330×10^{-10}	8.896×10^{-9}	2.962×10^{-9}	8.902×10^{-8}	1.069×10^{-9}	4.922×10^{-11}	1.423×10^{-8}	6.133×10^{-8}
	ε^t_{yy}	1.818×10^{-9}	1.064×10^{-9}	2.134×10^{-7}	2.658×10^{-9}	1.434×10^{-8}	6.562×10^{-9}	9.683×10^{-10}	2.272×10^{-9}	4.385×10^{-8}
	ε^p_{xx}	9.135×10^{-11}	5.298×10^{-11}	4.910×10^{-9}	1.033×10^{-10}	7.838×10^{-9}	5.041×10^{-9}	6.403×10^{-11}	5.983×10^{-10}	8.458×10^{-9}
	ε^p_{xy}	1.017×10^{-11}	4.605×10^{-12}	1.921×10^{-11}	6.723×10^{-12}	8.092×10^{-10}	2.014×10^{-10}	4.389×10^{-12}	1.955×10^{-10}	6.011×10^{-10}
	ε^p_{yy}	1.126×10^{-10}	5.894×10^{-11}	6.510×10^{-9}	5.889×10^{-10}	1.068×10^{-8}	7.262×10^{-9}	8.856×10^{-11}	7.876×10^{-10}	1.242×10^{-8}
	ε^p_{zz}	4.137×10^{-12}	1.456×10^{-12}	2.030×10^{-8}	6.220×10^{-10}	3.271×10^{-10}	2.126×10^{-10}	8.001×10^{-13}	1.840×10^{-11}	3.959×10^{-10}
	σ_{xx}	4.891	1.801	2.311	4.333×10^{-1}	3.219×10^{3}	4.183	4.173×10^{-1}	3.174×10^{2}	1.717×10^{3}
	σ_{xy}	1.582	7.091×10^{-1}	1.820×10^{1}	1.420×10^{1}	5.125×10^{2}	4.749	2.123×10^{-1}	8.043×10^{1}	3.589×10^{2}
	σ_{yy}	8.720×10^{1}	5.288×10^{1}	1.785×10^{4}	5.289×10^{2}	2.192×10^{2}	4.124×10^{2}	4.896×10^{1}	4.382×10^{1}	1.044×10^{3}
	σ_{zz}	8.678	5.215	3.603×10^{2}	1.305	2.061×10^{2}	2.859×10^{1}	4.174	3.387×10^{1}	2.660×10^{2}
	u	7.648×10^{-4}	4.353×10^{-5}	3.887×10^{-3}	5.607×10^{-4}	1.611×10^{-5}	6.987×10^{-4}	1.685×10^{-4}	3.679×10^{-4}	2.779×10^{-3}
	v	9.476×10^{-6}	2.243×10^{-6}	1.663×10^{-3}	3.645×10^{-5}	4.862×10^{-7}	1.622×10^{-5}	4.937×10^{-6}	1.937×10^{-6}	8.695×10^{-5}

Table A14. Detailed results for the bending beam use case Simulation 9.

		SIMULATION 9								
		MLP		PINN		SVR	GBDTR		KNNR	GPR
		mean	std	mean	std		mean	std		
R2	ε^t_{xx}	0.7851	3.018×10^{-3}	0.6358	6.166×10^{-4}	0.8656	0.8172	3.255×10^{-5}	0.7939	0.9395
	ε^t_{xy}	0.8255	4.803×10^{-3}	0.8392	2.690×10^{-2}	0.6478	0.9133	4.185×10^{-7}	0.7931	0.6226
	ε^t_{yy}	0.9627	3.529×10^{-6}	0.9126	3.213×10^{-4}	0.9776	0.9721	3.246×10^{-8}	0.9732	0.9432
	ε^p_{xx}	−984.3572	3.895×10^{4}	−2372.9520	1.322×10^{1}	−305.8183	−823.3812	3.067×10^{1}	-809.9266	−220.5149
	ε^p_{xy}	−13394.3542	1.990×10^{6}	−13885.9308	3.975×10^{2}	−604.7674	−8965.1296	4.317×10^{2}	−2397.3474	−4955.1166
	ε^p_{yy}	−821.5068	2.422×10^{4}	−343.2822	2.128	−282.6090	−698.7968	1.881×10^{-1}	−683.0748	−193.7274
	ε^p_{zz}	−359.9536	1.372×10^{3}	−382.8576	1.639×10^{1}	−214.6722	−322.9159	4.464×10^{-2}	−309.8423	−109.3652
	σ_{xx}	0.9529	1.611×10^{-3}	0.9893	1.808×10^{-3}	0.0275	0.9932	2.982×10^{-9}	0.5629	0.2630
	σ_{xy}	0.9120	1.109×10^{-3}	0.9260	1.163×10^{-2}	0.4862	0.9627	3.386×10^{-7}	0.6904	0.5129
	σ_{yy}	0.9672	5.183×10^{-6}	0.8105	3.138×10^{-3}	0.9569	0.9763	4.897×10^{-8}	0.9745	0.8841
	σ_{zz}	0.9825	1.315×10^{-5}	0.9681	8.932×10^{-5}	0.9738	0.9887	2.130×10^{-7}	0.9855	0.9184
	u	0.9947	1.871×10^{-5}	0.9985	4.685×10^{-4}	0.9950	0.9980	2.820×10^{-9}	0.9978	0.9581
	v	0.9933	2.244×10^{-5}	0.9493	1.716×10^{-3}	0.9960	0.9980	2.141×10^{-9}	0.9982	0.9443
mean		−1196.2920	6.166×10^{3}	−1305.9226	3.269×10^{1}	−107.7646	−830.8926	4.298	−322.4940	−420.9029
MSE	ε^t_{xx}	2.655×10^{-7}	4.608×10^{-15}	4.500×10^{-7}	7.618×10^{-10}	1.660×10^{-7}	2.177×10^{-7}	2.073×10^{-17}	2.546×10^{-7}	7.480×10^{-8}
	ε^t_{xy}	4.174×10^{-8}	2.747×10^{-16}	3.847×10^{-8}	6.433×10^{-9}	8.423×10^{-8}	2.062×10^{-8}	6.416×10^{-22}	4.949×10^{-8}	9.025×10^{-8}
	ε^t_{yy}	2.494×10^{-7}	1.581×10^{-16}	5.852×10^{-7}	2.150×10^{-9}	1.500×10^{-7}	1.973×10^{-7}	2.708×10^{-16}	1.793×10^{-7}	3.804×10^{-7}
	ε^p_{xx}	3.694×10^{-7}	5.475×10^{-15}	8.900×10^{-7}	4.957×10^{-9}	1.150×10^{-7}	3.109×10^{-7}	2.740×10^{-19}	3.040×10^{-7}	8.305×10^{-8}
	ε^p_{xy}	1.826×10^{-8}	3.699×10^{-18}	1.893×10^{-8}	5.419×10^{-10}	8.258×10^{-10}	1.224×10^{-8}	6.153×10^{-24}	3.270×10^{-9}	6.756×10^{-9}
	ε^p_{yy}	4.966×10^{-7}	8.827×10^{-15}	2.079×10^{-7}	1.285×10^{-9}	1.712×10^{-7}	4.229×10^{-7}	6.942×10^{-20}	4.130×10^{-7}	1.176×10^{-7}
	ε^p_{zz}	9.798×10^{-9}	1.011×10^{-18}	2.318×10^{-7}	9.893×10^{-9}	5.854×10^{-9}	8.779×10^{-9}	6.376×10^{-22}	8.438×10^{-9}	2.996×10^{-9}
	σ_{xx}	1.599×10^{2}	1.859×10^{4}	3.640×10^{1}	6.144	3.304×10^{3}	2.324×10^{1}	2.649×10^{-2}	1.485×10^{3}	2.504×10^{3}
	σ_{xy}	8.763×10^{1}	1.101×10^{3}	7.371×10^{1}	1.158×10^{1}	5.118×10^{2}	3.585×10^{1}	1.534	3.084×10^{2}	4.852×10^{2}
	σ_{yy}	1.180×10^{4}	6.690×10^{5}	6.809×10^{4}	1.127×10^{3}	1.547×10^{4}	8.593×10^{3}	1.887×10^{3}	9.171×10^{3}	4.165×10^{4}
	σ_{zz}	5.870×10^{2}	1.478×10^{4}	1.069×10^{3}	2.994	8.768×10^{2}	3.739×10^{2}	4.308×10^{2}	4.876×10^{2}	2.737×10^{3}
	u	1.288×10^{-2}	1.115×10^{-4}	3.569×10^{-3}	1.144×10^{-3}	1.217×10^{-2}	4.617×10^{-3}	6.444×10^{-9}	5.283×10^{-3}	1.023×10^{-1}
	v	2.259×10^{-4}	2.552×10^{-8}	1.711×10^{-3}	5.786×10^{-5}	1.339×10^{-4}	6.601×10^{-5}	2.616×10^{-12}	6.196×10^{-5}	1.880×10^{-3}

Table A15. Detailed results for the block compression use case Simulation 1.

		SIMULATION 1								
		MLP		PINN		SVR	GBDTR		KNNR	GPR
		mean	std	mean	std		mean	std		
R2	ε^t_{xx}	0.7303	1.285×10^{-1}	−3.1200	1.797	−1.0225	0.4661	1.606×10^{-3}	0.7233	0.0611
	ε^t_{xy}	0.5272	2.271×10^{-1}	−1.8630	3.719×10^{-1}	−0.1080	0.4285	3.330×10^{-4}	0.7319	0.0383
	ε^t_{yy}	0.7301	1.291×10^{-1}	−3.0260	2.761	−1.0094	0.4515	2.165×10^{-4}	0.7236	0.0549
	ε^p_{xx}	0.7267	1.309×10^{-1}	−2.6201	1.502	−0.9605	0.4632	1.165×10^{-3}	0.7213	0.0619
	ε^p_{xy}	0.5208	2.282×10^{-1}	−0.2147	6.512×10^{-1}	−0.0480	0.3885	1.938×10^{-2}	0.7295	0.0348
	ε^p_{yy}	0.7273	1.308×10^{-1}	−0.4142	7.218×10^{-1}	−0.9576	0.4298	1.476×10^{-3}	0.7217	0.0624
	ε^p_{zz}	0.3664	2.504×10^{-1}	0.2890	8.202×10^{-2}	−1.0694	0.2682	1.412×10^{-3}	0.6842	0.0806
	σ_{xx}	0.2825	2.839×10^{-1}	0.8531	3.238×10^{-2}	−0.5722	0.7233	2.191×10^{-4}	0.8244	0.1672
	σ_{xy}	0.2157	4.538×10^{-1}	0.8347	8.739×10^{-3}	0.0408	0.5910	4.449×10^{-4}	0.7997	0.1585
	σ_{yy}	0.2810	2.974×10^{-1}	0.7870	2.325×10^{-2}	0.7210	0.6808	3.475×10^{-4}	0.7726	−0.3058
	σ_{zz}	0.3929	3.008×10^{-1}	0.8048	3.480×10^{-3}	0.6143	0.6356	2.169×10^{-4}	0.8023	−0.1644
	u	0.8266	7.245×10^{-2}	0.9294	4.231×10^{-3}	0.4551	0.9157	4.560×10^{-5}	0.9360	0.5587
	v	0.9031	5.961×10^{-2}	0.9872	2.228×10^{-4}	0.7144	0.9619	1.290×10^{-5}	0.9805	0.5683
mean		0.5562	1.952×10^{-1}	−0.4441	6.066×10^{-1}	−0.2463	0.5695	1.665×10^{-3}	0.7808	0.1059
MSE	ε^t_{xx}	9.226×10^{-4}	4.397×10^{-4}	1.409×10^{-2}	6.146×10^{-3}	6.918×10^{-3}	1.826×10^{-3}	5.492×10^{-6}	9.465×10^{-4}	3.211×10^{-3}
	ε^t_{xy}	2.624×10^{-3}	1.261×10^{-3}	1.589×10^{-2}	2.065×10^{-3}	6.150×10^{-3}	3.172×10^{-3}	1.849×10^{-6}	1.488×10^{-3}	5.338×10^{-3}
	ε^t_{yy}	9.301×10^{-4}	4.449×10^{-4}	1.388×10^{-2}	9.514×10^{-3}	6.925×10^{-3}	1.890×10^{-3}	7.460×10^{-7}	9.526×10^{-4}	3.257×10^{-3}
	ε^p_{xx}	9.391×10^{-4}	4.498×10^{-4}	1.244×10^{-2}	5.160×10^{-3}	6.737×10^{-3}	1.845×10^{-3}	4.004×10^{-6}	9.577×10^{-4}	3.224×10^{-3}
	ε^p_{xy}	2.464×10^{-3}	1.173×10^{-3}	6.245×10^{-3}	3.348×10^{-3}	5.387×10^{-3}	3.143×10^{-3}	9.964×10^{-5}	1.391×10^{-3}	4.962×10^{-3}
	ε^p_{yy}	9.431×10^{-4}	4.524×10^{-4}	4.890×10^{-3}	2.496×10^{-3}	6.769×10^{-3}	1.972×10^{-3}	5.103×10^{-6}	9.625×10^{-4}	3.242×10^{-3}
	ε^p_{zz}	6.827×10^{-8}	2.698×10^{-8}	2.458×10^{-3}	2.836×10^{-4}	2.230×10^{-7}	7.885×10^{-8}	1.521×10^{-10}	3.403×10^{-8}	9.906×10^{-8}
	σ_{xx}	1.703×10^{4}	6.737×10^{3}	3.486×10^{3}	7.684×10^{2}	3.732×10^{4}	6.568×10^{3}	5.200	4.168×10^{3}	1.977×10^{4}
	σ_{xy}	1.182×10^{4}	6.839×10^{3}	2.491×10^{3}	1.317×10^{2}	1.446×10^{4}	6.164×10^{3}	6.705	3.019×10^{3}	1.268×10^{4}
	σ_{yy}	9.018×10^{4}	3.730×10^{4}	2.671×10^{4}	2.916×10^{3}	3.499×10^{4}	4.003×10^{4}	4.358×10^{1}	2.852×10^{4}	1.638×10^{5}
	σ_{zz}	1.872×10^{4}	9.276×10^{3}	6.021×10^{3}	1.073×10^{2}	1.189×10^{4}	1.124×10^{4}	6.688	6.098×10^{3}	3.591×10^{4}
	u	3.141×10^{-1}	1.312×10^{-1}	1.278×10^{-1}	7.662×10^{-3}	9.869×10^{-1}	1.526×10^{-1}	8.258×10^{-5}	1.159×10^{-1}	7.993×10^{-1}
	v	4.938×10^{-1}	3.039×10^{-1}	6.544×10^{-2}	1.136×10^{-3}	1.456	1.943×10^{-1}	6.575×10^{-5}	9.955×10^{-2}	2.201

Table A16. Detailed results for the block compression use case Simulation 2.

		SIMULATION 2								
		MLP		PINN		SVR	GBDTR		KNNR	GPR
		mean	std	mean	std		mean	std		
R2	ε^t_{xx}	0.7096	2.194×10^{-1}	−1.8843	1.366×10^{-1}	−0.2305	0.5318	2.513×10^{-3}	0.7129	0.0908
	ε^t_{xy}	0.5973	3.255×10^{-1}	−1.0111	5.049×10^{-1}	−0.1731	0.2782	2.402×10^{-3}	0.6317	0.0484
	ε^t_{yy}	0.7065	2.224×10^{-1}	0.4236	2.569×10^{-1}	−0.2238	0.5362	1.582×10^{-3}	0.7144	0.0893
	ε^p_{xx}	0.7009	2.318×10^{-1}	−1.1047	4.716×10^{-1}	−0.1968	0.5474	7.692×10^{-4}	0.7105	0.0898
	ε^p_{xy}	0.6101	3.059×10^{-1}	0.6086	2.773×10^{-2}	−0.0903	0.2472	3.168×10^{-3}	0.6283	0.0483
	ε^p_{yy}	0.7005	2.322×10^{-1}	−0.0181	3.360×10^{-2}	−0.2002	0.5343	8.637×10^{-4}	0.7113	0.0899
	ε^p_{zz}	−0.5742	9.762×10^{-1}	0.6019	7.576×10^{-2}	−1.0839	0.3289	7.350×10^{-4}	0.7009	0.0660
	σ_{xx}	0.3274	3.777×10^{-1}	0.8818	2.484×10^{-3}	−1.6968	0.6480	3.008×10^{-4}	0.8516	0.3220
	σ_{xy}	−0.4305	9.030×10^{-1}	0.5625	6.341×10^{-3}	−0.5689	0.0328	2.754×10^{-4}	0.5789	0.2049
	σ_{yy}	0.1829	4.317×10^{-1}	0.7002	3.199×10^{-3}	0.5536	0.6007	2.593×10^{-5}	0.7000	−0.1031
	σ_{zz}	−0.3596	8.129×10^{-1}	0.7370	1.084×10^{-2}	0.4829	0.5160	1.780×10^{-4}	0.7096	−0.1446
	u	0.8932	1.158×10^{-1}	0.9263	8.374×10^{-3}	0.3561	0.9418	7.221×10^{-5}	0.9579	0.4798
	v	0.8345	1.753×10^{-1}	0.9813	4.156×10^{-3}	0.7313	0.9507	2.445×10^{-5}	0.9677	0.5499
mean		0.3768	3.803×10^{-1}	0.1850	5.531×10^{-2}	−0.1800	0.5149	3.320×10^{-4}	0.7366	0.1409
MSE	ε^t_{xx}	1.744×10^{-3}	1.318×10^{-3}	1.732×10^{-2}	8.206×10^{-4}	7.390×10^{-3}	2.812×10^{-3}	1.509×10^{-5}	1.724×10^{-3}	5.461×10^{-3}
	ε^t_{xy}	2.154×10^{-3}	1.741×10^{-3}	1.076×10^{-2}	2.700×10^{-3}	6.274×10^{-3}	3.861×10^{-3}	1.285×10^{-5}	1.970×10^{-3}	5.090×10^{-3}
	ε^t_{yy}	1.772×10^{-3}	1.342×10^{-3}	3.479×10^{-3}	1.550×10^{-3}	7.387×10^{-3}	2.799×10^{-3}	9.547×10^{-6}	1.724×10^{-3}	5.496×10^{-3}
	ε^p_{xx}	1.811×10^{-3}	1.404×10^{-3}	1.275×10^{-2}	2.856×10^{-3}	7.248×10^{-3}	2.741×10^{-3}	4.658×10^{-6}	1.753×10^{-3}	5.512×10^{-3}
	ε^p_{xy}	1.859×10^{-3}	1.459×10^{-3}	1.866×10^{-3}	1.322×10^{-4}	5.199×10^{-3}	3.590×10^{-3}	1.511×10^{-5}	1.772×10^{-3}	4.538×10^{-3}
	ε^p_{yy}	1.824×10^{-3}	1.414×10^{-3}	6.200×10^{-3}	2.046×10^{-4}	7.309×10^{-3}	2.836×10^{-3}	5.260×10^{-6}	1.758×10^{-3}	5.542×10^{-3}
	ε^p_{zz}	1.408×10^{-7}	8.729×10^{-8}	2.424×10^{-3}	4.614×10^{-4}	1.863×10^{-7}	6.001×10^{-8}	6.573×10^{-11}	2.675×10^{-8}	8.352×10^{-8}
	σ_{xx}	1.019×10^{4}	5.721×10^{3}	1.791×10^{3}	3.762×10^{1}	4.085×10^{4}	5.332×10^{3}	4.557	2.249×10^{3}	1.027×10^{4}
	σ_{xy}	8.585×10^{3}	5.419×10^{3}	2.626×10^{3}	3.805×10^{1}	9.416×10^{3}	5.805×10^{3}	1.653	2.527×10^{3}	4.772×10^{3}
	σ_{yy}	8.018×10^{4}	4.236×10^{4}	2.942×10^{4}	3.139×10^{2}	4.381×10^{4}	3.919×10^{4}	2.545	2.944×10^{4}	1.083×10^{5}
	σ_{zz}	2.661×10^{4}	1.591×10^{4}	5.149×10^{3}	2.122×10^{2}	1.012×10^{4}	9.474×10^{3}	3.484	5.684×10^{3}	2.240×10^{4}
	u	4.301×10^{-1}	4.665×10^{-1}	2.971×10^{-1}	3.374×10^{-2}	2.594	2.343×10^{-1}	2.909×10^{-4}	1.697×10^{-1}	2.096
	v	8.877×10^{-1}	9.405×10^{-1}	1.001×10^{-1}	2.229×10^{-2}	1.442	2.647×10^{-1}	1.312×10^{-4}	1.735×10^{-1}	2.414

Table A17. Detailed results for the block compression use case Simulation 7.

		SIMULATION 7								
		MLP		PINN		SVR	GBDTR		KNNR	GPR
		mean	std	mean	std		mean	std		
R2	ε^t_{xx}	0.9991	2.601×10^{-4}	0.9881	6.498×10^{-3}	0.5368	0.9664	2.355×10^{-2}	0.9688	0.3910
	ε^t_{xy}	0.9987	6.959×10^{-4}	0.9623	2.560×10^{-2}	0.2573	0.9512	3.642×10^{-2}	0.9508	0.2373
	ε^t_{yy}	0.9991	2.628×10^{-4}	0.9696	1.661×10^{-2}	0.5397	0.9663	2.395×10^{-2}	0.9691	0.3940
	ε^p_{xx}	0.9991	2.756×10^{-4}	0.8726	4.550×10^{-3}	0.5257	0.9655	2.371×10^{-2}	0.9688	0.3834
	ε^p_{xy}	0.9986	7.165×10^{-4}	0.9616	3.809×10^{-3}	0.2496	0.9577	2.802×10^{-2}	0.9484	0.2319
	ε^p_{yy}	0.9991	2.774×10^{-4}	0.8969	4.043×10^{-3}	0.5281	0.9635	2.655×10^{-2}	0.9690	0.3846
	ε^p_{zz}	0.9968	4.519×10^{-5}	0.7795	9.273×10^{-3}	0.6862	0.9629	2.088×10^{-2}	0.9789	0.4545
	σ_{xx}	0.9965	6.357×10^{-4}	0.9883	6.623×10^{-4}	0.7980	0.9743	1.251×10^{-2}	0.9868	0.5950
	σ_{xy}	0.9962	9.610×10^{-4}	0.9775	1.717×10^{-3}	0.6131	0.9410	3.089×10^{-2}	0.9772	0.5081
	σ_{yy}	0.9915	3.103×10^{-3}	0.8713	1.406×10^{-3}	0.8900	0.9891	4.567×10^{-3}	0.9945	0.7272
	σ_{zz}	0.9943	2.068×10^{-3}	0.9874	3.501×10^{-5}	0.8345	0.9800	1.025×10^{-2}	0.9906	0.6242
	u	0.9996	3.315×10^{-5}	0.9851	1.746×10^{-3}	0.9386	0.9973	1.892×10^{-3}	0.9967	0.9069
	v	0.9997	2.460×10^{-5}	0.9923	2.524×10^{-4}	0.9415	0.9976	1.783×10^{-3}	0.9972	0.9224
mean		0.9976	8.258×10^{-5}	0.9410	4.310×10^{-3}	0.6415	0.9702	1.884×10^{-2}	0.9767	0.5200
MSE	ε^t_{xx}	3.776×10^{-6}	1.124×10^{-6}	5.131×10^{-5}	2.808×10^{-5}	2.001×10^{-3}	1.454×10^{-4}	1.018×10^{-4}	1.349×10^{-4}	2.631×10^{-3}
	ε^t_{xy}	7.595×10^{-6}	4.066×10^{-6}	2.200×10^{-4}	1.496×10^{-4}	4.339×10^{-3}	2.854×10^{-4}	2.128×10^{-4}	2.873×10^{-4}	4.456×10^{-3}
	ε^t_{yy}	3.777×10^{-6}	1.144×10^{-6}	1.324×10^{-4}	7.231×10^{-5}	2.004×10^{-3}	1.467×10^{-4}	1.043×10^{-4}	1.347×10^{-4}	2.638×10^{-3}
	ε^p_{xx}	3.881×10^{-6}	1.196×10^{-6}	5.530×10^{-4}	1.975×10^{-5}	2.059×10^{-3}	1.496×10^{-4}	1.029×10^{-4}	1.356×10^{-4}	2.676×10^{-3}
	ε^p_{xy}	7.212×10^{-6}	3.779×10^{-6}	2.027×10^{-4}	2.009×10^{-5}	3.957×10^{-3}	2.229×10^{-4}	1.478×10^{-4}	2.722×10^{-4}	4.051×10^{-3}
	ε^p_{yy}	3.894×10^{-6}	1.213×10^{-6}	4.511×10^{-4}	1.769×10^{-5}	2.065×10^{-3}	1.598×10^{-4}	1.162×10^{-4}	1.357×10^{-4}	2.692×10^{-3}
	ε^p_{zz}	5.595×10^{-10}	7.808×10^{-12}	9.648×10^{-4}	4.057×10^{-5}	5.421×10^{-8}	6.411×10^{-9}	3.608×10^{-9}	3.652×10^{-9}	9.424×10^{-8}
	σ_{xx}	1.240×10^{2}	2.269×10^{1}	4.180×10^{2}	2.364×10^{1}	7.209×10^{3}	9.186×10^{2}	4.463×10^{2}	4.708×10^{2}	1.445×10^{4}
	σ_{xy}	6.742×10^{1}	1.684×10^{1}	3.950×10^{2}	3.009×10^{1}	6.781×10^{3}	1.034×10^{3}	5.413×10^{2}	3.991×10^{2}	8.620×10^{3}
	σ_{yy}	1.877×10^{3}	6.894×10^{2}	2.858×10^{4}	3.124×10^{2}	2.443×10^{4}	2.421×10^{3}	1.015×10^{3}	1.226×10^{3}	6.060×10^{4}
	σ_{zz}	2.603×10^{2}	9.521×10^{1}	5.801×10^{2}	1.612	7.619×10^{3}	9.203×10^{2}	4.717×10^{2}	4.317×10^{2}	1.730×10^{4}
	u	1.251×10^{-3}	9.453×10^{-5}	4.255×10^{-2}	4.980×10^{-3}	1.752×10^{-1}	7.838×10^{-3}	5.395×10^{-3}	9.348×10^{-3}	2.655×10^{-1}
	v	1.509×10^{-3}	1.275×10^{-4}	4.010×10^{-2}	1.308×10^{-3}	3.030×10^{-1}	1.242×10^{-2}	9.241×10^{-3}	1.433×10^{-2}	4.024×10^{-1}

Table A18. Detailed results for the block compression use case Simulation 12

		SIMULATION 12								
		MLP		PINN		SVR	GBDTR		KNNR	GPR
		mean	std	mean	std		mean	std		
R2	ε^t_{xx}	0.6584	3.166×10^{-1}	−3.4145	2.071	−1.1624	0.4656	2.148×10^{-4}	0.7187	0.0506
	ε^t_{xy}	0.5324	3.411×10^{-1}	−1.7259	5.343×10^{-1}	−0.1194	0.4278	4.223×10^{-4}	0.7196	0.0415
	ε^t_{yy}	0.6563	3.215×10^{-1}	−2.9570	2.652	−1.1091	0.4534	8.915×10^{-4}	0.7196	0.0503
	ε^p_{xx}	0.6578	3.175×10^{-1}	−2.7614	1.762	−1.1139	0.4618	1.278×10^{-3}	0.7158	0.0478
	ε^p_{xy}	0.5168	3.478×10^{-1}	−0.1122	6.743×10^{-1}	−0.0789	0.4009	8.991×10^{-3}	0.7157	0.0401
	ε^p_{yy}	0.6591	3.172×10^{-1}	−0.5161	8.422×10^{-1}	−1.1014	0.4300	2.718×10^{-4}	0.7165	0.0495
	ε^p_{zz}	0.6064	3.871×10^{-1}	0.2723	1.236×10^{-1}	−0.4602	0.2658	7.615×10^{-4}	0.6808	0.0781
	σ_{xx}	0.6182	1.745×10^{-1}	0.8828	4.138×10^{-2}	0.0180	0.7218	3.966×10^{-5}	0.8423	0.1688
	σ_{xy}	0.4799	4.811×10^{-1}	0.8803	1.074×10^{-2}	0.3261	0.5912	8.127×10^{-4}	0.7735	0.1288
	σ_{yy}	0.5583	4.457×10^{-1}	0.8136	9.338×10^{-3}	0.3793	0.6803	2.977×10^{-4}	0.8285	0.1275
	σ_{zz}	0.5919	3.890×10^{-1}	0.9196	1.648×10^{-3}	0.3378	0.6356	1.603×10^{-4}	0.7979	0.0963
	u	0.7277	2.782×10^{-1}	0.9072	6.483×10^{-3}	0.4679	0.9157	7.408×10^{-5}	0.9269	0.5649
	v	0.9310	5.347×10^{-2}	0.9866	1.786×10^{-4}	0.7166	0.9621	2.432×10^{-6}	0.9800	0.5744
mean		0.6303	3.204×10^{-1}	−0.4480	6.652×10^{-1}	−0.2230	0.5702	6.266×10^{-4}	0.7797	0.1553
MSE	ε^t_{xx}	1.092×10^{-3}	1.012×10^{-3}	1.412×10^{-2}	6.623×10^{-3}	6.915×10^{-3}	1.828×10^{-3}	7.349×10^{-7}	8.997×10^{-4}	3.036×10^{-3}
	ε^t_{xy}	2.505×10^{-3}	1.827×10^{-3}	1.460×10^{-2}	2.863×10^{-3}	5.997×10^{-3}	3.176×10^{-3}	2.344×10^{-6}	1.502×10^{-3}	5.135×10^{-3}
	ε^t_{yy}	1.115×10^{-3}	1.043×10^{-3}	1.284×10^{-2}	8.604×10^{-3}	6.842×10^{-3}	1.884×10^{-3}	3.073×10^{-6}	9.098×10^{-4}	3.081×10^{-3}
	ε^p_{xx}	1.098×10^{-3}	1.019×10^{-3}	1.207×10^{-2}	5.655×10^{-3}	6.783×10^{-3}	1.849×10^{-3}	4.391×10^{-6}	9.120×10^{-4}	3.055×10^{-3}
	ε^p_{xy}	2.264×10^{-3}	1.629×10^{-3}	5.210×10^{-3}	3.159×10^{-3}	5.054×10^{-3}	3.080×10^{-3}	4.622×10^{-5}	1.332×10^{-3}	4.497×10^{-3}
	ε^p_{yy}	1.107×10^{-3}	1.030×10^{-3}	4.921×10^{-3}	2.734×10^{-3}	6.821×10^{-3}	1.971×10^{-3}	9.399×10^{-7}	9.202×10^{-4}	3.085×10^{-3}
	ε^p_{zz}	1.229×10^{-7}	1.208×10^{-7}	2.362×10^{-3}	4.012×10^{-4}	4.558×10^{-7}	7.910×10^{-8}	8.205×10^{-11}	9.964×10^{-8}	2.878×10^{-7}
	σ_{xx}	2.769×10^{4}	1.265×10^{4}	8.502×10^{3}	3.001×10^{3}	7.121×10^{4}	6.602×10^{3}	9.414×10^{-1}	1.144×10^{4}	6.028×10^{4}
	σ_{xy}	2.415×10^{4}	2.233×10^{4}	5.559×10^{3}	4.987×10^{2}	3.128×10^{4}	6.161×10^{3}	1.225×10^{1}	1.052×10^{4}	4.045×10^{4}
	σ_{yy}	1.760×10^{5}	1.776×10^{5}	7.428×10^{4}	3.722×10^{3}	2.474×10^{5}	4.010×10^{4}	3.734×10^{1}	6.835×10^{4}	3.478×10^{5}
	σ_{zz}	3.808×10^{4}	3.630×10^{4}	7.501×10^{3}	1.538×10^{2}	6.179×10^{4}	1.124×10^{4}	4.943	1.886×10^{4}	8.433×10^{4}
	u	4.605×10^{-1}	4.704×10^{-1}	1.569×10^{-1}	1.096×10^{-2}	8.999×10^{-1}	1.527×10^{-1}	1.342×10^{-4}	1.235×10^{-1}	7.358×10^{-1}
	v	3.434×10^{-1}	2.660×10^{-1}	6.647×10^{-2}	8.887×10^{-4}	1.410	1.934×10^{-1}	1.240×10^{-5}	9.955×10^{-2}	2.117

Table A19. Detailed results for the block compression use case Simulation 13.

		SIMULATION 13								
		MLP		PINN		SVR	GBDTR		KNNR	GPR
		mean	std	mean	std		mean	std		
R2	ε^t_{xx}	0.7511	2.314×10^{-1}	−4.3817	2.685	−0.2933	0.5204	9.591×10^{-4}	0.7164	0.0857
	ε^t_{xy}	0.5801	3.853×10^{-1}	−0.9114	1.065×10^{-1}	−0.1797	0.2623	2.829×10^{-3}	0.6314	0.0485
	ε^t_{yy}	0.7517	2.319×10^{-1}	0.1125	9.052×10^{-2}	−0.2646	0.5280	1.380×10^{-3}	0.7186	0.0873
	ε^p_{xx}	0.7547	2.273×10^{-1}	−2.1130	9.596×10^{-1}	−0.2740	0.5357	4.099×10^{-4}	0.7131	0.0826
	ε^p_{xy}	0.5616	4.021×10^{-1}	0.5356	2.972×10^{-2}	−0.1116	0.2250	7.643×10^{-3}	0.6277	0.0480
	ε^p_{yy}	0.7556	2.269×10^{-1}	−0.4174	3.678×10^{-1}	−0.2741	0.5217	8.630×10^{-4}	0.7142	0.0830
	ε^p_{zz}	0.6959	2.525×10^{-1}	0.4548	2.042×10^{-1}	−0.4013	0.4990	7.038×10^{-5}	0.7267	0.0912
	σ_{xx}	0.4251	5.887×10^{-1}	0.9077	3.046×10^{-3}	−0.4988	0.7847	3.553×10^{-4}	0.8620	0.2942
	σ_{xy}	−0.0088	8.836×10^{-1}	0.8314	3.051×10^{-3}	0.1802	0.4255	6.447×10^{-4}	0.6805	0.1513
	σ_{yy}	0.6042	4.473×10^{-1}	0.6678	8.365×10^{-4}	−0.0250	0.7608	5.706×10^{-6}	0.8418	0.1470
	σ_{zz}	0.6356	3.848×10^{-1}	0.9351	1.247×10^{-3}	0.0788	0.6999	1.686×10^{-4}	0.8282	0.1321
	u	0.9627	2.725×10^{-2}	0.9392	4.615×10^{-3}	0.3676	0.9467	1.345×10^{-4}	0.9625	0.4940
	v	0.9478	5.094×10^{-2}	0.9805	3.346×10^{-3}	0.7270	0.9521	1.989×10^{-5}	0.9697	0.5572
mean		0.6475	3.326×10^{-1}	−0.1122	3.265×10^{-1}	−0.0745	0.5894	1.579×10^{-4}	0.7687	0.1771
MSE	ε^t_{xx}	1.412×10^{-3}	1.313×10^{-3}	3.054×10^{-2}	1.523×10^{-2}	7.339×10^{-3}	2.721×10^{-3}	5.442×10^{-6}	1.609×10^{-3}	5.188×10^{-3}
	ε^t_{xy}	2.253×10^{-3}	2.067×10^{-3}	1.025×10^{-2}	5.715×10^{-4}	6.329×10^{-3}	3.958×10^{-3}	1.518×10^{-5}	1.977×10^{-3}	5.105×10^{-3}
	ε^t_{yy}	1.422×10^{-3}	1.328×10^{-3}	5.080×10^{-3}	5.182×10^{-4}	7.239×10^{-3}	2.702×10^{-3}	7.900×10^{-6}	1.611×10^{-3}	5.225×10^{-3}
	ε^p_{xx}	1.389×10^{-3}	1.288×10^{-3}	1.763×10^{-2}	5.436×10^{-3}	7.217×10^{-3}	2.630×10^{-3}	2.322×10^{-6}	1.625×10^{-3}	5.197×10^{-3}
	ε^p_{xy}	2.107×10^{-3}	1.933×10^{-3}	2.232×10^{-3}	1.428×10^{-4}	5.342×10^{-3}	3.724×10^{-3}	3.673×10^{-5}	1.789×10^{-3}	4.575×10^{-3}
	ε^p_{yy}	1.398×10^{-3}	1.299×10^{-3}	8.111×10^{-3}	2.104×10^{-3}	7.291×10^{-3}	2.737×10^{-3}	4.939×10^{-6}	1.635×10^{-3}	5.247×10^{-3}
	ε^p_{zz}	7.358×10^{-8}	6.111×10^{-8}	3.120×10^{-3}	1.168×10^{-3}	3.391×10^{-7}	1.212×10^{-7}	1.703×10^{-11}	6.614×10^{-8}	2.199×10^{-7}
	σ_{xx}	2.766×10^{4}	2.832×10^{4}	4.438×10^{3}	1.466×10^{2}	7.211×10^{4}	1.036×10^{4}	1.709×10^{1}	6.640×10^{3}	3.395×10^{4}
	σ_{xy}	1.954×10^{4}	1.711×10^{4}	3.266×10^{3}	5.908×10^{1}	1.588×10^{4}	1.113×10^{4}	1.249×10^{1}	6.187×10^{3}	1.644×10^{4}
	σ_{yy}	1.214×10^{5}	1.372×10^{5}	1.019×10^{5}	2.565×10^{2}	3.143×10^{5}	7.333×10^{4}	1.750	4.851×10^{4}	2.615×10^{5}
	σ_{zz}	2.089×10^{4}	2.205×10^{4}	3.720×10^{3}	7.147×10^{1}	5.280×10^{4}	1.720×10^{4}	9.664	9.844×10^{3}	4.974×10^{4}
	u	1.395×10^{-1}	1.019×10^{-1}	2.273×10^{-1}	1.726×10^{-2}	2.365	1.992×10^{-1}	5.028×10^{-4}	1.401×10^{-1}	1.892
	v	2.707×10^{-1}	2.644×10^{-1}	1.011×10^{-1}	1.737×10^{-2}	1.417	2.487×10^{-1}	1.032×10^{-4}	1.573×10^{-1}	2.299

References

1. Reddy, J.N. *Introduction to the Finite Element Method;* Mechanical Engineering; McGraw Hill Education: New York, NY, USA, 2019.
2. Yang, X.-S.; Koziel, S.; Leifsson, L. Computational Optimization, Modelling and Simulation: Recent Trends and Challenges. *Procedia Comput. Sci.* **2013**, *18*, 855–860. [CrossRef]
3. Roberts, S.M.; Kusiak, J.; Liu, Y.L.; Forcellese, A.; Withers, P.J. Prediction of damage evolution in forged aluminium metal matrix composites using a neural network approach. *J. Mater. Process. Technol.* **1998**, *80–81*, 507–512. [CrossRef]
4. Hans Raj, K.; Sharma, R.S.; Srivastava, S.; Patvardhan, C. Modeling of manufacturing processes with ANNs for intelligent manufacturing. *Int. J. Mach. Tools Manuf.* **2000**, *40*, 851–868. [CrossRef]
5. García-Crespo, A.; Ruiz-Mezcua, B.; Fernández-Fdz, D.; Zaera, R. Prediction of the response under impact of steel armours using a multilayer perceptron. *Neural Comput. Appl.* **2007**, *16*, 147–154. [CrossRef]
6. Nourbakhsh, M.; Irizarry, J.; Haymaker, J. Generalizable surrogate model features to approximate stress in 3D trusses. *Eng. Appl. Artif. Intell.* **2018**, *71*, 15–27. [CrossRef]
7. Chan, W.L.; Fu, M.W.; Lu, J. An integrated FEM and ANN methodology for metal-formed product design. *Eng. Appl. Artif. Intell.* **2008**, *21*, 1170–1181. [CrossRef]
8. D'Addona, D.M.; Antonelli, D. Neural Network Multiobjective Optimization of Hot Forging. *Procedia CIRP* **2018**, *67*, 498–503. [CrossRef]
9. Gudur, P.P.; Dixit, U.S. A neural network-assisted finite element analysis of cold flat rolling. *Eng. Appl. Artif. Intell.* **2008**, *21*, 43–52. [CrossRef]
10. Pellicer-Valero, O.J.; Rupérez, M.J.; Martínez-Sanchis, S.; Martín-Guerrero, J.D. Real-time biomechanical modeling of the liver using Machine Learning models trained on Finite Element Method simulations. *Expert Syst. Appl.* **2020**, *143*, 113083. [CrossRef]
11. Abueidda, D.W.; Almasri, M.; Ammourah, R.; Ravaioli, U.; Jasiuk, I.M.; Sobh, N.A. Prediction and optimization of mechanical properties of composites using convolutional neural networks. *Compos. Struct.* **2019**, *227*, 111264. [CrossRef]
12. Pfaff, T.; Fortunato, M.; Sanchez-Gonzalez, A.; Battaglia, P.W. Learning Mesh-Based Simulation with Graph Networks. In Proceedings of the International Conference on Learning Representations (ICLR), Addis Ababa, Ethiopia, 26 April–1 May 2020.
13. Loghin, A.; Ismonov, S. Augmenting Generic Fatigue Crack Growth Models Using 3D Finite Element Simulations and Gaussian Process Modeling. In Proceedings of the ASME 2019 Pressure Vessels & Piping Conference, San Antonio, TX, USA, 14–19 July 2019; Volume 2. [CrossRef]
14. Ming, W.; Zhang, G.; Li, H.; Guo, J.; Zhang, Z.; Huang, Y.; Chen, Z. A hybrid process model for EDM based on finite-element method and Gaussian process regression. *Int. J. Adv. Manuf. Technol.* **2014**, *74*, 1197–1211. [CrossRef]
15. Pan, F.; Zhu, P.; Zhang, Y. Metamodel-based lightweight design of B-pillar with TWB structure via support vector regression. *Comput. Struct.* **2010**, *88*, 36–44. [CrossRef]
16. Li, H.; Shi, M.; Liu, X.; Shi, Y. Uncertainty optimization of dental implant based on finite element method, global sensitivity analysis and support vector regression. Proceedings of the Institution of Mechanical Engineers. *Part H J. Eng. Med.* **2019**, *233*, 232–243. [CrossRef] [PubMed]
17. Hu, F.; Li, D. Modelling and Simulation of Milling Forces Using an Arbitrary Lagrangian–Eulerian Finite Element Method and Support Vector Regression. *J. Optim. Theory Appl.* **2012**, *153*, 461–484. [CrossRef]
18. Martínez-Martínez, F.; Rupérez-Moreno, M.J.; Martínez-Sober, M.; Solves-Llorens, J.A.; Lorente, D.; Serrano-López, A.J.; Martínez-Sanchis, S.; Monserrat, C.; Martín-Guerrero, J.D. A finite element-based machine learning approach for modeling the mechanical behavior of the breast tissues under compression in real-time. *Comput. Biol. Med.* **2017**, *90*, 116–124. [CrossRef] [PubMed]
19. Zhang, W.; Zhang, R.; Wu, C.; Goh, A.T.C.; Wang, L. Assessment of basal heave stability for braced excavations in anisotropic clay using extreme gradient boosting and random forest regression. *Undergr. Space* **2020**. [CrossRef]
20. Qi, Z.; Zhang, N.; Liu, Y.; Chen, W. Prediction of mechanical properties of carbon fiber based on cross-scale FEM and machine learning. *Compos. Struct.* **2019**, *212*, 199–206. [CrossRef]
21. Haghighat, E.; Juanes, R. SciANN: A Keras/TensorFlow wrapper for scientific computations and physics-informed deep learning using artificial neural networks. *Comput. Methods Appl. Mech. Eng.* **2021**, *373*, 113552. [CrossRef]
22. Haghighat, E.; Raissi, M.; Moure, A.; Gomez, H.; Juanes, R. A physics-informed deep learning framework for inversion and surrogate modeling in solid mechanics. *Comput. Methods Appl. Mech. Eng.* **2021**, *379*, 113741. [CrossRef]
23. Shin, Y. On the Convergence of Physics Informed Neural Networks for Linear Second-Order Elliptic and Parabolic Type PDEs. *CiCP* **2020**, *28*, 2042–2074. [CrossRef]
24. Yin, M.; Zheng, X.; Humphrey, J.D.; Em Karniadakis, G. Non-invasive Inference of Thrombus Material Properties with Physics-Informed Neural Networks. *Comput. Methods Appl. Mech. Eng.* **2021**, *375*, 113603. [CrossRef]
25. Arnold, F.; King, R. State–space modeling for control based on physics-informed neural networks. *Eng. Appl. Artif. Intell.* **2021**, *101*, 104195. [CrossRef]
26. Zobeiry, N.; Humfeld, K.D. A physics-informed machine learning approach for solving heat transfer equation in advanced manufacturing and engineering applications. *Eng. Appl. Artif. Intell.* **2021**, *101*, 104232. [CrossRef]
27. Nascimento, R.G.; Fricke, K.; Viana, F.A.C. A tutorial on solving ordinary differential equations using Python and hybrid physics-informed neural network. *Eng. Appl. Artif. Intell.* **2020**, *96*, 103996. [CrossRef]
28. Jin, X.; Cai, S.; Li, H.; Karniadakis, G.E. NSFnets (Navier-Stokes flow nets): Physics-informed neural networks for the incompressible Navier-Stokes equations. *J. Comput. Phys.* **2021**, *426*, 109951. [CrossRef]

29. Mao, Z.; Jagtap, A.D.; Karniadakis, G.E. Physics-informed neural networks for high-speed flows. *Comput. Methods Appl. Mech. Eng.* **2020**, *360*, 112789. [CrossRef]
30. Goan, E.; Fookes, C. Bayesian Neural Networks: An Introduction and Survey. In *Case Studies in Applied Bayesian Data Science*; CIRM Jean-Morlet Chair, Fall 2018; Lecture Notes in Mathematics; Mengersen, K.L., Pudlo, P., Robert, C.P., Eds.; Springer: Cham, Switzerland, 2020; Volume 2259, pp. 45–87.
31. Garnelo, M.; Rosenbaum, D.; Maddison, C.; Ramalho, T.; Saxton, D.; Shanahan, M.; Teh, Y.W.; Rezende, D.; Eslami, S.M.A. Conditional Neural Processes. In Proceedings of the 35th International Conference on Machine Learning, Stockholm, Sweden, 10–15 July 2018; Dy, J., Krause, A., Eds.; pp. 1704–1713.
32. Wang, S.; Teng, Y.; Perdikaris, P. Understanding and mitigating gradient pathologies in physics-informed neural networks. *arXiv* **2020**, arXiv:2001.04536.
33. Pang, G.; Lu, L.; Karniadakis, G.E. fPINNs: Fractional Physics-Informed Neural Networks. *SIAM J. Sci. Comput.* **2019**, *41*, A2603–A2626. [CrossRef]
34. Yang, Y.; Perdikaris, P. Adversarial uncertainty quantification in physics-informed neural networks. *J. Comput. Phys.* **2019**, *394*, 136–152. [CrossRef]
35. Raissi, M.; Perdikaris, P.; Karniadakis, G.E. Physics-informed neural networks: A deep learning framework for solving forward and inverse problems involving nonlinear partial differential equations. *J. Comput. Phys.* **2019**, *378*, 686–707. [CrossRef]
36. Ngatchou, P.; Zarei, A.; El-Sharkawi, A. Pareto Multi Objective Optimization. In Proceedings of the 13th International Conference on Intelligent Systems Application to Power Systems, Arlington, VA, USA, 6–10 November 2005; IEEE Service Center: Piscataway, NJ, USA, 2005; pp. 84–91. [CrossRef]
37. Pettit, C.L.; Wilson, D.K. A physics-informed neural network for sound propagation in the atmospheric boundary layer. In Proceedings of the 179th Meeting of the Acoustical Society of America, Acoustics Virtually Everywhere, 7–11 December 2020; p. 22002. [CrossRef]
38. Lihua, L. Simulation physics-informed deep neural network by adaptive Adam optimization method to perform a comparative study of the system. *Eng. Comput.* **2021**, 1–20. [CrossRef]
39. McClenny, L.; Braga-Neto, U. Self-Adaptive Physics-Informed Neural Networks using a Soft Attention Mechanism. *arXiv* **2020**, arXiv:2009.04544.
40. Jagtap, A.D.; Kawaguchi, K.; Karniadakis, G.E. Adaptive activation functions accelerate convergence in deep and physics-informed neural networks. *J. Comput. Phys.* **2020**, *404*, 109136. [CrossRef]
41. 1.11. Ensemble Methods—Scikit-Learn 0.24.1 Documentation (2021.000Z). Available online: https://scikit-learn.org/stable/modules/ensemble.html (accessed on 5 November 2020).
42. Friedman, J.H. Greedy Function Approximation: A Gradient Boosting Machine. *Ann. Stat.* **2001**, *29*, 1189–1232. [CrossRef]
43. 1.6. Nearest Neighbors—Scikit-Learn 0.24.1 Documentation (2021.000Z). Available online: https://scikit-learn.org/stable/modules/neighbors.html (accessed on 5 November 2020).
44. Schulz, E.; Speekenbrink, M.; Krause, A. A tutorial on Gaussian process regression: Modelling, exploring, and exploiting functions. *J. Math. Psychol.* **2018**, *85*, 1–16. [CrossRef]
45. Rasmussen, C.E.; Williams, C.K.I. *Gaussian Processes for Machine Learning*, 3rd ed.; Adaptive Computation and Machine Learning; MIT Press: Cambridge, MA, USA, 2008.
46. Awad, M.; Khanna, R. Support Vector Regression. In *Efficient Learning Machines. Theories, Concepts, and Applications for Engineers and System Designers*; The Expert's Voice in Machine Learning; Awad, M., Khanna, R., Eds.; Apress Open: New York, NY, USA, 2015; pp. 67–80.
47. Smola, A.J.; Schölkopf, B. A tutorial on support vector regression. *Stat. Comput.* **2004**, *14*, 199–222. [CrossRef]
48. Svozil, D.; Kvasnicka, V.; Pospichal, J. Introduction to multi-layer feed-forward neural networks. *Chemom. Intell. Lab. Syst.* **1997**, *39*, 43–62. [CrossRef]
49. Leshno, M.; Lin, V.Y.; Pinkus, A.; Schocken, S. Multilayer feedforward networks with a nonpolynomial activation function can approximate any function. *Neural Netw.* **1993**, *6*, 861–867. [CrossRef]
50. Lu, Z.; Pu, H.; Wang, F.; Hu, Z.; Wang, L. The Expressive Power of Neural Networks: A View from the Width. In Proceedings of the 31st International Conference on Neural Information Processing Systems, Long Beach, CA, USA, 4–9 December 2017.
51. Hornik, K.; Stinchcombe, M.; White, H. Multilayer feedforward networks are universal approximators. *Neural Netw.* **1989**, *2*, 359–366. [CrossRef]
52. Gardner, M.W.; Dorling, S.R. Artificial neural networks (the multilayer perceptron)—A review of applications in the atmospheric sciences. *Atmos. Environ.* **1998**, *32*, 2627–2636. [CrossRef]
53. Rumelhart, D.E.; Hinton, G.E.; Williams, R.J. Learning representations by back-propagating errors. *Nature* **1986**, *323*, 533–536. [CrossRef]
54. Lu, L.; Pestourie, R.; Yao, W.; Wang, Z.; Verdugo, F.; Johnson, S.G. Physics-informed neural networks with hard constraints for inverse design. *arXiv* **2021**, arXiv:2102.04626.

applied sciences

MDPI

Article

Topology Optimization of Deformable Bodies with Linear Dynamic Impact and Frictionless Contact Condition

Gil-Eon Jeong

Radwaste Transportation and Storage Research Team, Korea Atomic Energy Research Institute, Daejeon 34057, Korea; gejeong@kaeri.re.kr; Tel.: +82-10-3670-4177

Abstract: There has been an increasing demand for the design of an optimum topological layout in several engineering fields for a simple part, along with a system that considers the relative behaviors between adjacent parts. This paper presents a method of designing an optimum topological layout to achieve a linear dynamic impact and frictionless contact conditions in which relative behaviors can be observed between adjacent deformable parts. The solid isotropic method with penalization (SIMP) method is used with an appropriate filtering scheme to obtain an optimum topological layout. The condensed mortar method is used to handle the non-matching interface, which inevitably occurs in the impact and contact regions, since it can easily apply the existing well-known topology optimization approach even in the presence of a non-matching interface. The validity of the proposed method is verified through a numerical example. In the future, the proposed optimization approach will be applied to more general and highly nonlinear non-matching interface problems, such as friction contact and multi-physics problems.

Keywords: SIMP method; impact condition; non-matching interface; condensed mortar method

Citation: Jeong, G.-E. Topology Optimization of Deformable Bodies with Linear Dynamic Impact and Frictionless Contact Condition. *Appl. Sci.* **2021**, *11*, 10518. https://doi.org/10.3390/app112210518

Academic Editors: Ramin Rahmani and Nicola Pio Belfiore

Received: 27 August 2021
Accepted: 29 October 2021
Published: 9 November 2021

1. Introduction

The demand for creative and optimized designs that cannot be realized using conventional manufacturing techniques has been increasing in recent years owing to the development of additive manufacturing. The design of additive manufacturing (DfAM) has received widespread attention due to developments in the industrial field [1,2]. DfAM is an advanced technique compared to design for manufacturing (DfM) and can be used to overcome the limitations faced by the DfM process. DfAM can be used to produce more creative and high-quality designs since additive manufacturing can produce highly rigid and lightweight products with complicated internal structures. Among the various DfAM techniques, the most widely analyzed is topology optimization, which presents an excellent synergistic effect when combined with additive manufacturing [1–5].

Topology optimization is a scheme which determines the means of the distribution of materials in order to find an optimum topological layout based on the designer's purpose using various simulations [6–8]. The basic concept of topology optimization is to distinguish between the necessary and unnecessary parts in order to best meet the optimization formulation within a given design domain. It has been used in various industries and in research owing to its theoretical simplicity and ability to produce a concise, esthetic, and creative design [9–13]. Despite the several studies conducted on topology optimization, most have been focused on a single part [6–13]. However, in the practical engineering field, each part interacts with its adjacent parts, and most studies have replaced the interaction effects by imposing an equivalent interface condition to address this issue [14–17]. Unfortunately, it is incompatible and impossible in highly nonlinear systems such as contact and impact conditions.

In order to consider the effects of adjacent parts, Strömberg et al. suggested a topology optimization framework which simultaneously considers multiple deformable parts through a node-to-node contact condition [18]. However, this is only feasible in the matching interface meshes owing to the features of the node-to-node contact condition. Jeong et al. presented a new topology optimization technique based on a surface-to-surface contact condition to handle both the matching and non-matching interface meshes [19]. The surface-to-surface contact condition was treated by the condensed mortar method and a filtering scheme considering the size of the element was introduced to alleviate some discontinuities present on the non-matching interface. Recently, Fernandez et al. proposed a topology optimization technique to handle multiple deformable bodies in static and large deformation contacts based on a mortar formulation approach [20].

This paper presents a method to determine an optimum topological layout for the linear dynamic impact and frictionless contact conditions to present a more generalized application of topology optimization. Firstly, the condensed mortar method is used to handle the non-matching interface, which is inevitably caused by the relative motion near the impact region. The condensed mortar method presents several advantages when applied in topology optimization because it can be applied in the same manner as that of conventional structural analysis, even in the presence of a non-matching interface [19]. Since previous studies only considered a linear static problem [19], the proposed approach presents a way of applying the condensed mortar method in a linear dynamic topology optimization. Secondly, the solid isotropic method with penalization (SIMP) method is used to determine the optimum topological layout [13]. The mean dynamic compliance and mean squared displacement of a target DOF are considered as an objective function in this study to obtain an optimum topological layout for a linear dynamic system [13]. Furthermore, a filtering scheme that considers the size of an element is applied to handle discontinuities along the non-matching interface [19]. As a result, the approach proposed in this paper can be easily used in various applications including analysis and design because the non-matching interface are simply handled by applying the condensed localized mortar method to the well-known linear dynamic analysis methods.

The remainder of this paper is organized as follows. Section 2 briefly introduces the concept of the condensed mortar method, which is a scheme used to handle non-matching interfaces. The validity of the condensed mortar method is verified through a simple verification example for the linear dynamic analysis. Section 3 describes a method to determine an optimum topological layout in the linear dynamic problems. The condensed mortar method is applied to handle the non-matching interface on the impact regions, based on the conventional dynamic response topology optimization framework. Section 4 presents the verification of the validity of the proposed method using an impact problem in which a beam and a flying block are impacted by frictionless contact conditions. Lastly, Section 5 presents the concluding remarks.

2. Linear Dynamic Analysis with Condensed Localized Mortar Method

This section presents a linear dynamic analysis framework using the condensed localized mortar method to handle the impact and frictionless contact conditions. Figure 1 illustrates the overall analysis process.

When the analysis is initiated, it is necessary to define the system and to determine how the system is considered at the problem start step. Subsequently, the contact region between the adjoined parts is searched in the case of the impact and frictionless contact problem. A linear dynamic structural analysis considering the contact conditions is performed in the event of an impact or contact; otherwise, a linear dynamic structural analysis is performed for one part. In this paper, only the impact and frictionless contact conditions are considered among the various contact conditions (i.e., interface-tie, friction or frictionless contact, and impact conditions). Additionally, the contact conditions are handled using the condensed mortar method. A detailed explanation of this is presented in the following subsections.

The overall process ends when the time t is greater than the final time t_f. Otherwise, the variables including the time and information for node, displacement, velocity, and acceleration are updated, and the process is repeated, as shown in Figure 1.

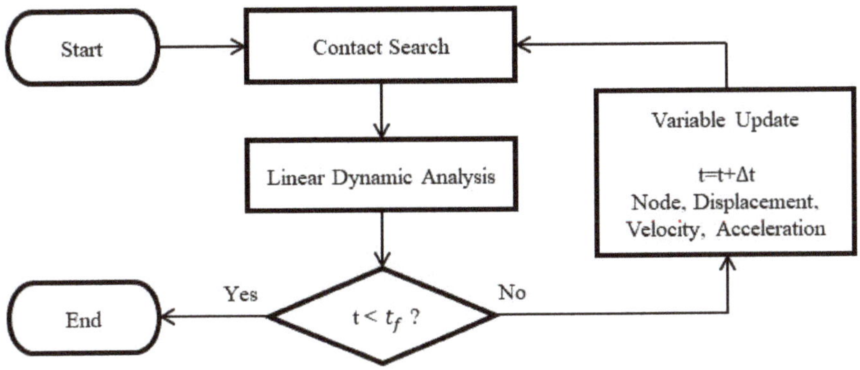

Figure 1. Flowchart for linear dynamic contact analysis.

2.1. Linear Dynamic Finite Element Analysis Considering Contact Condition

This subsection presents the basic concept of the 2D linear dynamic finite element analysis used to solve the impact and frictionless contact problems by employing the condensed mortar method. Figure 2 presents a schematic diagram of the 2D contact problem. Even though only two contacting parts are considered in Figure 2, it can be easily extended to multiple parts. The initial boundary value problem of the general structural mechanics comprises a set of combined second-order partial differential equations and a given set of initial and boundary conditions [21].

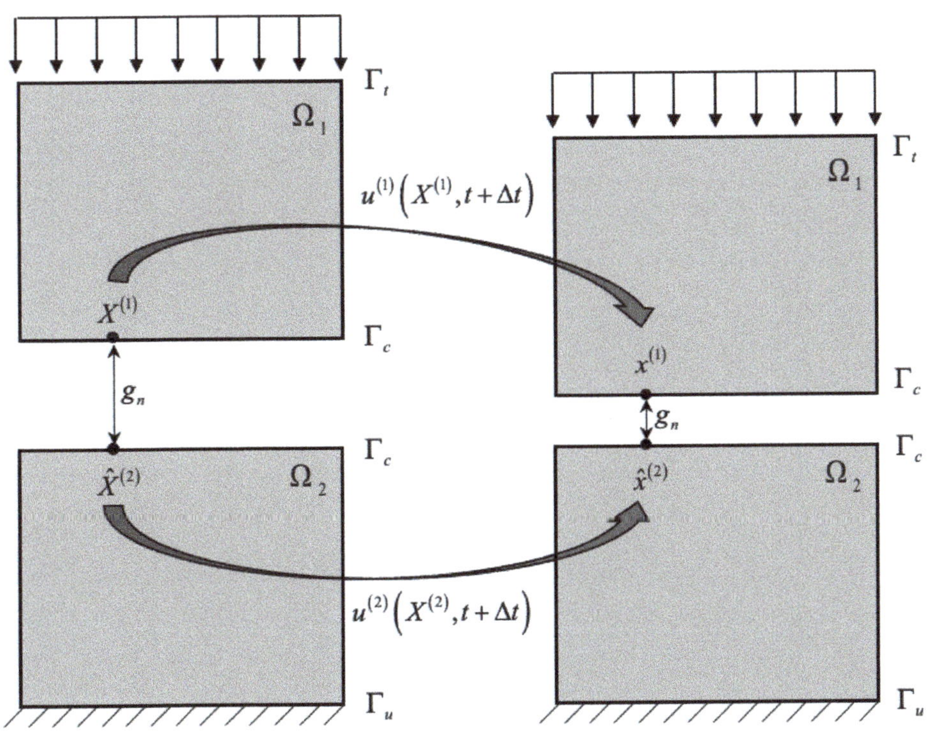

Figure 2. Schematic description for contact problem.

The initial boundary value problem for the linear dynamic structural mechanics is given as:

$$\begin{aligned} \mathrm{Div}\mathbf{P} + \underline{\mathbf{b}} &= \rho\ddot{\mathbf{u}} \text{ in } \Omega \times [0,\ \mathrm{T}] \\ \mathbf{u} &= \underline{\mathbf{u}} \text{ on } \Gamma_u \times [0,\ \mathrm{T}] \\ \mathbf{P}\cdot\mathbf{N} &= \mathbf{t} \text{ on } \Gamma_t \times [0,\ \mathrm{T}] \end{aligned} \tag{1}$$

where Ω is the domain occupied by all the material points, Γ_u and Γ_t denote the Dirichlet and Neumann boundary, where displacements $\bar{\mathbf{u}}$ and tractions $\bar{\mathbf{t}}$ are prescribed. $\mathbf{P}$ represents the first Piola-Kirchhoff stress tensor, $\mathbf{b}$ represents an external body force vector, ρ is the density, $\mathbf{u}$ is the displacement vector, and the overdot indicates the time derivative.

In order to address the impact condition with the frictionless contact between adjoined parts, the Karush-Kuhn-Tucker conditions must be satisfied in the normal direction

$$g_n \geq 0,\ \tau_n \leq 0,\ \tau_n g_n = 0 \tag{2}$$

where τ_n represents the normal component of the contact traction and g_n indicates a gap function in the normal direction, which is defined by:

$$g_n = -\mathbf{n}\,\mathbf{x}^{(1)}\left[\mathbf{x}^{(1)} - \tilde{\mathbf{x}}^{(2)}\right] \tag{3}$$

where $\tilde{\mathbf{x}}^{(2)}$ is defined by projecting the point $\mathbf{x}^{(1)}$ onto the surface Γ_c in the domain Ω_2 along the normal direction $\mathbf{n}$. The projected point $\tilde{\mathbf{x}}^{(2)}$ can be obtained by using a well-known contact search algorithm (e.g., [22]).

When the finite element method is employed to perform structural analysis based on Equations (1)–(3), the momentum equation under an external load can be given by:

$$\mathbf{M}\ddot{\mathbf{u}} + \mathbf{f}_{\mathrm{int}}(\mathbf{u}) + \mathbf{f}_c(\mathbf{u}, \boldsymbol{\lambda}) - \mathbf{f}_{ext} = 0 \tag{4}$$

where $\mathbf{M}$ is the global mass matrix and $\mathbf{f}_{\mathrm{int}}$, $\mathbf{f}_{\mathrm{ext}}$ and $\mathbf{f}_c$ represent the global internal, external, and contact force vectors, respectively. Here, the mass matrix $\mathbf{M}$ is assumed to be constant.

In order to obtain the discrete single unknown $\mathbf{u}_{n+1}$ in single step 'n + 1', Newmark approximation [23] is used to describe other quantities as a function of $\mathbf{u}_{n+1}$.

$$\begin{aligned} \dot{\mathbf{u}}_{n+1} &= \tfrac{\gamma}{\beta\Delta t}(\mathbf{u}_{n+1} - \mathbf{u}_n) - \tfrac{\gamma-\beta}{\beta}\dot{\mathbf{u}}_n - \tfrac{\gamma-2\beta}{2\beta}\Delta t\ddot{\mathbf{u}}_n \\ \ddot{\mathbf{u}}_{n+1} &= \tfrac{\gamma}{\beta\Delta t^2}(\mathbf{u}_{n+1} - \mathbf{u}_n) - \tfrac{1}{\beta\Delta t}\dot{\mathbf{u}}_n - \tfrac{1-2\beta}{2\beta}\ddot{\mathbf{u}}_n \end{aligned} \tag{5}$$

where β and γ represent the algorithmic parameters for the Newmark method, typically $2\beta = \gamma = 1/2$.

2.2. Application of Condensed Localized Mortar Method in Linear Dynamic Analysis

Based on the basic concept and framework of the linear dynamic analysis considering frictionless contact conditions, this subsection presents a method to apply the condensed localized mortar method [19] in the linear dynamic analysis to derive the contact force vector $\mathbf{f}_c$ in Equation (4).

The mortar method is the most widely used approach to solve the interface, contact, impact, and multi-physics problems by introducing Lagrange multipliers to impose the interface compatibility condition in a weak sense [24–27]. The mortar methods can be divided into classical and localized versions depending on how the interface compatibility conditions and Lagrange multipliers are defined. The classical mortar method is the most widely used approach for treating the interface conditions by introducing Lagrange multipliers as the interface pressure, as shown in Figure 3b. A fictitious frame is introduced in the localized mortar method, and the interface compatibility condition is imposed through the frame domain, as shown in Figure 3c [19,28–32].

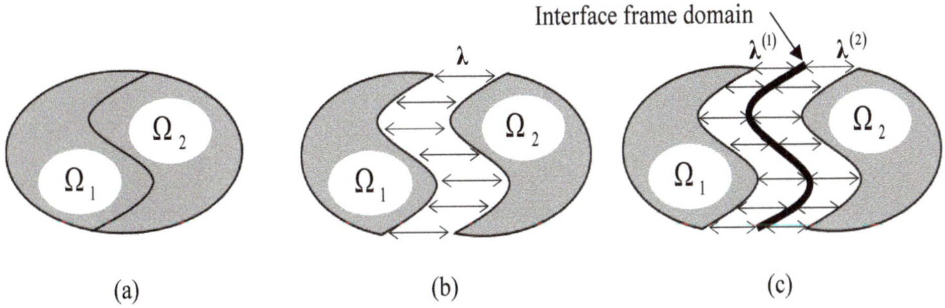

Figure 3. Simple patch test problem: (**a**) a continuum structure composed of two subdomains, and each subdomain is connected using the (**b**) classical and (**c**) localized version.

This study employs the localized mortar method to impose the interface compatibility conditions because it is successfully utilized in various problems by uniquely defining the interface compatibility conditions through the frame domain [19,28–32]. In the case of the localized mortar method, the interface compatibility condition between the newly introduced frame domain and the adjacent two parts, as shown in Figure 3c, is given by

$$\mathbf{u}^{(1)} - \mathbf{u}^{(2)} = 0 \Rightarrow \begin{cases} \mathbf{u}^{(1)} - \mathbf{u}^{(f)} = 0 \\ \mathbf{u}^{(2)} - \mathbf{u}^{(f)} = 0 \end{cases} \tag{6}$$

Subsequently, two sets of Lagrange multipliers, i.e., $\boldsymbol{\lambda}^{(1)}$ and $\boldsymbol{\lambda}^{(2)}$, are defined along the interface to impose the interface capability condition (6) in a weak sense. By substituting the interface condition (6) and two sets of Lagrange multipliers into the momentum equation of the mortar method (4) and organizing the matrix form, the total system equation of the localized mortar method in a linear dynamic condition is given by:

$$\begin{bmatrix} \mathbf{M}^{(1)} & 0 & 0 & 0 & 0 \\ 0 & \mathbf{M}^{(2)} & 0 & 0 & 0 \\ 0 & 0 & 0 & 0 & 0 \\ 0 & 0 & 0 & 0 & 0 \\ 0 & 0 & 0 & 0 & 0 \end{bmatrix} \begin{bmatrix} \ddot{\mathbf{u}}^{(1)} \\ \ddot{\mathbf{u}}^{(2)} \\ \ddot{\boldsymbol{\lambda}}^{(1)} \\ \ddot{\boldsymbol{\lambda}}^{(2)} \\ \ddot{\mathbf{u}}^{(f)} \end{bmatrix} + \begin{bmatrix} \mathbf{K}^{(1)} & 0 & \mathbf{B}_1\mathbf{Q}_1 & 0 & 0 \\ 0 & \mathbf{K}^{(2)} & 0 & \mathbf{B}_2\mathbf{Q}_2 & 0 \\ \mathbf{Q}_1^T\mathbf{B}_1^T & 0 & 0 & 0 & -\mathbf{W}_1 \\ 0 & \mathbf{Q}_2^T\mathbf{B}_2^T & 0 & 0 & -\mathbf{W}_2 \\ 0 & 0 & -\mathbf{W}_1^T & -\mathbf{W}_2^T & 0 \end{bmatrix} \begin{bmatrix} \mathbf{u}^{(1)} \\ \mathbf{u}^{(2)} \\ \boldsymbol{\lambda}^{(1)} \\ \boldsymbol{\lambda}^{(2)} \\ \mathbf{u}^{(f)} \end{bmatrix} = \begin{bmatrix} \mathbf{f}^{(1)} \\ \mathbf{f}^{(2)} \\ \mathbf{G}^{(1)} \\ \mathbf{G}^{(2)} \\ 0 \end{bmatrix}, \tag{7}$$

$$\mathbf{Q}_1 = \int_{\Gamma_I} \mathbf{N}_{u_{\Gamma 1}}^T \mathbf{N}_{\lambda 1} d\Gamma, \quad \mathbf{Q}_2 = \int_{\Gamma_I} \mathbf{N}_{u_{\Gamma 2}}^T \mathbf{N}_{\lambda 2} d\Gamma, \quad \mathbf{W}_1 = \int_{\Gamma_I} \mathbf{N}_{\lambda 1}^T \mathbf{N}_{uf} d\Gamma, \quad \mathbf{W}_2 = \int_{\Gamma_I} \mathbf{N}_{\lambda 2}^T \mathbf{N}_{uf} d\Gamma$$
$$\mathbf{G}^{(1)} = \int_{\Gamma_I} \mathbf{N}_{\lambda 1}^T g_{1f} d\Gamma, \quad \mathbf{G}^{(2)} = \int_{\Gamma_I} \mathbf{N}_{\lambda 2}^T g_{2f} d\Gamma$$

where $\mathbf{B}_1$ and $\mathbf{B}_2$ are the Boolean matrices representing the boundary displacement components, and $\mathbf{Q}_1$, $\mathbf{Q}_2$, $\mathbf{W}_1$, and $\mathbf{W}_2$ are the projection matrices, which represent the interface condition between the interface displacements and frame displacement. Additionally, $\mathbf{G}^{(1)}$ and $\mathbf{G}^{(2)}$ indicate the non-penetration condition caused by the KKT condition (2) and the interface compatibility condition (6)

The condensed localized mortar method, which was developed to be easily applied in topology optimization algorithms even in the presence of a non-matching interface, is employed [19]. This can be obtained by condensing the Lagrange multipliers and overlapping interface displacements using the localized mortar method (7).

When the total system equation is condensed, the numerical efficiency is increased drastically, and the total system equation can be used universally because it can be considered equivalent to the framework of the general structural finite element analysis [19].

Before proceeding with the condensed localized mortar method, the displacement $\mathbf{u}^{(n)}$ at each part must be divided into the internal component $\mathbf{u}_i^{(n)}$ and the interface component

$\mathbf{u}_b^{(n)}$ as shown in Figure 4. When the displacements are decomposed, the total system Equation (7) for the localized mortar method is rewritten as:

$$\begin{bmatrix} \mathbf{M}_{ii}^{(1)} & \mathbf{M}_{ib}^{(1)} & 0 & 0 & 0 & 0 & 0 \\ \mathbf{M}_{bi}^{(1)} & \mathbf{M}_{bb}^{(1)} & 0 & 0 & 0 & 0 & 0 \\ 0 & 0 & \mathbf{M}_{ii}^{(2)} & \mathbf{M}_{ib}^{(2)} & 0 & 0 & 0 \\ 0 & 0 & \mathbf{M}_{bi}^{(2)} & \mathbf{M}_{bb}^{(2)} & 0 & 0 & 0 \\ 0 & 0 & 0 & 0 & 0 & 0 & 0 \\ 0 & 0 & 0 & 0 & 0 & 0 & 0 \\ 0 & 0 & 0 & 0 & 0 & 0 & 0 \end{bmatrix} \begin{bmatrix} \ddot{\mathbf{u}}_i^{(1)} \\ \ddot{\mathbf{u}}_b^{(1)} \\ \ddot{\mathbf{u}}_i^{(2)} \\ \ddot{\mathbf{u}}_b^{(2)} \\ \ddot{\boldsymbol{\lambda}}^{(1)} \\ \ddot{\boldsymbol{\lambda}}^{(2)} \\ \ddot{\mathbf{u}}^{(f)} \end{bmatrix} + \begin{bmatrix} \mathbf{K}_{ii}^{(1)} & \mathbf{K}_{ib}^{(1)} & 0 & 0 & 0 & 0 & 0 \\ \mathbf{K}_{bi}^{(1)} & \mathbf{K}_{bb}^{(1)} & 0 & 0 & \mathbf{Q}^{(1)} & 0 & 0 \\ 0 & 0 & \mathbf{K}_{ii}^{(2)} & \mathbf{K}_{ib}^{(2)} & 0 & 0 & 0 \\ 0 & 0 & \mathbf{K}_{bi}^{(2)} & \mathbf{K}_{bb}^{(2)} & 0 & \mathbf{Q}^{(2)} & 0 \\ 0 & \mathbf{Q}^{(1)T} & 0 & 0 & 0 & 0 & -\mathbf{W}^{(1)} \\ 0 & 0 & 0 & \mathbf{Q}^{(2)T} & 0 & 0 & -\mathbf{W}^{(1)} \\ 0 & 0 & 0 & 0 & -\mathbf{W}^{(1)T} & -\mathbf{W}^{(1)T} & 0 \end{bmatrix} \begin{bmatrix} \mathbf{u}_i^{(1)} \\ \mathbf{u}_b^{(1)} \\ \mathbf{u}_i^{(2)} \\ \mathbf{u}_b^{(2)} \\ \boldsymbol{\lambda}^{(1)} \\ \boldsymbol{\lambda}^{(2)} \\ \mathbf{u}^{(f)} \end{bmatrix} = \begin{Bmatrix} \mathbf{f}_i^{(1)} \\ \mathbf{f}_b^{(1)} \\ \mathbf{f}_i^{(2)} \\ \mathbf{f}_b^{(2)} \\ \mathbf{G}^{(1)} \\ \mathbf{G}^{(2)} \\ 0 \end{Bmatrix} \tag{8}$$

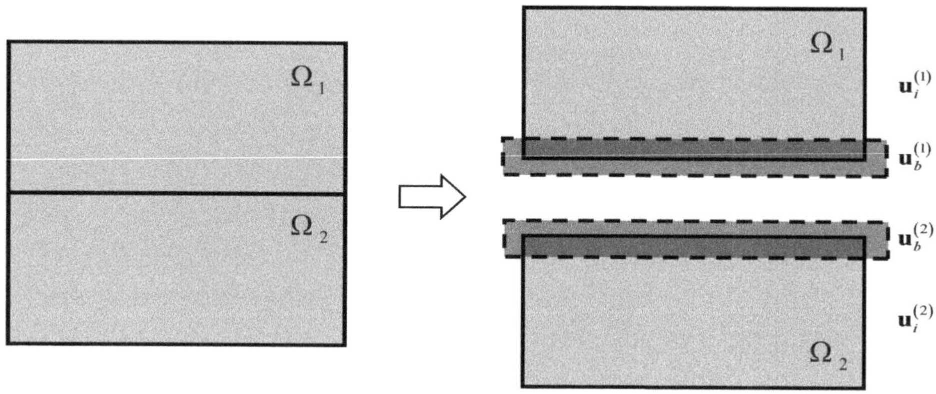

Figure 4. Division of the total displacements into the internal and interface components.

The following two steps must be performed to derive the condensed localized mortar method. Firstly, two relations must be obtained through the fifth and sixth rows of the total system Equation (8), to condense the boundary displacements $\mathbf{u}_b^{(1)}$, $\mathbf{u}_b^{(2)}$ into the frame displacement $\mathbf{u}^{(f)}$.

$$\begin{aligned} &\text{5th– row}: \mathbf{Q}^{(1)T}\mathbf{u}_b^{(1)} - \mathbf{W}^{(1)}\mathbf{u}^{(f)} = 0 \rightarrow \mathbf{u}_b^{(1)} = \left(\mathbf{Q}^{(1)T}\right)^{-1}\mathbf{W}^{(1)}\mathbf{u}^{(f)} = \mathbf{C}_{1f}\mathbf{u}^{(f)} \\ &\text{6th– row}: \mathbf{Q}^{(2)T}\mathbf{u}_b^{(2)} - \mathbf{W}^{(2)}\mathbf{u}^{(f)} = 0 \rightarrow \mathbf{u}_b^{(2)} = \left(\mathbf{Q}^{(2)T}\right)^{-1}\mathbf{W}^{(2)}\mathbf{u}^{(f)} = \mathbf{C}_{2f}\mathbf{u}^{(f)} \end{aligned} \tag{9}$$

Secondly, the second and fourth rows of the total system Equation (8) are used as follows to eliminate the Lagrange multipliers, $\boldsymbol{\lambda}^{(1)}$ and $\boldsymbol{\lambda}^{(2)}$:

$$\begin{aligned} &\text{2nd– row}: \mathbf{M}_{bi}^{(1)}\ddot{\mathbf{u}}_i^{(1)} + \mathbf{K}_{bi}^{(1)}\mathbf{u}_i^{(1)} + \mathbf{M}_{bb}^{(1)}\ddot{\mathbf{u}}_b^{(1)} + \mathbf{K}_{bb}^{(1)}\mathbf{u}_b^{(1)} + \mathbf{Q}^{(1)}\boldsymbol{\lambda}^{(1)} = \mathbf{f}_b^{(1)} \\ &\rightarrow \boldsymbol{\lambda}^{(1)} = \left(\mathbf{Q}^{(1)}\right)^{-1}\left(\mathbf{f}_b^{(1)} - \mathbf{M}_{bi}^{(1)}\ddot{\mathbf{u}}_i^{(1)} - \mathbf{K}_{bi}^{(1)}\mathbf{u}_i^{(1)} - \mathbf{M}_{bb}^{(1)}\ddot{\mathbf{u}}_b^{(1)} - \mathbf{K}_{bb}^{(1)}\mathbf{C}_{1f}\mathbf{u}^{(f)}\right) \\ &\text{4th– row}: \mathbf{M}_{bi}^{(2)}\ddot{\mathbf{u}}_i^{(2)} + \mathbf{K}_{bi}^{(2)}\mathbf{u}_i^{(2)} + \mathbf{M}_{bb}^{(2)}\ddot{\mathbf{u}}_b^{(2)} + \mathbf{K}_{bb}^{(2)}\mathbf{u}_b^{(2)} + \mathbf{Q}^{(2)}\boldsymbol{\lambda}^{(2)} = \mathbf{f}_b^{(2)} \\ &\rightarrow \boldsymbol{\lambda}^{(2)} = \left(\mathbf{Q}^{(2)}\right)^{-1}\left(\mathbf{f}_b^{(2)} - \mathbf{M}_{bi}^{(2)}\ddot{\mathbf{u}}_i^{(2)} - \mathbf{K}_{bi}^{(2)}\mathbf{u}_i^{(2)} - \mathbf{M}_{bb}^{(2)}\ddot{\mathbf{u}}_b^{(2)} - \mathbf{K}_{bb}^{(2)}\mathbf{C}_{2f}\mathbf{u}^{(f)}\right) \end{aligned} \tag{10}$$

When the relations (9) and (10) are substituted into the total system Equation (8), the Lagrange multipliers $\boldsymbol{\lambda}^{(1)}$ and $\boldsymbol{\lambda}^{(2)}$ and the overlapped boundary displacements $\mathbf{u}_b^{(1)}$ and $\mathbf{u}_b^{(2)}$ are condensed by the internal displacements $\mathbf{u}_i^{(1)}$ and $\mathbf{u}_i^{(2)}$ and the frame displacement $\mathbf{u}^{(f)}$. Lastly, the matrix form of the total system equation can be obtained as follows by summarizing and re-organizing Equation (8):

$$\hat{\mathbf{M}}\ddot{\hat{\mathbf{u}}} + \hat{\mathbf{K}}\hat{\mathbf{u}} = \hat{\mathbf{F}} \text{ where } \hat{\mathbf{M}} = \begin{bmatrix} \mathbf{M}_{ii}^{(1)} & 0 & \mathbf{M}_{ib}^{(1)}\mathbf{C}_{1f} \\ 0 & \mathbf{M}_{ii}^{(2)} & \mathbf{M}_{ib}^{(2)}\mathbf{C}_{2f} \\ \mathbf{C}_{1f}^T\mathbf{M}_{bi}^{(1)} & \mathbf{C}_{2f}^T\mathbf{M}_{bi}^{(2)} & \mathbf{C}_{1f}^T\mathbf{M}_{bb}^{(1)}\mathbf{C}_{1f} + \mathbf{C}_{2f}^T\mathbf{M}_{bb}^{(2)}\mathbf{C}_{2f} \end{bmatrix}, \hat{\mathbf{u}} = \begin{bmatrix} \mathbf{u}_i^{(1)} \\ \mathbf{u}_i^{(2)} \\ \mathbf{u}^{(f)} \end{bmatrix}, \ddot{\hat{\mathbf{u}}} = \begin{bmatrix} \ddot{\mathbf{u}}_i^{(1)} \\ \ddot{\mathbf{u}}_i^{(2)} \\ \ddot{\mathbf{u}}^{(f)} \end{bmatrix}$$
$$\hat{\mathbf{K}} = \begin{bmatrix} \mathbf{K}_{ii}^{(1)} & 0 & \mathbf{K}_{ib}^{(1)}\mathbf{C}_{1f} \\ 0 & \mathbf{K}_{ii}^{(2)} & \mathbf{K}_{ib}^{(2)}\mathbf{C}_{2f} \\ \mathbf{C}_{1f}^T\mathbf{K}_{bi}^{(1)} & \mathbf{C}_{2f}^T\mathbf{K}_{bi}^{(2)} & \mathbf{C}_{1f}^T\mathbf{K}_{bb}^{(1)}\mathbf{C}_{1f} + \mathbf{C}_{2f}^T\mathbf{K}_{bb}^{(2)}\mathbf{C}_{2f} \end{bmatrix}, \hat{\mathbf{F}} = \begin{bmatrix} \mathbf{f}_i^{(1)} \\ \mathbf{f}_i^{(2)} \\ \mathbf{C}_{1f}^T\mathbf{f}_b^{(1)} + \mathbf{C}_{2f}^T\mathbf{f}_b^{(2)} \end{bmatrix} \quad (11)$$

The internal displacements $\mathbf{u}_i^{(1)}$ and $\mathbf{u}_i^{(2)}$ and the frame displacement $\mathbf{u}^{(f)}$ can be obtained directly by solving the total system equation of the condensed localized mortar method (11). The condensed variables $\mathbf{u}_b^{(1)}$, $\mathbf{u}_b^{(2)}$, $\boldsymbol{\lambda}^{(1)}$ and $\boldsymbol{\lambda}^{(2)}$ can then be calculated using the relation (9) and (10), if required.

2.3. Verification Example for Linear Dynamic Impact and Frictionless Contact Problem

The problem of ring contacts to the rigid support beam [33] is considered to ensure that the condensed localized mortar method works well in the linear dynamic impact and frictionless contact conditions. Figure 5a presents the problem definition.

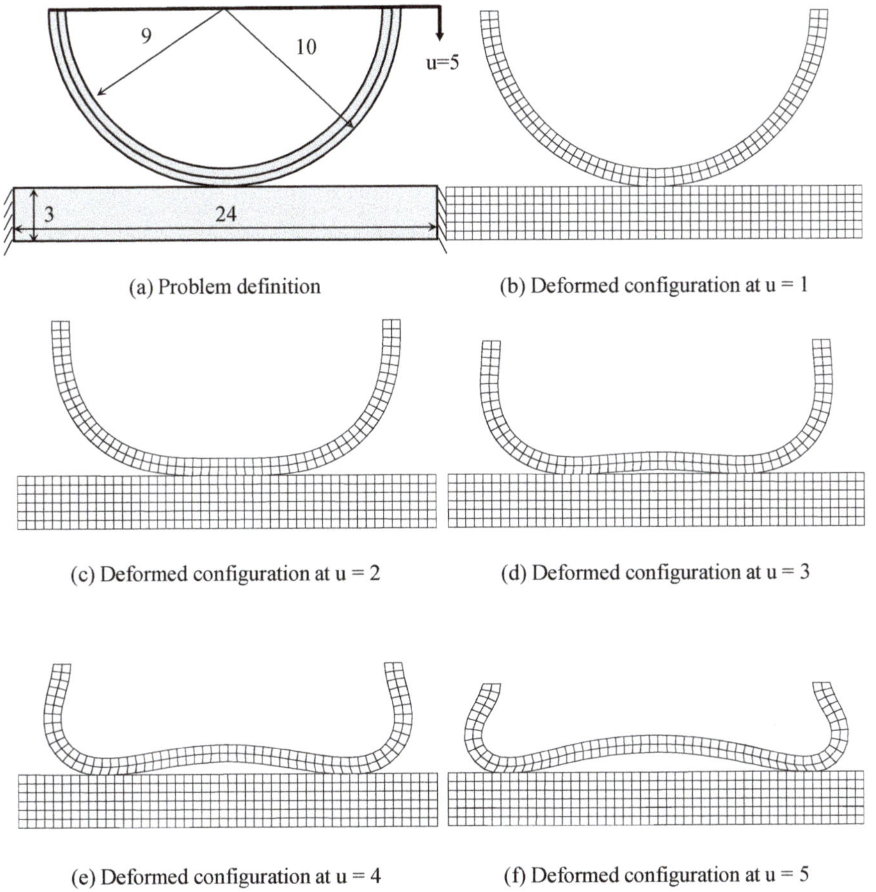

Figure 5. Problem definition and deformed configurations of a ring with rigid support.

As shown in Figure 5a, both side of the rigid support with a size of 24 by 3 is completely fixed and the displacement boundary condition is imposed on the top surface of the ring so that the ring goes down constantly. All the parts are modeled with a 2D linear quadrilateral element for both parts. In the verification example, the elastic ring consists of an inner and outer ring with the same thickness and different materials [33]. In order to ensure that the inner ring is stiffer than the outer ring, the material properties of the inner and outer rings are defined as follows:

$$\begin{aligned} E_{\text{inner}} &= 1E+05, \nu = 0.3 \\ E_{\text{outer}} &= 1E+03, \nu = 0.3 \end{aligned} \tag{12}$$

By imposing the condition of pressing on the top side of the ring, as shown in Figure 5a, the finite deformable ring is in contact with a rigid support, and a non-matching interface is then observed on the interface. The contact condition is solved by using the condensed localized mortar method (11).

A linear dynamic finite element analysis is conducted in 120 steps to solve the verification example. The deformed configurations are presented in Figure 5b–f. The contact area is changed from one spot to one region and then expanded to two regions to distribute the pressure by increasing the contact surface in order to support a given boundary condition as the analysis proceeds. These analysis results demonstrate identical appearances as shown in reference [33].

This result indicates that the condensed localized mortar method used in this study works well even in the linear dynamic impact and frictionless contact conditions, and that the contact regions are changed. Therefore, the interface is treated by the condensed localized mortar method to handle the topology optimization with the impact and frictionless contact conditions.

3. Dynamic Response Topology Optimization with Condensed Localized Mortar Method

Topology optimization is widely applied as a conceptual design that can be used to derive well-matched structures without any intervention from the designer [6–8]. The SIMP method is the most widely used approach among the various topology optimization techniques [9–12]; it is used to gradually find an optimum layout by determining the material existence or absence of each element in a given design domain. The SIMP method uses a relative density function to determine the existence of a material η.

$$\eta(x) = \begin{cases} 1 \text{ if } x \in \Omega \\ 0 \text{ if } x \notin \Omega \end{cases} \tag{13}$$

where x indicates the location of the finite element in the design domain, Ω. From Equation (13), it can be inferred that the material exists when the relative density function η is 1, and the material does not exist when the relative density function η, is 0. In the modified SIMP method [13], material properties such as Young's modulus and density of each element are defined based on the function of relative density η as given below:

$$\begin{aligned} E_e &= E_0\left(\alpha\eta_e^p + (1-\alpha)\eta_e\right) \\ \rho_e &= \rho_0\eta_e \end{aligned} \tag{14}$$

where E_0 and ρ_0 are the Young's modulus and density of the material, respectively, α is a positive constant that is typically defined as 15/16, and p is a penalization parameter that is typically set as 3.

3.1. Optimization Formulation with Condensed Localized Mortar Method

The optimization formulation used to obtain an optimum topological layout in a linear dynamic system [13,34,35] is represented by Equation (15), where the condensed localized mortar method described in the previous section is adopted by the matrices $\hat{\mathbf{M}}$, $\hat{\mathbf{K}}$, and vectors $\hat{\mathbf{u}}(t)$, $\ddot{\hat{\mathbf{u}}}(t)$ and $\hat{\mathbf{F}}(t)$ to handle the impact problem with the non-matching interface.

$$\begin{aligned}&\underset{\eta}{\text{minimize}}\ J_d\\&\text{subject to}\begin{cases}\hat{\mathbf{M}}\ddot{\hat{\mathbf{u}}}(t)+\hat{\mathbf{K}}\hat{\mathbf{u}}(t)=\hat{\mathbf{F}}(t)\ ,t\in[0,t_f]\\ g(\eta)=V(\eta)-V_{req}=\sum\limits_{e=1}^{N}\eta_e v_e-V_{req}\le 0\\ 0<\eta_{\min}\le\eta\le 1\end{cases}\end{aligned} \tag{15}$$

The optimization formulation (15) is organized to minimize the objective function J_d using the momentum equation and volume constraint, $g(n)$ [13]. In this study, the objective function J_d is defined for mean dynamic compliance to find a stiff structure as follows:

$$J_d=\frac{1}{t_f}\int_0^{t_f}c(\hat{\mathbf{u}}(t),\boldsymbol{\eta})\mathrm{d}t, c(\hat{\mathbf{u}}(t),\boldsymbol{\eta})=\hat{\mathbf{F}}^T(t)\hat{\mathbf{u}}(t)\text{ where }\frac{\partial c}{\partial\eta_e}=0,\frac{\partial c}{\partial\hat{\mathbf{u}}}=\hat{\mathbf{F}}(t) \tag{16}$$

Additionally, for the mean squared displacement of a target, DOF is given as:

$$J_d=\frac{1}{t_f}\int_0^{t_f}c(\hat{\mathbf{u}}(t),\boldsymbol{\eta})\mathrm{d}t, c(\hat{\mathbf{u}}(t),\boldsymbol{\eta})=(\mathbf{L}\hat{\mathbf{u}}(t))^2\text{ where }\frac{\partial c}{\partial\eta_e}=0,\frac{\partial c}{\partial\hat{\mathbf{u}}}=2\hat{\mathbf{u}}(t)\mathbf{L} \tag{17}$$

where $\mathbf{L}$ represents the role in selecting the target DOF from the displacement $\hat{\mathbf{u}}(t)$.

The optimum topological layout that minimizes the objective function can be determined, which includes the mean dynamic compliance and mean squared displacement of a target degree-of-freedom while satisfying the constraints, by performing the optimization formulation defined in Equations (15)–(17).

The sensitivity of the optimization formulation (15) must be calculated when a gradient-based optimization algorithm is used. Since the topology optimization presents several design variables by defining the relative density function η, for each element, the adjoint method is more appropriate than the direct method [6,7,13]. The adjoint variables λ_{adj} must be introduced to simplify the objective function in order to use the adjoint method to calculate the sensitivities of the objective function. When the adjoint variable λ_{adj} is inserted with the additional adjoint equation $\hat{\mathbf{M}}\ddot{\hat{\mathbf{u}}}+\hat{\mathbf{K}}\hat{\mathbf{u}}-\hat{\mathbf{F}}$ the objective function J_d can be rewritten as:

$$J_d\cong J_d+\boldsymbol{\lambda}^T\int_0^{t_f}(\hat{\mathbf{M}}\ddot{\hat{\mathbf{u}}}+\hat{\mathbf{K}}\hat{\mathbf{u}}-\hat{\mathbf{F}})\mathrm{d}t \tag{18}$$

Since the additional adjoint equation $\hat{\mathbf{M}}\ddot{\hat{\mathbf{u}}}+\hat{\mathbf{K}}\hat{\mathbf{u}}-\hat{\mathbf{F}}$ is always zero, as illustrated in Equation (11), it can be confirmed that the objective function J_d does not change even when the adjoint variable and the adjoint equation are added.

The sensitivity of the objective function is calculated by using the adjoint method, as follows:

$$\frac{\partial J_d}{\partial\eta_e}=\frac{\partial J_d}{\partial\eta_e}+\int_0^{t_f}\boldsymbol{\lambda}^T\frac{\partial}{\partial\eta_e}(\hat{\mathbf{M}}\ddot{\hat{\mathbf{u}}}+\hat{\mathbf{K}}\hat{\mathbf{u}}-\hat{\mathbf{F}})\mathrm{d}t=\int_0^{t_f}\left[\left\{\frac{\partial c}{\partial\eta_e}+\frac{\partial\hat{\mathbf{u}}}{\partial\eta_e}^T\frac{\partial c}{\partial\hat{\mathbf{u}}}\right\}+\left\{\boldsymbol{\lambda}^T\frac{\partial}{\partial\eta_e}(\hat{\mathbf{M}}\ddot{\hat{\mathbf{u}}}+\hat{\mathbf{K}}\hat{\mathbf{u}})\right\}\right]\mathrm{d}t \tag{19}$$

Subsequently, by integrating-by-parts and re-organizing, Equation (19) yields:

$$\frac{\partial J_d}{\partial\eta_e}=\int_0^{t_f}\left[\boldsymbol{\lambda}^T\frac{\partial\hat{\mathbf{M}}}{\partial\eta_e}\ddot{\hat{\mathbf{u}}}+\boldsymbol{\lambda}^T\frac{\partial\hat{\mathbf{K}}}{\partial\eta_e}\hat{\mathbf{u}}+\frac{\partial c}{\partial\eta_e}\right]\mathrm{d}t+\int_0^{t_f}\left[\frac{\partial\hat{\mathbf{u}}}{\partial\eta_e}^T\left(\frac{\partial c}{\partial\hat{\mathbf{u}}}+\hat{\mathbf{M}}\ddot{\boldsymbol{\lambda}}+\hat{\mathbf{K}}\boldsymbol{\lambda}\right)\right]\mathrm{d}t+\left[\frac{\partial\dot{\hat{\mathbf{u}}}}{\partial\eta_e}^T\hat{\mathbf{M}}\boldsymbol{\lambda}-\frac{\partial\hat{\mathbf{u}}}{\partial\eta_e}^T\hat{\mathbf{M}}\dot{\boldsymbol{\lambda}}\right]\Bigg|_{t=t_f} \tag{20}$$

In Equation (20), to obtain the adjoint variable $\boldsymbol{\lambda}_{\text{adj}}$ that is arbitrarily inserted to simplify the sensitivity process, the following differential equation must be solved:

$$\hat{\mathbf{M}}\ddot{\boldsymbol{\lambda}} + \hat{\mathbf{K}}\boldsymbol{\lambda} = -\frac{\partial c}{\partial \hat{\mathbf{u}}},\ \mathrm{t} \in \left[0, t_f\right] \text{ where } \boldsymbol{\lambda}_{t_f} = 0, \dot{\boldsymbol{\lambda}}_{t_f} = 0 \tag{21}$$

where Equation (21) is a similar form of the equation of motion (11), which is widely known. Therefore, the adjoint variable is transformed by $\Lambda(s) = \boldsymbol{\lambda}_{\text{adj}}(t_f - s)$ in order to make an equivalent form, and Equation (21) is re-written as:

$$\hat{\mathbf{M}}\ddot{\Lambda}(s) + \hat{\mathbf{K}}\Lambda(s) = -\left.\frac{\partial c}{\partial \hat{\mathbf{u}}}\right|_{t_f - s},\ \mathrm{s} \in \left[0, t_f\right] \text{ where } \Lambda(0) = 0, \dot{\Lambda}(0) = 0 \tag{22}$$

Since the form of Equation (22) is exactly identical to that of Equation (11), the transformed adjoint variable $\Lambda(s)$ can be easily calculated, along with the adjoint variable $\boldsymbol{\lambda}_{\text{adj}}$ through an inverse transformation.

3.2. Verification Example for Dynamic Response Topology Optimization

A Michell-type structure problem is used to validate the optimization formulation (15) for the linear dynamic system. The geometry and boundary conditions are shown in Figure 6a,b. The material properties of the entire design domain are assumed to be those of structural steel ($E = 2 \times 10^{11}$, Poisson's ratio is 0.3, and density $\rho = 7890$). The total time step is considered as 50.

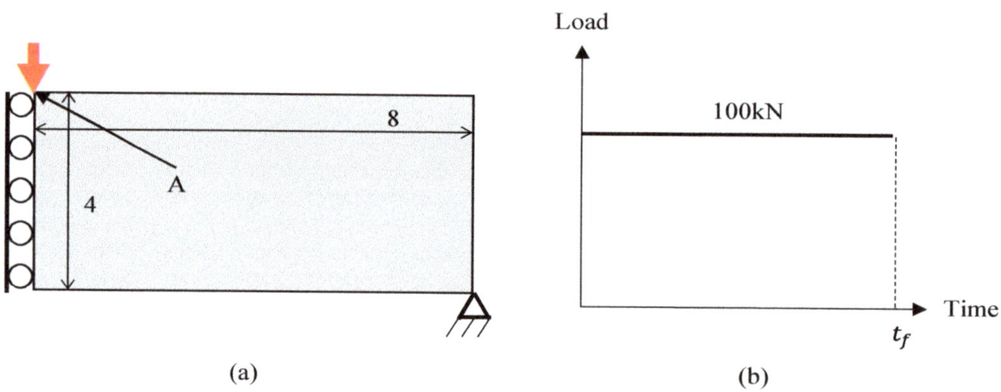

Figure 6. Problem definition for Michell-type structure problem: (**a**) schematic description (**b**) load condition.

Topology optimization is performed to obtain the stiffest design while satisfying 50% of the volume using the optimization formulation (15) and (16). The adjoint method is used for sensitivity analysis. In order to identify the differences and effectiveness of implementation in the linear dynamic system before performing the dynamic response topology optimization, the static response topology optimization under the same conditions and optimization formulations is performed as shown in Figure 7.

In order to perform topology optimization, it is necessary to confirm whether the dynamic analysis is performed adequately due to which, the Michell-type structure is analyzed first. Figure 8 shows the vertical directional displacements at point "A". The red color represents the result of the static analysis, and the other results indicate the displacement of the dynamic analysis corresponding to the final time t_f. In the verification example, analysis and optimization is conducted with the final time t_f set to 0.01, 0.05, 0.1, and 0.5 to check the response in the topology optimization for the dynamic system.

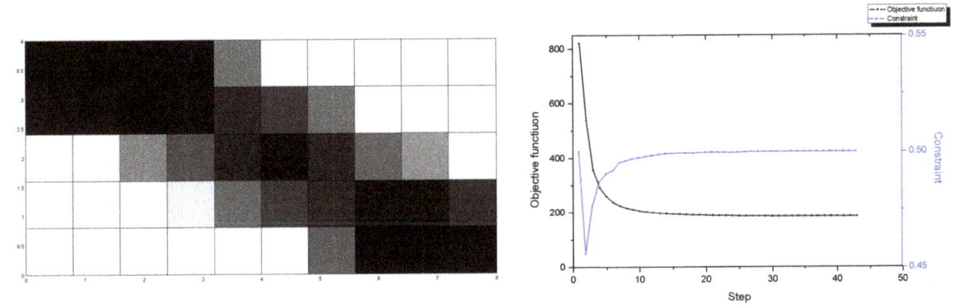

(a) (b)

Figure 7. Optimized results for Michell-type structure with static analysis: (**a**) Optimized layout (**b**) convergence history.

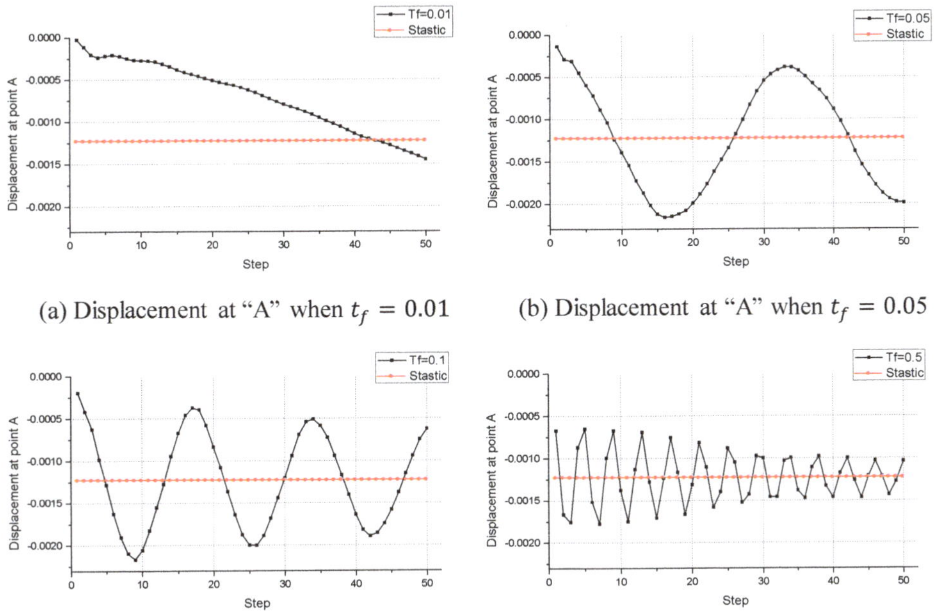

(a) Displacement at "A" when $t_f = 0.01$ (b) Displacement at "A" when $t_f = 0.05$

(c) Displacement at "A" when $t_f = 0.1$ (d) Displacement at "A" when $t_f = 0.5$

Figure 8. Displacement histories of the tip point for Michell-type structure.

As shown in Figure 8, the results of the displacement at point "A" are converged to the quasi-static results as the final time t_f increases. Figure 8a shows the displacement at point "A" when the final time t_f is 0.01 s, and the structure acts as an impact state because the load is applied for a very short period of time. This indicates that the dynamic effect produced by the impact phenomenon is essential in the state of Figure 8a with a final time t_f of 0.01 s.

In the case of the impact state as shown in Figure 9a, the material is concentrated near the point "A" where the load is applied due to the effect of inertia. Conversely, in the case of quasi-static cases, as shown in Figure 9b–d, the optimized topological layout is identical to the result of the static analysis presented in Figure 7 because the structure acts as a quasi-static state when the final time t_f is greater than 0.05.

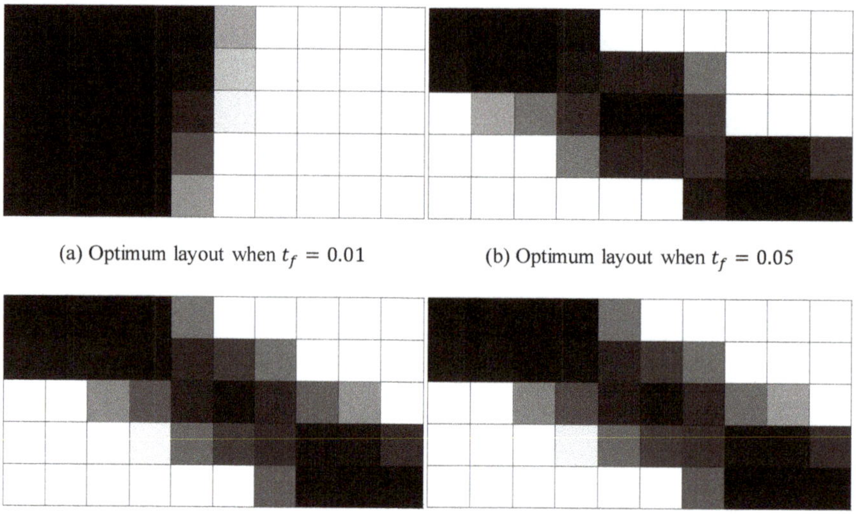

(a) Optimum layout when $t_f = 0.01$ (b) Optimum layout when $t_f = 0.05$

(c) Optimum layout when $t_f = 0.1$ (d) Optimum layout when $t_f = 0.5$

Figure 9. Optimized results for Michell-type structure with dynamic analysis.

The verification example is used to verify that the dynamic response topology optimization adequately derives the optimized layout when the dynamic effect is critical. As shown in Figure 9, the material is concentrated near the impact region due to the inertia effect in the impact state. However, in a quasi-static state, the optimized topological layout is obtained as the result of the static analysis as the final time t_f increases. This indicates that the dynamic response topology optimization is applied well in both the impact and quasi-static states when the formulation (15) is used.

4. Numerical Example: Impact between Fixed Beam and Flying Block

Hitherto, two methods have been employed to find an optimum topological layout in the linear dynamic impact problem: the condensed localized mortar method and the dynamic response topology optimization approach. The condensed localized mortar method used to handle the non-matching interface at the impact and frictionless contact region was described in Section 2, and the simple frictionless contact problem is solved to verify its effectiveness in the impact state. The dynamic response topology optimization was introduced in Section 3, and an optimization formulation with the condensed mortar method was presented. Additionally, by performing topology optimization on the verification example, the effectiveness of the dynamic response topology optimization is confirmed when the beam is in the impact state.

In this section, topology optimization is performed by assuming a situation in which two independently behaving objects are in the impact and frictionless contact conditions to examine the applicability of the proposed approach to the linear dynamic impact problem. The problem definition is presented in Figure 10. A block is separated at 0.05 intervals on a lower beam which is completely fixed on both sides, as shown in Figure 10a. A constant pressure is applied to the block for 0.12 s as depicted in Figure 10.

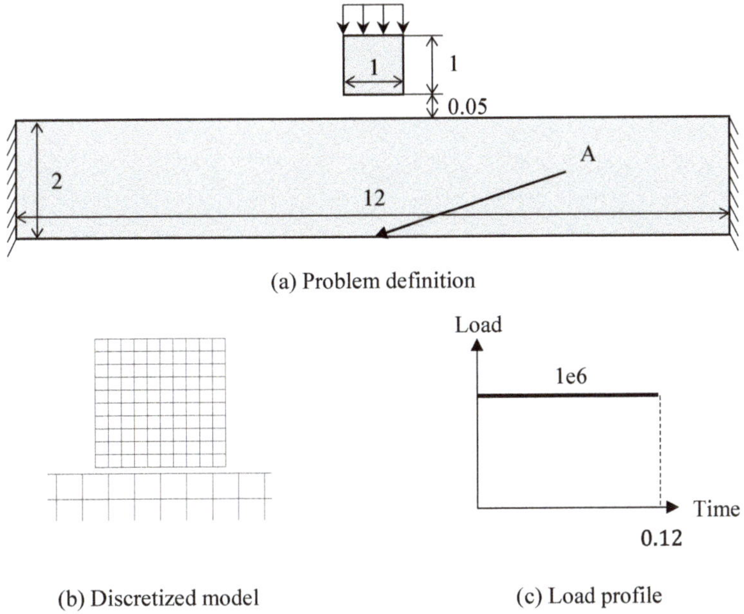

Figure 10. Problem definition of beam with a flying block.

The frictionless contact condition (2) is imposed to allow for minimal sliding in the tangential direction after the occurrence of the impact. Therefore, a situation in which the configurations on the non-matching interface inevitably occur is considered. The material properties for both the structures are assumed to be those of the structural steel ($E = 2 \times 10^{11}$, Poisson's ratio is 0.3, and density $\rho = 7890$). The mean squared displacement of a point "A" is used as an objective function (17) and the volume is limited to 50% for the topology optimization process (15).

The optimization was performed considering only the lower beam and not the upper block. A dynamic analysis is preferentially conducted to confirm the contact kinematics before performing topology optimization. Figure 11 illustrates the deformed shapes obtained by conducting the dynamic analysis, where it is observed that the impact occurs after 0.06 s, and then the frictionless contact occurs between two parts.

The topology optimization results based on the analysis are presented in Figure 12. In the numerical example, the optimization is performed for three cases: no contact (t = 0.04), impact (t = 0.08) and steady state (t = 0.12). The topological change in the design domain does not occur since there is no contact until 0.06 s, as shown in Figure 12a. The optimized layouts in the impact (t = 0.08) and steady state (t = 0.12) are derived as shown in Figure 12b,c, after 0.06 s to minimize the beam deflection. A large portion of the materials are distributed in the position where the beam is impacted when the design domain is in the impact state (t = 0.08), since the force is not completely transmitted to the beam, as shown in Figure 12b. However, when the analysis is performed over a period of 0.12 s, a shape in which the material is spread throughout the design domain is observed since the load is distributed throughout the beam, as shown in Figure 12c. The result shown in Figure 12c is a truss-type layout similar to the optimized results that can be derived under static analysis with the same boundary conditions.

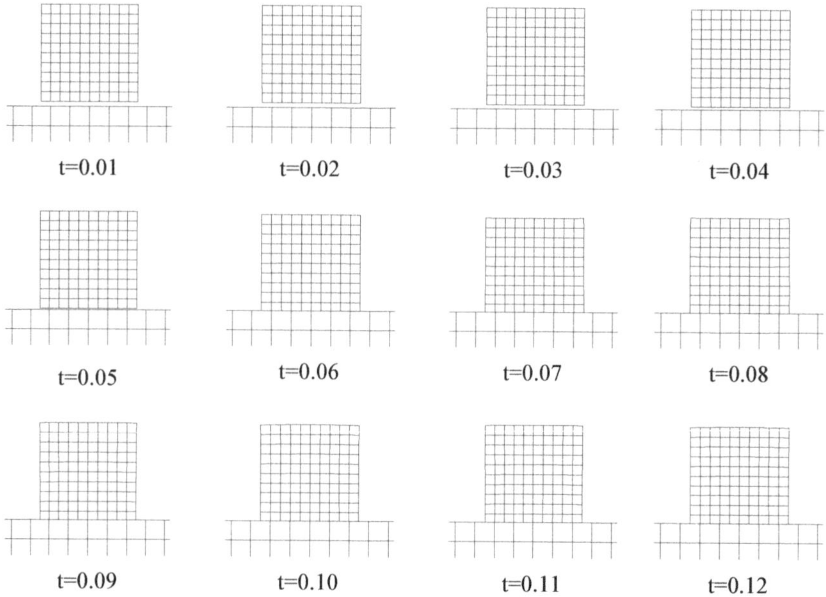

Figure 11. Deformed configuration for beam with flying block.

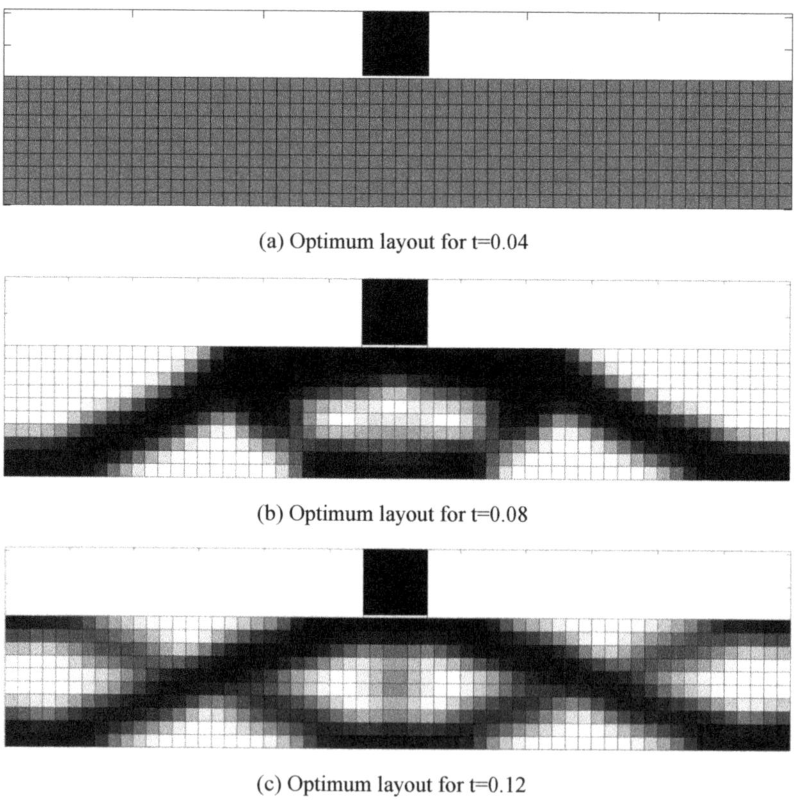

(a) Optimum layout for t=0.04

(b) Optimum layout for t=0.08

(c) Optimum layout for t=0.12

Figure 12. Optimized for beam with flying block using condensed localized mortar method.

As a result, it is possible to obtain the optimum topological layout to which the dynamic effect is applied in the impact state in which the non-matching interface inevitably occurred. In addition, after the impact state, it is confirmed that the identical layout with the quasi-static state can be obtained when the load by the frictionless contact condition is transmitted to the entire design domain.

It is confirmed through this example that the proposed approach can derive an acceptable result in the case of dynamic analysis where impact and frictionless contact are observed. Consequently, it is concluded that this approach can be used to perform topology optimization for various linear dynamic impact and frictionless contact problems.

5. Conclusions

This study presents an approach to apply the treatment of the non-matching interface to the dynamic response topology optimization for determining an optimum topological layout in the problem where the non-matching interface inevitably exists owing to the linear dynamic impact and frictionless contact conditions. First, the condensed localized mortar method is used to deal with linear dynamic impact and frictionless contact conditions which must occur in the non-matching interface during the analysis. The detailed explanation is in Section 2, and its usefulness is verified through the verification examples in Section 2.3. Second, the SIMP method, the most widely used topology optimization approach, is applied to treat the linear dynamic problems with the non-matching interface by using the condensed localized mortar method. The detailed explanation is in Section 3, and its effectiveness is verified through the verification examples in Section 3.2. Since the non-matching interface could be considered as the general structural analysis in the same framework by the features of the condensed localized mortar method, it can be easily applied to the well-known and well-established linear dynamic optimization framework. The effectiveness of the proposed approach is verified through a numerical example in Section 4. From Section 4, it is confirmed that physically reasonable results are obtained when the presented approach is applied in the dynamic response topology optimization in which the non-matching interface inevitably exist during the impact and frictionless contact conditions. Consequently, the problem of relative behavior including the impact and frictionless contact conditions along the adjacent parts could be adequately designed using the proposed method.

This study focuses on the linear dynamic impact and frictionless contact conditions to determine an optimum topological layout. In the future, the proposed optimization approach will be applied to more general and highly nonlinear non-matching interface problems, such as friction contact and multi-physics problems.

Funding: This work was supported by the Institute for Korea Spent Nuclear Fuel (iKSNF) and National Research Foundation of Korea (NRF) grant funded by the Korean government (Ministry of Science and ICT, MSIT) (No. 2021M2E1A1085229).

Institutional Review Board Statement: Not applicable.

Informed Consent Statement: Not applicable.

Data Availability Statement: Not applicable.

Acknowledgments: The author specially would like to thanks Sung-Kie Youn (Department of Mechanical Engineering, Korea Advanced Institute of Science and Technology) for useful discussion and suggestion and constant academic supports.

Conflicts of Interest: The author declares no conflict of interest.

Abbreviations/Nomenclature:

Ω_i	Domain for domain "i"
Γ	Boundary (Subscript, u: Dirichlet, t: Neumann, c: Contact)
$X^{(i)}$	Position for un-deformed domain "i"
$x^{(i)}$	Position for deformed domain "i"
ρ	Density
E	Young's modulus
$\mathbf{P}$	First Piola-Kirchhoff stress tensor
$\mathbf{b}$	External body force
$\mathbf{u}^{(i)}$	Displacement for domain "i"
g_n	Gap function in normal direction
τ_n	Contact traction in normal direction
α, β	Algorithmic parameter for Newmark method
λ	Lagrange multiplier
$\mathbf{f}_{\text{int}}$	Internal force vector
$\mathbf{f}_{\text{c}}$	Contact force vector
$\mathbf{f}_{\text{ext}}$	External force vector
η	Design variable for topology optimization
J_d	Objective function for topology optimization
V	Volume constraint for topology optimization
λ_{adj}	Adjoint variable for topology optimization
Λ	Transformed adjoint variable λ_{adj}

References

1. Vaneker, T.; Bernard, A.; Moroni, G.; Gibson, I.; Zhang, Y. Design for Additive Manufacturing: Framework and Methodology. *CIRP Ann. Manuf. Technol.* **2020**, *69*, 578–599. [CrossRef]
2. Kathryn, M.; Moroni, G.; Vaneker, T.; Fadel, G.; Campbell, I.; Gibson, I.; Bernard, A.; Schulz, J.; Graf, P.; Ahuja, B.; et al. Design for Additive Manufacturing: Trends, Opportunities, Considerations, and Constraints. *CIRP Ann. Manuf. Technol.* **2016**, *65*, 737–760. [CrossRef]
3. Meng, L.; Zhang, W.; Quan, D.; Shi, G.; Tang, L.; Hou, Y.; Breitkopf, P.; Zhu, J.; Gao, T. From Topology Optimization Design to Additive Manufacturing: Today's Success and Tomorrow's Roadmap. *Arch. Comput. Methods Eng.* **2020**, *27*, 805–830. [CrossRef]
4. Liu, J.; Gaynor, A.T.; Chen, S.; Kang, Z.; Suresh, K.; Takezawa, A.; Li, L.; Kato, J.; Tang, J.; Wang, C.C.L.; et al. Current and Future Trends in Topology Optimization for Additive Manufacturing. *Struct. Multidiscip. Optim.* **2018**, *57*, 2457–2483. [CrossRef]
5. Guo, X.; Zhou, J.; Zhang, W.; Du, Z.; Liu, C.; Liu, Y. Self-Supporting Structure Design in Additive Manufacturing Through Explicit Topology Optimization. *Comput. Methods Appl. Mech. Eng.* **2017**, *323*, 27–63. [CrossRef]
6. Bendsøe, M.P.; Sigmund, O. *Topology Optimization-Theory, Methods and Applications*; Springer: Berlin/Heidelberg, Germany, 2002.
7. Sigmund, O.; Maute, K. Topology Optimization Approaches. *Struct. Multidiscip. Optim.* **2013**, *48*, 1031–1055. [CrossRef]
8. Bendsøe, M.P.; Kikuchi, N. Generating Optimal Topologies in Structural Design Using a Homogenization Method. *Comput. Methods Appl. Mech. Eng.* **1988**, *71*, 197–224. [CrossRef]
9. Andreassen, E.; Clausen, A.; Schevenels, M.; Lazarov, B.S.; Sigmund, O. Efficient Topology Optimization in MATLAB Using 88 Lines of Code. *Struct. Multidiscip. Optim.* **2011**, *43*, 811–825. [CrossRef]
10. Sigmund, O. A 99 Line Topology Optimization Code Written in MATLAB. *Struct. Multidiscip. Optim.* **2001**, *21*, 120–127. [CrossRef]
11. Deaton, J.D.; Grandhi, R.V. A Survey of Structural and Multidisciplinary Continuum Topology Optimization: Post 2000. *Struct. Multidiscip. Optim.* **2014**, *49*, 1–38. [CrossRef]
12. Sigmund, O. Morphology-Based Black and White Filters for Topology Optimization. *Struct. Multidiscip. Optim.* **2007**, *33*, 401–424. [CrossRef]
13. Zhao, J.; Wang, C. Dynamic Response Topology Optimization in the Time Domain Using Model Reduction Method. *Struct. Multidiscip. Optim.* **2016**, *53*, 101–114. [CrossRef]
14. Mankame, N.D.; Ananthasuresh, G.K. Topology Optimization for Synthesis of Contact-Aided Compliant Mechanisms Using Regularized Contact Modeling. *Comput. Struct.* **2004**, *82*, 1267–1290. [CrossRef]
15. Hilding, D.; Klarbring, A. Optimization of Structures in Frictional Contact. *Comput. Methods Appl. Mech. Eng.* **2012**, *205–208*, 83–90. [CrossRef]
16. Luo, Y.; Li, M.; Kang, Z. Topology Optimization of Hyperelastic Structures with Frictionless Contact Supports. *Int. J. Solids Struct.* **2016**, *81*, 373–382. [CrossRef]
17. Strömberg, N.; Klarbring, A. Topology Optimization of Structures in Unilateral Contact. *Struct. Multidiscip. Optim.* **2010**, *41*, 57–64. [CrossRef]
18. Strömberg, N. Topology Optimization of Two Linear Elastic Bodies in Unilateral Contact. In Proceedings of the 2nd International Conference Engenharia Optimisation, Lisbon, Portugal, 6–9 September 2010; pp. 1–8.

19. Jeong, G.E.; Youn, S.K.; Park, K.C. Topology Optimization of Deformable Bodies with Dissimilar Interfaces. *Comput. Struct.* **2018**, *198*, 1–11. [CrossRef]
20. Fernandez, F.; Puso, M.A.; Solberg, J.; Tortorelli, D.A. Topology Optimization of Multiple Deformable Bodies in Contact with Large Deformations. *Comput. Methods Appl. Mech. Eng.* **2020**, *371*, 113288. [CrossRef]
21. Zienkiewicz, O.C.; Taylor, R.L. The Finite Element Method. *J. Solid Mech.* **2000**.
22. Fujun, W.; Jiangang, C.; Zhenhan, Y. A Contact Searching Algorithm for Contact-Impact Problems. *Acta Mech. Sin.* **2000**, *16*, 374–382. [CrossRef]
23. Newmark, N.M. A Method of Computation for Structural Dynamics. *J. Eng. Mech. Div.* **1959**, *85*, 67–94. [CrossRef]
24. Fischer, K.A.; Wriggers, P. Frictionless 2D Contact Formulations for Finite Deformations Based on the Mortar Method. *Comput. Mech.* **2005**, *36*, 226–244. [CrossRef]
25. Hesch, C.; Betsch, P. A Mortar Method for Energy-Momentum Conserving Schemes in Frictionless Dynamic Contact Problems. *Int. J. Numer. Methods Eng.* **2009**, *77*, 1468–1500. [CrossRef]
26. Flemisch, B.; Puso, M.A.; Wohlmuth, B.I. A New Dual Mortar Method for Curved Interfaces: 2D Elasticity. *Int. J. Numer. Methods Eng.* **2005**, *63*, 813–832. [CrossRef]
27. Puso, M.A. A 3D Mortar Method for Solid Mechanics. *Int. J. Numer. Meth. Eng.* **2004**, *59*, 315–336. [CrossRef]
28. Park, K.C.; Felippa, C.A. A Variational Principle for the Formulation of Partitioned Structural Systems. *Int. J. Numer. Meth. Eng.* **2000**, *47*, 395–418. [CrossRef]
29. Park, K.C.; Felippa, C.A.; Gumaste, U.A. A Localized Version of the Method of Lagrange Multipliers and Its Applications. *Comput. Mech.* **2000**, *24*, 476–490. [CrossRef]
30. Park, K.C.; Felippa, C.A.; Rebel, G. A Simple Algorithm for Localized Construction of Non-Matching Structural Interfaces. *Int. J. Numer. Meth. Eng.* **2002**, *53*, 2117–2142. [CrossRef]
31. Jeong, G.E.; Song, Y.U.; Youn, S.K.; Park, K.C. A New Approach for Nonmatching Interface Construction by the Method of Localized Lagrange Multipliers. *Comput. Methods Appl. Mech. Eng.* **2020**, *361*, 112728. [CrossRef]
32. Song, Y.; Jeong, G.; Youn, S.; Park, K.C. Virtual Tetrahedral Gap Element to Connect Three-Dimensional Non-Coincident Interfaces. *Finite Elem. Anal. Des.* **2018**, *152*, 18–26. [CrossRef]
33. Wriggers, P. *Computational Contact Mechanics*, 2nd ed.; Springer: Berlin/Heidelberg, Germany, 2006.
34. Min, S.; Kikuchi, N.; Park, Y.C.; Kim, S.; Chang, S. Optimal Topology Design of Structures Under Dynamic Loads. *Struct. Optim.* **1999**, *17*, 208–218. [CrossRef]
35. Rong, J.H.; Xie, Y.M.; Yang, X.Y.; Liang, Q.Q. Topology Optimization of Structures Under Dynamic Response Constraints. *J. Sound Vib.* **2000**, *234*, 177–189. [CrossRef]

applied sciences

MDPI

Article

Free and Forced Wave Motion in a Two-Dimensional Plate with Radial Periodicity

Elisabetta Manconi [1,*], Sergey V. Sorokin [2], Rinaldo Garziera [1] and Matheus Mikael Quartaroli [3]

1 Dipartimento di Ingegneria e Architettura, Università di Parma, 43124 Parma, Italy; rinaldo.garziera@unipr.it
2 Department of Materials and Production, Aalborg University, DK-9220 Aalborg East, Denmark; svs@m-tech.aau.dk
3 Department of Mechanical Engineering, UNESP-FEB, Bauru 17033-360, Brazil; matheus.quartaroli@unesp.br
* Correspondence: elisabetta.manconi@unipr.it

Abstract: In many practical engineering situations, a source of vibrations may excite a large and flexible structure such as a ship's deck, an aeroplane fuselage, a satellite antenna, a wall panel. To avoid transmission of the vibration and structure-borne sound, radial or polar periodicity may be used. In these cases, numerical approaches to study free and forced wave propagation close to the excitation source in polar coordinates are desirable. This is the paper's aim, where a numerical method based on Floquet-theory and the FE discretisation of a finite slice of the radial periodic structure is presented and verified. Only a small slice of the structure is analysed, which is approximated using piecewise Cartesian segments. Wave characteristics in each segment are obtained by the theory of wave propagation in periodic Cartesian structures and Finite Element analysis, while wave amplitude change due to the changes in the geometry of the slice is accommodated in the model assuming that the energy flow through the segments is the same. Forced response of the structure is then evaluated in the wave domain. Results are verified for an infinite isotropic thin plate excited by a point harmonic force. A plate with a periodic radial change of thickness is then studied. Free waves propagation are shown, and the forced response in the nearfield is evaluated, showing the validity of the method and the computational advantage compared to FE harmonic analysis for infinite structures.

Keywords: periodic structures; polar coordinates; wave propagation; forced response of plates and shells; finite element analysis; unbounded structures

Citation: Manconi, E.; Sorokin, S.V.; Garziera, R.; Quartaroli, M.M. Free and Forced Wave Motion in a Two-Dimensional Plate with Radial Periodicity. *Appl. Sci.* **2021**, *11*, 10948. https://doi.org/10.3390/app112210948

Academic Editor: Jin-Gyun Kim

Received: 21 September 2021
Accepted: 16 November 2021
Published: 19 November 2021

1. Introduction

Starting from the milestone book written by Leon Brillouin [1], wave propagation in periodic media has been a subject extensively studied. In particular, periodicity effects in electromagnetic wave motion have been largely investigated, and they have found applications in many optical and electromagnetic devices. The ability of periodic configuration of creating electronic/photonic band-gaps in semiconductors and crystals is similar to structural/acoustical band-gaps in elastic media. The subject has recently found renewed interest due to its new and potential applications in vibroacoustic isolation, e.g., [2], noise suppression devices, e.g., [3], mitigation of seismic waves, e.g., [4,5], elastic/acoustic metamaterials [6]. A comprehensive review of the research in periodic materials and structures has been presented in [7]. In [7] the authors gave an overview of the numerical and experimental research in periodic structures, phononic crystals, and acoustic/elastic metamaterials up to 2014, showing some of the recent progress and the growth in academic and applied research interests in these fields. Experimental studies on periodic structures were presented recently in many papers, and periodicity effects such as structural and acoustic band-gaps, attenuation, directional energy flow, were verified in many specific cases, e.g., [7,8]. While these phenomena are well known, most of the literature on periodic engineering structures has been devoted to developing theoretical and numerical approaches due to the importance of understanding and characterising wave propagation behaviour.

Mead [9] and his coworkers at the University of Southampton [10] gave a substantial contribution to methods for predicting and analysing wave motion in periodic engineering structures. Among the numerical approaches proposed, those based on Finite Element theory have shown large versatility and applicability to modelling physical structures. Two different main approaches are typically applied. The firsts implement absorbing boundary conditions/layers that mimick the infinity domains through artificial non-reflective boundaries [11,12]; the advantage of this approach relies on the capability of standard FEA to model waveguides as a part of more complex structures. However, the computational cost remains a common issue in such an analysis, especially at high frequencies, when the mesh must be refined to reach sufficient accuracy in studies of the performance of a waveguide. It implies a substantial increase in the number of DOFs of the model, which leads to prohibitive computational time at higher frequency.

The seconds are numerical methods based on FE analysis of a unit cell of the structures and the theory of wave propagation in periodic structures, e.g., [13–15]. Compared to FE modelling with absorbing boundary conditions/layers, these numerical methods allow high accuracy up to high frequency at very low computation cost, and they can be the preferable choice when the requirement is the prediction of wave motion in 1-dimensional and 2-dimensional single waveguides, e.g., beam-like structures, panels and shells, cylindrical waveguides, etc. Some FE commercial software have recently implemented new specific modules for studying Cartesian periodic structures exploiting Bloch-Floquet theory. Although their use can be advantageous due to their ability to tackle complicated geometries, they involve a high computational cost in many cases.

A common feature of these methods is Cartesian spatial periodicity. The Bloch-Floquet theorem, which was applied in 1946 by Brillouin to solve the wave equation, relies, in fact, on the translational invariance of the problem formulation and is not rigorously applicable to polar coordinates systems [16]. In many practical engineering situations, a source of vibrations may excite a large and flexible structure such as a ship's deck, an aeroplane fuselage, a satellite antenna, a wall panel. In these cases, radial periodicity (e.g., as a sequence of annuli with alternating properties) may be used to reduce transmission of the vibration and structure-borne sound, and numerical approaches to study free and forced wave propagation in polar coordinates are desirable. The problem of wave propagation in radially periodic structures has been mainly formulated in optics within the theory of Bragg fibre [17]. Leaving aside numerous purely numerical studies (see, for instance, [18–20]), we notice that two approaches have been used so far to "adjust" the Bloch theorem for a cylindrically symmetric Bragg fibre. The first one is based on the use of far-field approximation of Bessel functions [21,22], and the second one implies special radial varying of material properties [23–25]. Some recent works presented an approximation of the formulation of the Floquet theory for radially periodic membrane and plates [26]. However, studies of wave propagation in a polar periodic configuration in structural mechanics are rare.

The paper aims to present a numerical method based on Floquet-theory and the FE discretisation of a small finite slice of a radial periodic structure. The method relies on adapting the Wave Finite Element (WFE) approach [14,15] to radial periodic waveguides. In this method, a unit cell of the waveguide is discretised using standard FE elements. The FE mass and stiffness matrices are reduced using wave propagation theory in periodic structures, and wave characteristics are numerically evaluated from an eigenvalue problem. This allows a very substantial reduction in the number of DOFs involved in the computation, with a dramatic reduction of the computational time compared to other FE approaches. Forced wave propagation is then evaluated in the wave domain as described in [27,28]. Amongst the numerical methods that could be used to investigate wave propagation problems, the WFE technique has several desirable features: it can be applied both to continuous and periodic structures; it exploits Bloch-Floquet theory and the versatility of standard FE analysis of a very small part of the structure; it allows the study of waveguides with complex cross-sectional characteristics in a systematic manner, up to high frequency and with a low-computation cost. Applications of the technique to periodic, axisymmetric, helical, and slowly varying waveguides were presented in the literature, and the method has been validated through many benchmark cases, e.g., [29–31]. The approach rigorously assumes wave propagation in structures posed in Cartesian coordinates, and application to forced response in two dimensions was only possible by a semi-analytical computation via contour integration and the residue theorem [32].

Following a previous study by the same authors [33], this paper presents a simplified adaptation of this WFE technique to structures in polar coordinates exhibiting periodicity in the radial directions. Cylindrical wave propagation is thus estimated exploiting the Floquet theory formulation for an infinite periodic structure in one dimension. The approximation is achieved by taking only a very small slice of the structure and discretising the slice through piecewise Cartesian segments. Wave characteristics in each segment are obtained by the WFE approach, while wave amplitudes change due to the changes in the geometry of the slice are accommodated in the model assuming that the energy flow through each segment is the same. In this paper, the forced response of a flexible periodic plate excited by a transverse source of vibration force is considered. Nearfield response and low-frequency harmonic excitation, far enough from the first stop-band, is assumed. The aim is to quantify the response level of the structure close to the excitation point and in the frequency regime where the fundamental modes of the structures are excited.

The paper is organised as follows. In Section 2 the method is described. Section 3 contains a numerical example of a thin flexible plate, for which analytical solutions are available [34]. Section 4 deals with a plate with a periodic change of thickness in the radial direction for which an analytical solution is not available. The numerical results were verified through a standard Finite Element model of the plate with perfectly matched layers (PMLs) in COMSOL Multiphysics®. The last section is devoted to conclusions.

2. Stepwise WFE Approximation of a Radially Periodic Plate

A lossless and linear elastic plate with radial periodicity is considered. The plate consists of an infinite repetition of a sequence of annuli of the same width with alternating properties, as shown in Figure 1a. The lattice constant, or the characteristic length of the unit cell, is here defined in the radial direction and denoted by $\overline{R}$.

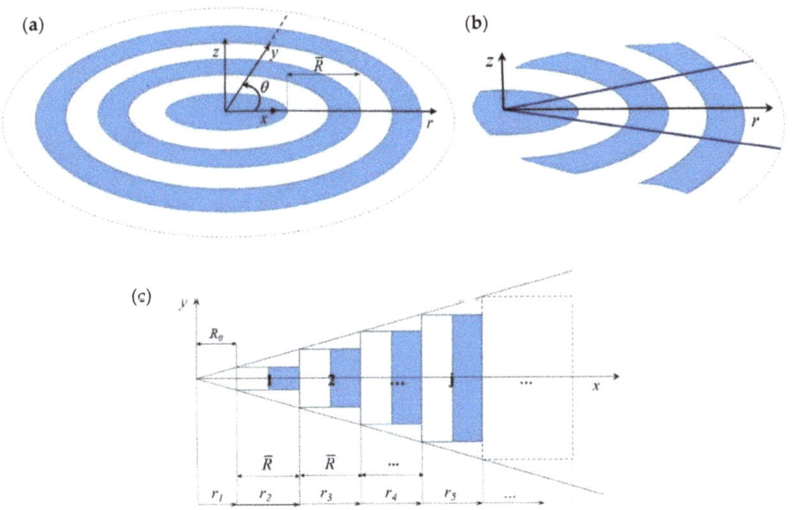

Figure 1. (**a**) radially periodic structure; (**b**) slice of the radially periodic structure; (**c**) approximation in piecewise Cartesian periodic waveguides.

In order to apply the WFE method, a slice of the plate is taken, as shown in Figure 1b. Since the periodicity is in the radial direction, the angle of the slice can be arbitrarily small. The slice is approximated as a piecewise rectangular waveguide in Cartesian coordinates, as shown in Figure 1c. A finite number of segments is assumed. These segments are numbered increasingly in the radial direction from 1 to n, where n is the cell distant $\bar{r}$ from the centre. Wave characteristics of each of these rectangular segments are obtained using the WFE approach as described in Section 2.1. In the model, the left nodes of the first cell are shifted at an arbitrarily small distance R_0 from the origin of the coordinates; the inner part, $0 < r < R_0$, is uniform and homogenous.

2.1. Free Wave Propagation Characteristics

Wave characteristics of each segment (cell) in Figure 1c are evaluated using the WFE method as described in [14,28]. The waves are numerically represented by the dispersion curves, $(\mathbf{k}, \omega)$, which give the information about the wave vector $\mathbf{k}$ available for each frequency ω, and the corresponding FE nodal displacements $\mathbf{\Phi_q}$ and nodal forces $\mathbf{\Phi_f}$, which occur under the passage of a wave. In the WFE, only the unit periodic cell is analysed. Figure 2 shows a finite element mesh of a unit cell using 4-noded rectangular elements and the correspondent WFE model. The degrees of freedom $\mathbf{q}$ in the dicrestised equation of motion, Equation (1), are ordered and condensed as $\mathbf{q} = \begin{bmatrix} \mathbf{q}_L^T & \mathbf{q}_R^T & \mathbf{q}_I^T \end{bmatrix}^T$, where the subscripts L, R and I are associated with the right $\mathbf{q}_R = \begin{bmatrix} \mathbf{q}_1^T & \mathbf{q}_2^T \end{bmatrix}^T$, left $\mathbf{q}_L = \begin{bmatrix} \mathbf{q}_5^T & \mathbf{q}_6^T \end{bmatrix}^T$ and interior $\mathbf{q}_I = \begin{bmatrix} \mathbf{q}_3^T & \mathbf{q}_4^T \end{bmatrix}^T$ nodal degrees of freedom of the cell, with a similar expression for the nodal forces $\mathbf{f} = \begin{bmatrix} \mathbf{f}_L^T & \mathbf{f}_R^T & \mathbf{f}_I^T \end{bmatrix}^T$.

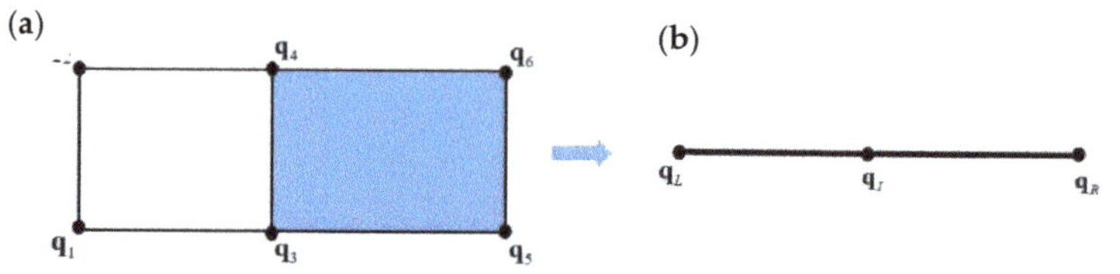

Figure 2. (**a**) Finite element mesh of a unit cell using 4-noded rectangular elements and (**b**) WFE nodes condensation.

The discretised equation of motion, $(\mathbf{K} - \omega^2\mathbf{M})\mathbf{q} = \mathbf{D}\mathbf{q} = \mathbf{f}$, is obtained using conventional FEA and it is partitioned as

$$\begin{bmatrix} \mathbf{D}_{LL} & \mathbf{D}_{LR} & \mathbf{D}_{RI} \\ \mathbf{D}_{RL} & \mathbf{D}_{RR} & \mathbf{D}_{RI} \\ \mathbf{D}_{IL} & \mathbf{D}_{IR} & \mathbf{D}_{II} \end{bmatrix} \begin{Bmatrix} \mathbf{q}_L \\ \mathbf{q}_R \\ \mathbf{q}_I \end{Bmatrix} = \begin{Bmatrix} \mathbf{f}_L \\ \mathbf{f}_R \\ \mathbf{f}_I \end{Bmatrix} \tag{1}$$

To apply the WFE method, the internal Dofs of the unit cell must be reduced. In this paper, a standard dynamic condensation is applied as

$$\mathbf{q}_I = -\mathbf{D}_{II}{}^{-1}(\mathbf{D}_{IL}\mathbf{q}_L + \mathbf{D}_{IR}\mathbf{q}_R) \tag{2}$$

However, according to the size of the model, more efficient methods for condensing the inner DOFs and speeding up the resolution of the WFE eigenvalue problem can be applied, e.g., [35,36].

The resulting equation of motion is

$$\begin{bmatrix} \mathbf{D}_{LL} & \mathbf{D}_{LR} \\ \mathbf{D}_{RL} & \mathbf{D}_{RR} \end{bmatrix} \begin{Bmatrix} \mathbf{q}_L \\ \mathbf{q}_R \end{Bmatrix} = \begin{Bmatrix} \mathbf{f}_L \\ \mathbf{f}_R \end{Bmatrix} \tag{3}$$

The inclusion of stress, temperature effects and damping in the model can be accommodated, if necessary, as described in [37]. Periodicity conditions are applied so that

$$\mathbf{q}_R = \lambda \mathbf{q}_L; \quad \mathbf{f}_R = -\lambda \mathbf{f}_L \tag{4}$$

where $\lambda = e^{-ikL}$ and k is the wavenumber. Combining Equation (4) with Equation (3), the equation of free wave motion takes the form of a quadratic eigenvalue problem, as

$$\left[\lambda^2 \mathbf{D}_{\mathrm{LR}} + \lambda(\mathbf{D}_{\mathrm{LL}} + \mathbf{D}_{\mathrm{RR}}) + \mathbf{D}_{\mathrm{RL}}\right]\mathbf{q}_{\mathrm{L}} = 0 \tag{5}$$

whose solutions yield the relationship between the wavenumber and frequency (dispersion curves) and the displacement $\mathbf{q}_L$ of the cross-section due to wave motion (wave mode shapes). Equation (5) can be further recast as a standard linear eigenvalue problem as $(\mathbf{A} - \gamma\mathbf{I})\mathbf{Z} = 0$,

where $\mathbf{A} = \begin{bmatrix} 0 & \mathbf{I} \\ -\mathbf{A}_2^{-1}\mathbf{A}_0 & -\mathbf{A}_2^{-1}\mathbf{A}_1 \end{bmatrix}$ $\mathbf{A}_2 = \mathbf{D}_{LR}$, $\mathbf{A}_1 = \mathbf{D}_{LL} + \mathbf{D}_{RR}$, $\mathbf{A}_0 = \mathbf{D}_{RL}$, and $\mathbf{Z} = [\mathbf{q}_L^T \quad \lambda\mathbf{q}_L^T]^T$.

The time average energy flow in each segment can be then obtained from

$$\Pi = \frac{1}{2}\mathrm{Re}\left[\mathbf{f}^H i\omega \mathbf{q}\right] \tag{6}$$

where the superscript H denotes the Hermitian transpose and $\mathbf{f}$ and $\mathbf{q}$ are recovered using Equations (3) and (4).

2.2. Forced Wave Amplitude

Forced wave propagation and structural response are evaluated following the theory presented in [27,28]. Wave properties are grouped into positive and negative going waves, which can be described by the two sets of results $(\mathbf{k}^+, \boldsymbol{\Phi}^+, \mathbf{a}^+)$ and $(\mathbf{k}^-, \boldsymbol{\Phi}^-, \mathbf{a}^-)$. Here $\mathbf{k}^+, \mathbf{k}^-$ and $\mathbf{a}^+, \mathbf{a}^-$ are wavenumbers and waves' amplitudes travelling in the positive and negative direction, while $\boldsymbol{\Phi}^+ = [\boldsymbol{\Phi}_\mathbf{q}^+, \boldsymbol{\Phi}_\mathbf{f}^+]^T$ and $\boldsymbol{\Phi}^- = [\boldsymbol{\Phi}_\mathbf{q}^-, \boldsymbol{\Phi}_\mathbf{f}^-]^T$ are the corresponding nodal displacements and nodal forces, that is the FE discretisation of wavemodes. For evaluating the forced response, it is advantageous to obtain also the left eigenvectors of the WFE eigenvalue problem in Equation (5). These are partitioned in the same manner as the wavemodes, that is $\boldsymbol{\Psi}^+ = [\boldsymbol{\Psi}_\mathbf{f}^+, \boldsymbol{\Psi}_\mathbf{q}^+]$ and $\boldsymbol{\Psi}^- = [\boldsymbol{\Psi}_\mathbf{f}^-, \boldsymbol{\Psi}_\mathbf{q}^-]$. Left and right eigenvectors are orthogonal and they can be normalised so that

$$\boldsymbol{\Psi}^+\boldsymbol{\Phi}^+ = \mathbf{I}, \boldsymbol{\Psi}^-\boldsymbol{\Phi}^- = \mathbf{I} \tag{7}$$

In an analogous manner to the modal analysis, the total displacement and force at the junction of a cell is described by the sum of the positive and negative wavemodes so that

$$\begin{bmatrix} \mathbf{q} \\ \mathbf{f} \end{bmatrix} = \begin{bmatrix} \boldsymbol{\Phi}_\mathbf{q}^+ & \boldsymbol{\Phi}_\mathbf{q}^- \\ \boldsymbol{\Phi}_\mathbf{f}^+ & \boldsymbol{\Phi}_\mathbf{f}^- \end{bmatrix} \begin{bmatrix} \mathbf{a}^+ \\ \mathbf{a}^- \end{bmatrix}. \tag{8}$$

Equation (8) defines the transformation between the physical domain, where the motion is described in terms of nodal displacements and forces, and the wave domain, where the motion is described in terms of waves of amplitudes $\mathbf{a}^+, \mathbf{a}^-$ travelling in the positive and negative directions. A point force $\mathbf{f}_e$ will generate excited positive and negative going waves, propagating away from the excitation point. Compared to the Cartesian periodicity, in the case of a radially periodic plate, the position of this point force is an independent parameter. Here, we assume a transverse harmonic point force exciting the structure at the left nodes of the first cell in Figure 1c. With reference to Figure 1, continuity and equilibrium equations can be written at the left side of the first cell, and excited wave amplitudes can be recovered from Equation (9)

$$\begin{bmatrix} \mathbf{\Phi}^+_{q,1} & \mathbf{\Phi}^-_{q,1} \\ \mathbf{\Phi}^+_{f,1} & \mathbf{\Phi}^-_{f,1} \end{bmatrix} \begin{bmatrix} \mathbf{a}_1^+(r_1) \\ \mathbf{a}_1^-(r_1) \end{bmatrix} = \begin{bmatrix} 0 \\ \mathbf{f}_e \end{bmatrix} \tag{9}$$

In practice, as in modal analysis, only m pairs of (positive- and negative-going) waves are retained (a reduced wave basis is assumed in most cases). Moreover, the number of assumed modes can be different at each frequency. It is noteworthy to mention that all the waves propagating in the structure (real wavenumbers) should be considered. Evanescent and attenuating waves (corresponding to pure imaginary or complex wavenumbers) should be also assumed since they play an important role in scattering and forcing problems. Consequently, the wavemodes' matrix is typically rectangular and the use of standard pseudoinverse operation can lead to numerical errors. To overcome this issue, the orthogonality between right and left eigenvectors in Equation (7) can be exploited, and the excited wave amplitudes at r_1 can be obtained from

$$\begin{bmatrix} \mathbf{a}_1^+(r_1) \\ \mathbf{a}_1^-(r_1) \end{bmatrix} = \begin{bmatrix} \mathbf{\Psi}^+_{f,1} & \mathbf{\Psi}^+_{q,1} \\ \mathbf{\Psi}^-_{f,1} & \mathbf{\Psi}^-_{q,1} \end{bmatrix} \begin{bmatrix} 0 \\ \mathbf{f}_e \end{bmatrix} \tag{10}$$

2.3. Coupling of the Segments and Wave Amplitude Decay

Compared to the corresponding Cartesian waveguides, wave propagation is not translational invariant in the radially periodic structures, and amplitude attenuation occurs as the waves travel from the centre in the radial direction. Since the energy flowing along the slice is constant, wave amplitude change due to changes in the geometry can be accommodated in the model according to an energy balance principle as in [31]. In this section, the procedure is presented for the first two cells. The same passages must be applied up to the cell $j = n$, including the point at which the response must be evaluated.

In the Cartesian cell, amplitudes are related at two points r_1 and $r_2 = r_1 + \overline{R}$, by $\mathbf{a}_1{}^+(r_2) = \mathbf{T}^+(\overline{R})\mathbf{a}_1^+(r_1)$ and $\mathbf{a}_1^-(r_2) = \mathbf{T}_1^-(\overline{R})\mathbf{a}_1^-(r_1)$, where $\mathbf{T}_1^+(\overline{R}) = \mathrm{diag}\left[\exp(-i\mathbf{k}_1^+\overline{R})\right]$ and $\mathbf{T}_1^-(\overline{R}) = \mathrm{diag}\left[\exp(-i\mathbf{k}_1^-\overline{R})\right]$. Therefore, nodal displacements and nodal forces at the interface between cell 1 and cell 2 are

$$\begin{bmatrix} \mathbf{q}_1 \\ \mathbf{f}_1 \end{bmatrix} = \begin{bmatrix} \mathbf{\Phi}^+_{q,1} & \mathbf{\Phi}^-_{q,1} \\ \mathbf{\Phi}^+_{f,1} & \mathbf{\Phi}^-_{f,1} \end{bmatrix} \begin{bmatrix} \mathbf{T}_1{}^+(\overline{R}) & 0 \\ 0 & \mathbf{T}_1{}^-(\overline{R}) \end{bmatrix} \begin{bmatrix} \mathbf{a}_1{}^+(r_1) \\ \mathbf{a}_1{}^-(r_1) \end{bmatrix} \tag{11}$$

Using the left eigenvectors as in Equation (10), the first attempt to find the excited wave amplitude in cell 2 gives

$$\begin{bmatrix} \tilde{\mathbf{a}}_2^+ \\ \tilde{\mathbf{a}}_2^- \end{bmatrix} = \begin{bmatrix} \mathbf{\Psi}^+_{f,2} & \mathbf{\Psi}^+_{q,2} \\ \mathbf{\Psi}^-_{f,2} & \mathbf{\Psi}^-_{q,2} \end{bmatrix} \begin{bmatrix} \mathbf{q}_1 \\ \mathbf{f}_1 \end{bmatrix} \tag{12}$$

Nodal displacements and nodal forces at cell 2 can be obtained using the same expression of Equation (10)

$$\begin{bmatrix} \mathbf{q}_2 \\ \mathbf{f}_2 \end{bmatrix} = \begin{bmatrix} \mathbf{\Phi}^+_{q,2} & \mathbf{\Phi}^-_{q,2} \\ \mathbf{\Phi}^+_{f,2} & \mathbf{\Phi}^-_{f,2} \end{bmatrix} \begin{bmatrix} \mathbf{T}_2{}^+(\overline{R}) & 0 \\ 0 & \mathbf{T}_2{}^-(\overline{R}) \end{bmatrix} \begin{bmatrix} \tilde{\mathbf{a}}_2{}^+ \\ \tilde{\mathbf{a}}_2{}^- \end{bmatrix} \tag{13}$$

and the time-averaged energy flows, Equation (6), of cell 1 and cell 2 can be evaluated using nodal displacement and nodal forces from Equations (11) and (13)

$$\Pi_1 = \frac{1}{2}\mathrm{Re}\left\{\mathbf{f}_1{}^H i\omega\mathbf{q}_1\right\}, \quad \Pi_2 = \frac{1}{2}\mathrm{Re}\left\{\mathbf{f}_2{}^H i\omega\mathbf{q}_2\right\} \tag{14}$$

The ratio between Π_1 and Π_2 gives the amplitude decay ξ_2 in waveguide 2, which is evaluated as in [32]

$$\xi_2{}^2 = \frac{\mathrm{Re}\{\mathbf{f}_1{}^H i\omega\mathbf{q}_1\}}{\mathrm{Re}\{\mathbf{f}_2{}^H i\omega\mathbf{q}_2\}} \tag{15}$$

The wave and amplitudes in waveguide 2 are therefore approximated to

$$\begin{bmatrix} \mathbf{a}_2{}^+ \\ \mathbf{a}_2{}^- \end{bmatrix} = \xi_2 \begin{bmatrix} \tilde{\mathbf{a}}_2{}^+ \\ \tilde{\mathbf{a}}_2{}^- \end{bmatrix} \tag{16}$$

Equations (13)–(16) can be repeated iteratively until ξ^2 converges to one, and the final value of the wave amplitudes reach the required approximation. One passage was sufficient to converge to a useful approximation with a very low computational cost in the numerical cases studied. Excited wave amplitudes decay in the next cells up to cell n are evaluated following the same passages.

3. Numerical Examples

This section shows two numerical examples: the first concerns a thin isotropic plate excited by a central transverse harmonic force, while the second deals with a plate with a periodic radial change of thickness. In both cases, the WFE model is obtained using 4-noded plane elements in bending having three degrees of freedom per node: translation in the z-direction and rotations around the x- and y-directions. To verify the results, analytical solutions were compared in the first case, while results for the radially periodic plate were verified through comparison with those obtained by an FE model with PMLs. This model was realised in COMSOL Mutiphysics® using the Structural Mechanics Module. The plate was discretised by shell elements having six degrees of freedom per node (translations and rotations in the z, x, and y directions), while a second-order polynomial stretching function was chosen for the PMLs, see the COMSOL Multiphysics Reference Manual for further information. The FE models were realised both in Cartesian and Polar coordinates systems with similar results, and a convergence test was performed by refining the mesh in the PML domain.

The non-dimensional frequency

$$\Omega = \omega\overline{R}^2\sqrt{\frac{\rho h}{D}} \tag{17}$$

is introduced, where ρ, h and D are respectively the density, thickness and bending stiffness of the plate, ω is the frequency in radiant, and $\overline{R}$ is the period of the structure as in Figure 1.

3.1. Numerical Verification. Infinite Plate Subjected to a Transverse Harmonic Force

The literature has largely studied the problem of a thin plate subjected to harmonic loads, and closed-form solutions can be found in many classical books on vibrations, e.g., [34]. This numerical example is introduced here to verify the method described in Section 2 and show its applicability to continuous structures.

An aluminium plate of thickness $h = 1$ mm is considered. Flexural waves in the plate are excited by a central transverse harmonic force of magnitude $F/(Eh^2) = 1.4 \cdot 10^{-3}$, and the complex out of plane displacement is evaluated at a distance $\bar{r}/h = 300$ from the plate centre, viz. origin of the polar coordinates.

In this case, the structure is continuous and uniform, and it can be studied as a periodic structure with arbitrary radial and circumferential periodicity: the choice of the period, viz. length $\overline{R}$ of the unit cell, is arbitrary under standard FE assumptions to avoid element distortion and dispersion errors [38]. Therefore, the number of segments that are used to approximate the slice from the centre to $\bar{r}$ is arbitrarily chosen. The discretisation must be refined until it does not produce a negligible change in the solution. In this example, only three segments have been found sufficient for convergence.

In order to simplify the WFE model, a small circle around the point force is removed and the left nodes of the first cell are shifted at a distance $R_0/h = r_1 = 10$ from the point force. In this example, the WFE approximation is obtained for a slice of angle θ = 6°. Waves are induced only in the positive radial direction and therefore $\mathbf{a}_1^- = 0$ is assumed. Periodicity in each segment is further exploited. Here, the number of "sub-period" in each segment is chosen by a simple algorithm that optimises the mesh according to the slice's dimension and the number of segments. No significant differences are noticed decreasing the distance R_0 further.

Figure 3 shows the absolute value and the phase of the transverse displacement of the plate. A comparison between the numerical and analytical solutions is shown. The latter is evaluated using the analytical equation given in [34]: $w = \frac{-iF}{8Dk^2}\left(H_0^2(kr) - i\frac{2}{\pi}K_0(kr)\right)$, where $k = \sqrt[4]{\omega^2 \rho h/D}$ is the flexural wavenumber, $H_0^2(kr)$ is the zero-order Hankel function of the second kind and $K_0(kr)$ is the zero-order modified Bessel function of the second kind.

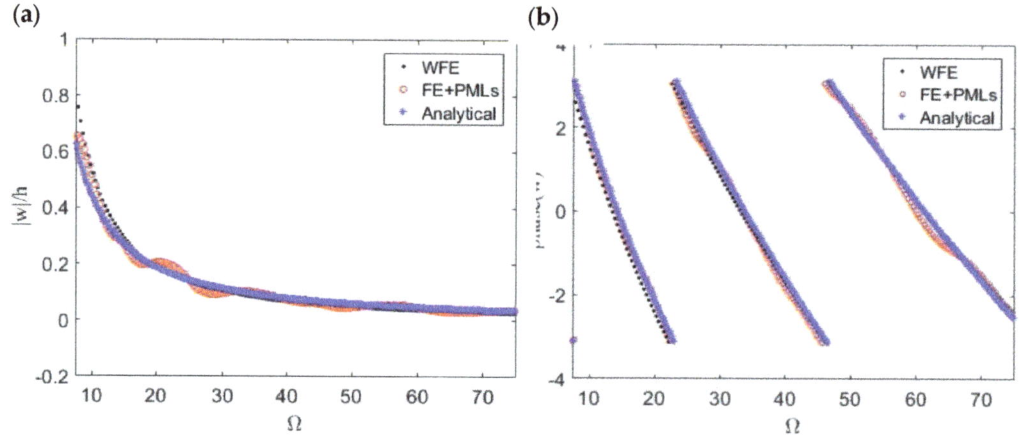

Figure 3. Response of a thin plate excited by a harmonic transverse point force, (**a**) absolute value; (**b**) phase. Comparison between the approximated WFE, FE model with PMLs and analytical results.

Results obtained using the FE model of the plate with PMLs are, as seen in Figure 3, in excellent agreement with the WFE results (small discrepancies are due to the differences in the FE models). However, this numerical study also demonstrates the computational superiority of the WFE method over the standard FE + PML analysis: the CPU time used by Matlab® to solve the WFE model of the plate was approximately four hundred times less than the CPU time used by COMSOL Mutiphysics® to solve the correspondent FE + PML model. One of the main reasons for such a difference in the computational cost is the very small size of the WFE matrices compared to the number of DOFs involved in standard FE analysis. The WFE model for this numerical example was realised using one shell element, resulting in 6 DOFs after the WFE reduction, Equation (5), while the size of the FE + PML model was 676,422 DOFs.

3.2. Radially Periodic Plate

In this section, we consider a plate in polar coordinates with radial periodicity. Figure 4 depicts a schematic figure of the plate and its unit cell (period). In the following, the characteristic length $\overline{R}$ and the thickness h_1 are used to define the dimensionless parameters:

$$\beta = h_2/h_1 = 2.7, \ \gamma = \overline{R}/R_1 = 2.7, \ \delta = \overline{R}/h_1 = 27$$

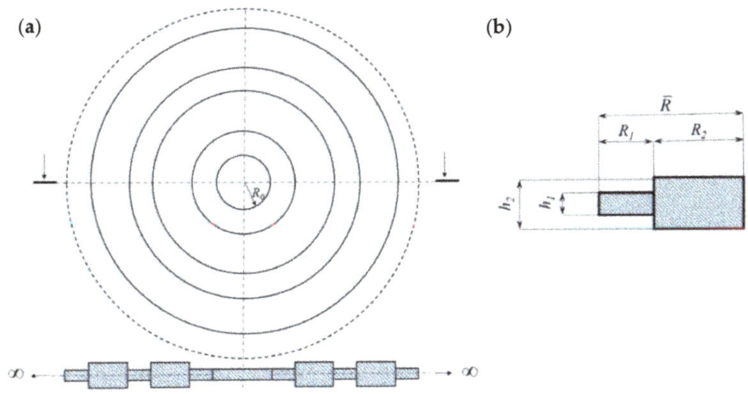

Figure 4. (**a**) radially periodic infinite plate with an internal hole of radius R_0; (**b**) period, or unit cell.

The WFE model is obtained considering an arbitrary circumferential periodicity of angle $\theta = 9^{\circ}$. The external harmonic force of magnitude $F/(2\pi R_0) = 140$ N/m is applied at the internal edge of the plate and included in the WFE by two equivalent transverse nodal forces exciting the left nodes of the first unit cell. Waves are induced only in the positive radial direction and therefore $\mathbf{a}_1^{-} = 0$ is assumed. The structure exhibits stop-bands due to its periodicity as shown in [26]. These stop-bands can be predicted with accuracy by the WFE up to high frequency. However, in this paper, the plate response is evaluated in the nearfield and at a low-frequency. Therefore, only the wave modes below the first stop-band are considered in the analysis. Figure 5 shows the dispersion curves of flexural waves propagating in the positive direction. Analytical flexural waves propagating in the corresponding homogeneous plates of thickness h_1 and h_2 are also shown for comparison.

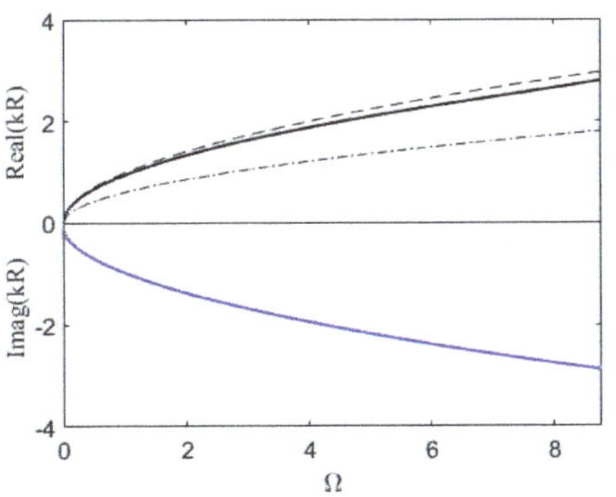

Figure 5. Dispersion curves for flexural waves in the radially periodic plate. Propagating flexural waves in the corresponding homogeneous plates of thickness h_1, - - - -, and h_2, - . - . -, are shown for comparison.

The transverse response of the plate is evaluated after 4 unit cells, at a distance $r = 5\bar{R}$ from the centre of the plate (or equally at a distance $r = 4\bar{R}$ from the border of the internal hole). Figure 6 shows the displacements in terms of absolute and phase values. Results obtained using the FE model with PMLs are also presented. Although some small discrepancies were expected due to the very different FE discretisation and a very different approach to the problem, it can be noticed that the results obtained by the WFE method are in good agreement with those obtained by the finite element harmonic analysis. The WFE model for each segment of the plate slice was set up with three shell elements, resulting in six DOFs after the WFE reduction, while the FE + PML had 794,448 DOFs. The computational time used by Matlab® to solve the WFE model of this radially periodic plate was almost two thousand times less than the CPU time used by COMSOL Mutiphysics® to solve the correspondent FE + PML model.

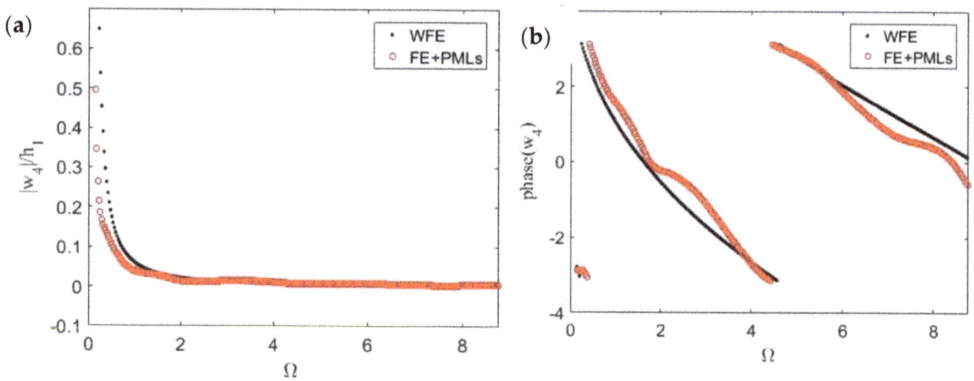

Figure 6. Response at the 4th cell of the radial periodic plate excited by harmonic transverse point forces at the border of the central hole, (**a**) absolute value; (**b**) phase. Comparison between the approximated WFE results FEM results.

4. Conclusions

In this paper, free and forced wave propagation in a radially periodic plate was studied using an adaptation of the Wave Finite Element method to structures showing radial periodicities. Cylindrical waves propagating were approximated exploiting the Floquet theory formulation for an infinite periodic structure in one dimension. This approximation was achieved by taking only a very small slice of the structure, which was discretised through piecewise Cartesian segments. Wave characteristics in each segment were obtained by the WFE method, while wave amplitudes change was accommodated in the model assuming an energy balance principle. The forced response of the structure was then evaluated in the wave domain. The paper's main goal was to verify the approach for predicting the forced response in the nearfield of a radial periodic structures. Two numerical examples were presented: an isotropic thin plate excited by a transverse harmonic force, for which analytical solutions are available, and a plate with a periodic change of thickness in the radial direction. In the latter, the numerical results were verified through a standard Finite Element model of the plate with perfectly matched layers (PMLs). In both cases, it was found that the numerical results were in good agreement with the analytical and numerical FE results, showing the advantages of the approach in terms of computational ime, approximation controllability, and modelling efficiency.

Author Contributions: Conceptualization, E.M., S.V.S. and R.G.; methodology, E.M. and S.V.S.; software, E.M. and M.M.Q.; validation, R.G. and M.M.Q.; formal analysis, E.M.; investigation, E.M. and M.M.Q.; data curation, E.M.; writing—original draft preparation, E.M. and S.V.S.; writing—review and editing, E.M. and S.V.S. All authors have read and agreed to the published version of the manuscript.

Funding: This research received no external funding.

Conflicts of Interest: The authors declare no conflict of interest.

References

1. Brillouin, L. *Wave Propagation in Periodic Structures*, 1st ed.; McGraw-Hill Book Co., Inc.: New York, NY, USA; London, UK, 1946.
2. Droz, C.; Zergoune, Z.; Boukadia, R.; Bareille, O.; Ichchou, M.N. Vibro-acoustic optimisation of sandwich panels using the wave/finite element method. *Compos. Struct.* **2016**, *156*, 108–114. [CrossRef]

3. Casadei, F.; Dozio, L.; Ruzzene, M.; Cunefare, K.A. Periodic shunted arrays for the control of noise radiation in an enclosure. *J. Sound Vib.* **2010**, *329*, 3623–3646. [CrossRef]
4. Andersen, L.V.; Persson, P.; Peplow, A.T. Mitigation of ground vibration from pile driving by circular arrays of rigid blocks placed on the ground surface. In Proceedings of the Seventh International Conference on Structural Dynamics, COMPDYN 2019, Crete, Greece, 24–26 June 2019; pp. 3422–3447.
5. Palermo, A.; Vitali, M.; Marzani, A. Metabarriers with multi-mass locally resonating units for broad band Rayleigh waves attenuation. *Soil Dyn. Earthq. Eng.* **2018**, *113*, 265–277. [CrossRef]
6. Bastawrous, M.V.; Hussein, M.I. Theoretical band-gap bounds and coupling sensitivity for a waveguide with periodically attached resonating branches. *J. Sound Vib.* **2021**, 116428. [CrossRef]
7. Hussein, M.I.; Leamy, M.J.; Ruzzene, M. Dynamics of Phononic Materials and Structures: Historical Origins, Recent Progress, and Future Outlook. *Appl. Mech. Rev.* **2014**, *66*, 040802. [CrossRef]
8. Zouari, S.; Brocail, J.; Génevaux, J.-M. Flexural wave band gaps in metamaterial plates: A numerical and experimental study from infinite to finite models. *J. Sound Vib.* **2018**, *435*, 246–263. [CrossRef]
9. Mead, D.J. Vibration Response and Wave Propagation in Periodic Structures. *ASME J. Eng. Ind.* **1971**, *93*, 783–792. [CrossRef]
10. Mead, D.M. Wave propagation in continuous periodic structures: Research contributions from Southampton, 1964–1995. *J. Sound Vib.* **1996**, *190*, 495–524. [CrossRef]
11. Engquist, B.; Majda, A. Absorbing boundary conditions for numerical simulation of waves. *Proc. Natl. Acad. Sci. USA* **1977**, *74*, 1765–1766. [CrossRef]
12. Bérenger, J.P. A Perfectly Matched Layer for the Absorption of Electromagnetic Waves. *J. Comput. Phys.* **1994**, *114*, 185–200. [CrossRef]
13. Orris, R.M.; Petyt, M. A finite element study of harmonic wave propagation in periodic structures. *J. Sound Vib.* **1974**, *33*, 223–236. [CrossRef]
14. Mace, B.R.; Duhamel, D.; Brennan, M.J.; Hinke, L. Finite element prediction of wave motion in structural waveguides. *J. Acoust. Soc. Am.* **2005**, *117*, 2835–2843. [CrossRef] [PubMed]
15. Mace, B.R.; Manconi, E. Modelling wave propagation in two-dimensional structures using finite element analysis. *J. Sound Vib.* **2008**, *318*, 884–902. [CrossRef]
16. Wilcox, C.H. Theory of Bloch waves. *J. Anal. Math.* **1978**, *33*, 146–167. [CrossRef]
17. Yeh, P.; Yariv, A.; Marom, E. Theory of Bragg fiber. *J. Opt. Soc. Am.* **1978**, *68*, 1196–1201. [CrossRef]
18. Li, Y.; Chen, T.; Wang, X.; Yu, K.; Chen, W. Propagation of lamb waves in one-dimensional radial phononic crystal plates with periodic corrugations. *J. Appl. Phys.* **2014**, *115*, 054907. [CrossRef]
19. Ma, T.; Chen, T.; Wang, X.; Li, Y.; Wang, P. Band structures of bilayer radial phononic crystal plate with crystal gliding. *J. Appl. Phys.* **2014**, *116*, 104505. [CrossRef]
20. Shi, X.; Shu, H.; Zhu, J.; Wang, X.; Dong, L.; Zhao, L.; Liang, S.; Liu, R. Research on wave band-gaps in a circular plate of radial phononic crystal. *Int. J. Mod. Phys. B* **2016**, *30*, 1650162. [CrossRef]
21. Xu, Y.; Ouyang, G.X.; Lee, R.K.; Yariv, A. Asymptotic matrix theory of Bragg fibers. *J. Lightwave Technol.* **2002**, *20*, 428–440.
22. Kitagawa, A.; Sakai, J. Bloch theorem in cylindrical coordinates and its application to a Bragg fiber. *Phys. Rev. A* **2009**, *80*, 033802. [CrossRef]
23. Torrent, D.; Sanchez-Dehesa, J. Radial wave crystals: Radially periodic structures from anisotropic metamaterials for engineering acoustic or electromagnetic waves. *Phys. Rev. Lett.* **2009**, *102*, 064301. [CrossRef] [PubMed]
24. Torrent, D.; Sanchez-Dehesa, J. Acoustic resonances in two-dimensional radial sonic crystal shells. *New J. Phys.* **2010**, *12*, 073034. [CrossRef]
25. Arretche, I.; Matlack, K.H. Effective phononic crystals for non-Cartesian elastic wave propagation. *Phys. Rev. B* **2020**, *102*, 134308. [CrossRef]
26. Hvatov, A.; Sorokin, S. On application of the Floquet theory for radially periodic membranes and plates. *J. Sound Vib.* **2018**, *414*, 15–30. [CrossRef]
27. Waki, Y.; Mace, B.R.; Brennan, M.J. Free and forced vibrations of a tyre using a wave/finite element approach. *J. Sound Vib.* **2009**, *323*, 737–756. [CrossRef]
28. Renno, J.M.; Manconi, E.; Mace, B.R. A Finite Element Method for Modelling Waves in Laminated Structures. *Adv. Struct. Eng.* **2013**, *16*, 61–75. [CrossRef]
29. Manconi, E.; Mace, B. Wave characterisation of cylindrical and curved panels using a finite element method. *J. Acoust. Soc. Am.* **2009**, *125*, 154–163. [CrossRef]
30. Sorokin, S.; Manconi, E.; Ledet, L.; Garziera, R. Wave propagation in helically orthotropic elastic cylindrical shells and lattices. *Int. J. Solids Struct.* **2019**, *170*, 11–21. [CrossRef]
31. Fabro, A.T.; Ferguson, N.S.; Mace, B.R. Wave propagation in slowly varying waveguides using a finite element approach. *J. Sound Vib.* **2019**, *442*, 308–329. [CrossRef]
32. Renno, J.M.; Mace, B.R. Calculating the forced response of two-dimensional homogeneous media using the wave and finite element method. *J. Sound Vib.* **2011**, *330*, 5913–5927. [CrossRef]
33. Manconi, E.; Sorokin, S.; Garziere, R. Wave propagation in polar periodic structures using Floquet theory and finite element analysis. In Proceedings of the Compdyn 2019, Crete, Greece, 24–26 June 2019.

34. Cremer, L.; Heckl, M.; Ungar, E.E. *Structure-Borne Sound*; Springer: Berlin/Heidelberg, Germany; New York, NY, USA, 1988.
35. Zhou, C.W.; Lainé, J.P.; Ichchou, M.N.; Zine, A.M. Multi-scale modelling for two-dimensional periodic structures using a combined mode/wave based approach. *Comput. Struct.* **2015**, *154*, 145–162. [CrossRef]
36. Boukadia, R.F.; Droz, C.; Ichchou, M.N.; Desmet, W.A. Bloch wave reduction scheme for ultrafast band diagram and dynamic response computation in periodic structures. *Finite Elem. Anal. Des.* **2018**, *148*, 1–12. [CrossRef]
37. Manconi, E.; Mace, B.R.; Garziera, R. The loss-factor of pre-stressed laminated curved panels and cylinders using a wave and finite element method. *J. Sound Vib.* **2013**, *332*, 1704–1711. [CrossRef]
38. Waki, B.; Mace, B.R.; Brennan, M.J. Numerical issues concerning the wave and finite element method for free and forced vibrations of waveguides. *J. Sound Vib.* **2009**, *327*, 92–108. [CrossRef]

applied sciences

MDPI

Article

Modeling and Validation of a Passive Truss-Link Mechanism for Deployable Structures Considering Friction Compensation with Response Surface Methods

Han-Sol Choi [1], Dong-Yeon Kim [2], Jeong-Hoon Park [1], Jae Hyuk Lim [1,*] and Tae Seong Jang [3]

1 Department of Mechanical Engineering, Jeonbuk National University, 567 Baekje-daero, Deokjin-gu, Jeonju-si 54896, Jeollabuk-do, Korea; paul2599@jbnu.ac.kr (H.-S.C.); 201513315@jbnu.ac.kr (J.-H.P.)
2 Hanwha System, 491-23, Gyeonggidong-ro, Namsa-eup, Cheoin-gu, Yongin-si 17121, Gyeonggi-do, Korea; dongyeon.kim@hanwha.com
3 SaTReC (Satellite Technology Research Center), KAIST, 291, Daehak-ro, Yuseong-gu, Daejeon 34141, Korea; tsjang@kaist.ac.kr
* Correspondence: jaehyuklim@jbnu.ac.kr

Abstract: In this study, a passive truss-link mechanism applicable to large-scale deployable structures was designed to achieve successful deployment in space. First, we simplified the selected truss-link mechanisms to the two-dimensional geometry and calculated the degrees of freedom (DOF) to determine whether a kinematic over-constraint occurs. The dimensions of the truss-link structure were determined through a deployment kinematic analysis. Second, a deployment simulation with the truss-link was conducted using multibody dynamics (MBD) software. Finally, a deployment test was performed considering gravity compensation, and the results were compared with those of MBD simulation. The results of the deployment simulations were confirmed to be slightly faster than those of the deployment test due to friction effects existing in the joints and gravity compensation devices. To address this issue, inverse identification of the equivalent frictional torque (EFT) at the revolute joints in the deployment test was conducted through response surface methods (RSM) combined with the central composite design technique. As a result, we confirmed that the deployment angle history of the deployment simulation was similar to that of the deployment test.

Keywords: deployment dynamics; truss-link mechanism; synthetic aperture radar; friction compensation

Citation: Choi, H.-S.; Kim, D.-Y.; Park, J.-H.; Lim, J.H.; Jang, T.S. Modeling and Validation of a Passive Truss-Link Mechanism for Deployable Structures Considering Friction Compensation with Response Surface Methods. *Appl. Sci.* **2022**, *12*, 451. https://doi.org/10.3390/app12010451

Academic Editors: Adel Razek

Received: 13 November 2021
Accepted: 1 January 2022
Published: 4 January 2022

1. Introduction

Synthetic aperture radar (SAR), which can provide high-resolution earth images regardless of weather conditions or time of day has recently been used in various fields of domestic and international observation satellites [1,2]. In the case of SAR antennas, to maximize resolution and power gain in acquiring high-resolution images, a large deployable panel is accommodated. However, to deploy such a large space deployment structure in orbit, it is essential to use specific mechanisms that allow the structure to be appropriately folded and stored inside the launch-vehicle fairing and fully deployed in orbit [1–6].

Furthermore, when a satellite performs a mission, line-of-sight (LOS) pointing stability must be secured. During ground imaging, if the relatively large antenna is shaken by external or internal disturbances, the quality of the captured images will be degraded. Therefore, the stiffness of the rotational spring hinges of deployable panels is often increased in order to obtain high deployment stability. This mechanism shortens the deployment time but leads to a high impact load when fully deployed and latched. This deployment impact load could cause damage to structures, payloads, and solar cells. By contrast, decreased stiffness of the rotational spring hinges reduces the deployment impact load by increasing the deployment time, but cannot guarantee successful full deployment due to harness resistance and mechanical friction of deployment devices at low temperatures [7].

In response to these issues, a variety of methods have been proposed to allow stable deployment and reduce deployment impact loads, such as a torsion spring-latching hinge [8], tape spring hinge [9], and shape memory alloy (SMA)-based hinge [10]. However, these deployment mechanisms still have limitations including low deployment stiffness of the relatively large deployable structures. To address this, truss-link mechanisms [11] have been applied to many satellite missions to increase the deployment stiffness: ERS-1 [12], RADARSAT-1 [13], RADARSAT-2 [14], RADARSAT-Constellation [15], ALOS-2 [16], SEASAT [17], WorldView-4 [18], and Sentinel-5P [19]. Among these, the RADARSAT mission employed the truss-link mechanism to increase deployment stiffness and attempted to reduce the number of truss-links as the follow-up mission progressed to reduce weight. Comparing the number of truss-links in satellites, RADARSAT-1 (1995) has 30 links, RADARSAT-2 (2007) has 22 links, and RADARSAT-Constellation (2019) has four links. The reduction in the number of links is the result of a trend to build up small satellites for constellation missions. In addition, the applications of the truss-link mechanisms have been diversified. Wang et al. proposed a pyramid deployable truss structure (PDTS) design for deployable SAR antennas through an analytical approach based on weighted graphs and kinematic chain techniques [20]. They also proposed a modular deployable truss structure (MDTS) for large SAR antennas [13] with a scaled model. A difference was found between the deployment test and the simulation, however, due to uneven and indeterminate ground friction as well as joint friction. Furthermore, to eliminate interference during the deployment process of the truss structure, the optimal motor driving torque was determined through the design of experiments (DOE) method. Han et al. proposed a large-scale deployable ring truss that composes a space antenna by complementing a rope-based actuation and cable net system [21]. Based on the absolute node coordinate formulation (ANCF), a method for simulating the rigid body motion of the antenna support arm was proposed, and the friction and motion equations of the flexible cable net were presented using the Lagrange equation.

Most of the aforementioned works essentially assume that an active driving mechanism such as motors with cables is employed for large-scale SAR antenna structures. However, such an active mechanism is hardly employed for small satellites due to deployment costs and complexity. In addition, the friction effects cannot be neglected as they generate a difference between deployment tests and simulations.

The goal of this work is to propose a passive truss-link mechanism for large deployable structures of small satellites. In contrast to the many truss-link mechanisms [11–17], we do not accommodate any active driving mechanisms because they are not simple and not sufficiently reliable. As driving mechanisms, only conventional torsion spring hinges with proper latching mechanisms were employed in this work. To achieve successful deployment, an inverse identification technique for equivalent friction torque (EFT) was proposed based on the results of the response surface method combined with the central composite design technique. Finally, a torque margin analysis was conducted to predict whether the deployment would be successful.

This works consisted of three sections: first, we discussed the configuration design from conceptual design to detailed modeling considering deployment kinematics. Second, deployment dynamics simulations and tests were conducted. To minimize the discrepancy between the test and simulation, the friction-compensation technique was introduced, and results were verified by a torque margin analysis. Finally, we provided concluding remarks.

2. Configuration Design of Truss-Link Mechanism

2.1. Concept Design

For full deployment of deployable structures with truss-links, the total degrees of freedom (*DOF*) has to be checked to determine whether the structure is over-constrained. If the *DOF* of the deployable structure becomes negative, the deployment will fail due to over-constraint problems. First, we analyzed the *DOF* by referring to RADARSAT-1 and RADARSAT-2 satellites using truss-links [13,14]. Two-dimensional (2D) designs of one payload wing of the satellite were considered, Type 1 (RADARSAT-1) and Type 2 (RADARSAT-2), to analyze the *DOF*, as seen in Figure 1. We employed Gruebler's equation, as shown in Equation (1) to evaluate the total *DOF* [22].

$$SysDOF = 3(L-1) - 2J_1 - J_2 \quad (1)$$

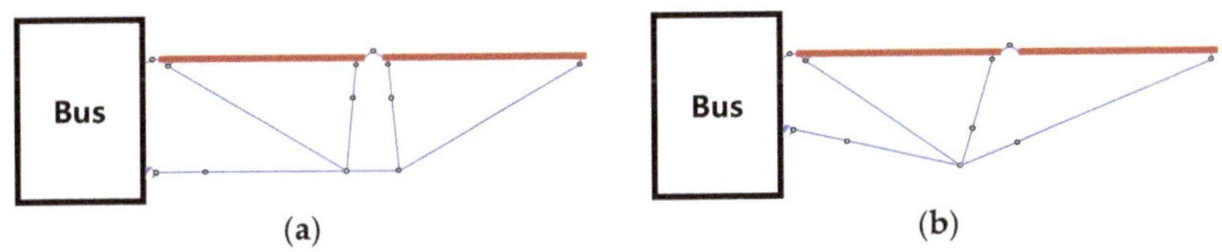

Figure 1. Configuration of 2-D truss-link mechanism: (**a**) Type 1 (RADARSAT-1) (**b**) Type 2 (RADARSAT-2).

Here, *SysDOF* represents the total *DOF* of the system, *L* represents the number of bodies, J_1 represents the number of joints with 1 *DOF* such as a revolute joint, and J_2 represents the number of joints with 2 *DOF* such as a universal joint. As a result, Type 1 has nine *DOF*, and Type 2 has seven *DOF*. The greater number of *DOF* in Type 1 compared with Type 2 is advantageous for preventing over-constraint but has the disadvantage of making the structure heavier owing to a greater number of links than Type 2. Thus, the deployment simulation model is designed based on Type 2 because it is lightweight due to the small number of truss-links and short total length. The total *DOF* calculation results are summarized in Table 1.

Table 1. Evaluation of total DOF.

	Type 1 (RADARSAT-1)	**Type 2 (RADARSAT-2)**
Body (*L*)	Bus (1) + Links (11) = 12	Bus (1) + Links (9) = 10
J_1, J_2	$J_1 = 12, J_2 = 0$	$J_1 = 10, J_2 = 0$
Degrees of freedom	3 × (12 − 1) − 2 × (12) = 9	3 × (10 − 1) − 2 × (10) = 7

2.2. Configuration Design

The configuration is simplified to a two-dimensional domain to obtain the truss-link's dimensions in the stowed configuration (Figure 2). By simplifying to a two-dimensional geometry, it is possible to reduce the *DOF* and variables used to obtain dimensions and coordinates of the body compared to traditional three-dimensional (3D) methods. The dimensions and coordinates were defined in the formula in the reference [23], which was derived using geometric construction methods and bar-groups methods. Upon verification of the formula's results, it was found that there were some errors, which were fixed accordingly in this study as presented in Equations (2)–(10).

With the panel folded, the length and angle of each truss-link are derived from input variables in Tables 2 and 3. Input variables including φ_1, φ_4, φ_6, L_1, L_4, L_6, X_A, X_C, X_D, X_F, and X_G are defined by the mechanical designer considering the size of the panel according to the design requirements. Through Equations (2)–(10), output variables, including φ_2, φ_3, φ_5, φ_7, L_2, L_3, L_5, L_7, X_B, X_E, and X_H, were determined.

Joint O corresponds to the origin of absolute coordinates as the rotation center of the hinge installed between the bus and the inner panel; Joint I is the rotation center between the inner panel and outer panel; Joint F is the intersection point of all truss-links. Furthermore, to keep the deployment state, three latching points, B, E, and H, were considered, as shown in Figure 2b.

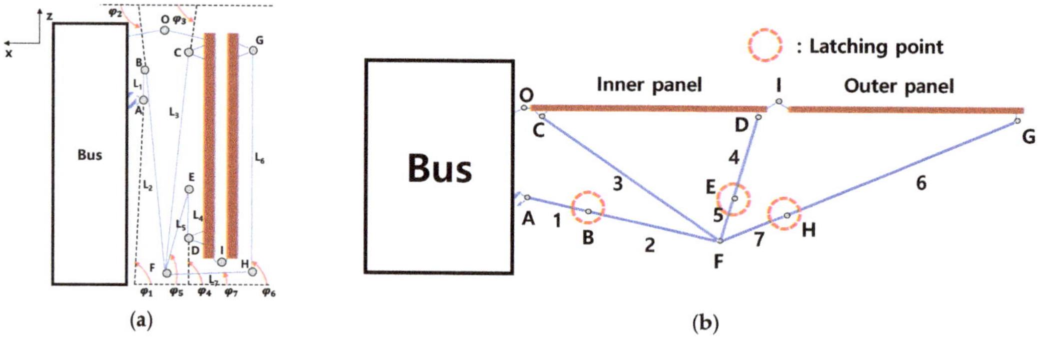

Figure 2. Configuration design of truss-link mechanism in (**a**) stowed configuration (**b**) deployed configuration with latching points: B, E, H.

Table 2. Design variables of the truss-link mechanism.

	Input	Output
Angle variables	$\varphi_1, \varphi_4, \varphi_6$	$\varphi_2, \varphi_3, \varphi_5, \varphi_7$
Length variables	L_1, L_4, L_6	L_2, L_3, L_5, L_7
Absolute coordinate variables	A, C, D, F, G	B, E, H

Table 3. Values of the input and output design variables of truss-link mechanism.

Input Variables	Value	Output Variables	Value
φ_1	91.55°	φ_2	93.11°
φ_4	129.18°	φ_3	91.64°
φ_6	92.06°	φ_5	104.20°
L_1	140.00 mm	φ_7	27.77°
L_4	86.00 mm	L_2	726.28 mm
L_6	661.34 mm	L_3	753.58 mm
X_A	(57.00, −288.00)	L_5	309.38 mm
X_C	(43.00, −120.00)	L_7	129.54 mm
X_D	(43.00, −640.00)	X_B	(60.79, −148.05)
X_F	(21.44, −873.27)	X_E	(97.34, −573.34)
X_G	(−69.40, −152.00)	X_H	(−93.18, −812.91)
X_O	(−13.00, 11.00)	X_I	(−13.20, −813.00)

$$L_3 = \sqrt[2]{(x_F - x_C)^2 + (z_F - z_C)^2} \tag{2}$$

$$(x_H, z_H) = (x_G + L_6 \cos\varphi_6, z_G - L_6 \sin\varphi_6) \tag{3}$$

$$L_7 = \sqrt[2]{(x_F - (x_G + L_6\cos\varphi_6))^2 + (z_F - (z_G - L_6\sin\varphi_6))^2} \tag{4}$$

$$(x_B, z_B) = (x_A - L_1\cos\varphi_1, z_A + L_1\sin\varphi_1) \tag{5}$$

$$L_2 = \sqrt[2]{(x_A - L_1\cos\varphi_1) - x_F)^2 + (z_A + L_1\sin\varphi_1) - z_F)^2} \tag{6}$$

$$\varphi_2 = \cos^{-1}(\frac{x_B - x_F}{L_2}) \tag{7}$$

$$(x_E, z_E) = (x_D - L_4\cos\varphi_4, z_D + L_4\sin\varphi_4) \tag{8}$$

$$L_5 = \sqrt[2]{(x_D - L_4\cos\varphi_4) - x_F)^2 + (z_D + L_4\sin\varphi_4) - z_F)^2} \tag{9}$$

$$\varphi_3 = \cos^{-1}(\frac{x_F - x_C}{L_3}) \tag{10}$$

$$\varphi_5 = \cos^{-1}(\frac{x_F - x_E}{L_5}) \tag{11}$$

$$\varphi_7 = \cos^{-1}(\frac{x_F - x_H}{L_7}) \tag{12}$$

For practical design, a 3D configuration design was conducted by CATIA. The 3D model consists of an inner panel, an outer panel, and 12 truss-links, with dimensions of 1620 mm × 800 mm in the fully deployed configuration (see Figure 3) and a total mass of 11.45 kg as shown in Table 4. The mass of hinges is included in the mass of the panel, and the mass of the bracket is included in the mass of each link. The truss-links connection was finally realized as illustrated in Figure 4a, with several connection angles, as shown in Figure 4b. The connection angles according to the truss-links are summarized in Table 5. The material of the truss-link is aluminum 6061, which has a Young's modulus of 68.9 GPa, a Poisson's ratio of 0.33, and density of 2700 kg/m^3 [24]. The cross-section of the truss was assumed as a tube-type with an outer diameter of 10 mm and an inner diameter of 8 mm.

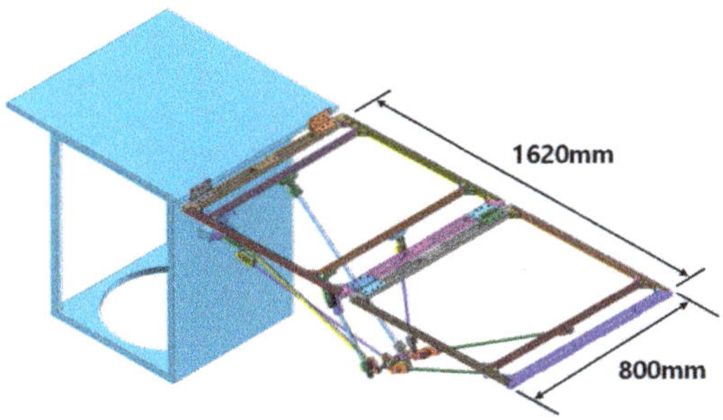

Figure 3. Configuration of a deployment analysis model.

Table 4. Length and weight of truss-link mechanism.

ID	Quantity [ea]	Length [mm]	Mass [g]
Link 1	1	140.00	173.60
Link 2	1	726.28	176.86
Link 3	2	753.58	370.41
Link 4	2	86.00	321.96
Link 5	2	309.38	379.38
Link 6	2	661.34	437.19
Link 7	2	129.54	342.14
Panel	2	800 × 800	8505.47
Panel joints	7	-	447.99
Shaft	10	-	291.31
Total	-	-	11,446.31

Table 5. Connection angle of truss-links.

Link	Connection Angle [°]	Link	Connection Angle [°]
Link 1	180	Link 5	168
Link 2	180	Link 6	170
Link 3	170.5	Link 7	170
Link 4	168		

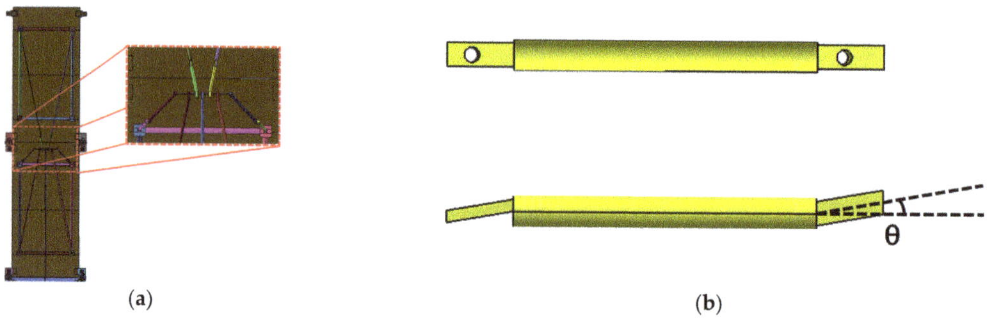

Figure 4. Configuration of truss-link mechanism: (**a**) 3-D truss, (**b**) truss-links with connection angles θ.

2.3. Latching Mechanism

Once the truss-links are deployed, they must keep their positions and angles through proper latching mechanisms, which are the most critical factors in maintaining the deployment configuration. As shown in Figure 5, the latching structure consists of a neodymium magnet, a ball plunger, and a stopping protrusion. The neodymium magnet and the ball plunger are installed on the lower bracket, and the stopping protrusion is installed on the upper bracket. If the truss-links rotate, the stopping protrusion meets the ball plunger, slides over the ball plunger and locks. In addition, neodymium magnets were installed on the wall to prevent the stopping protrusion from bouncing by the reaction force.

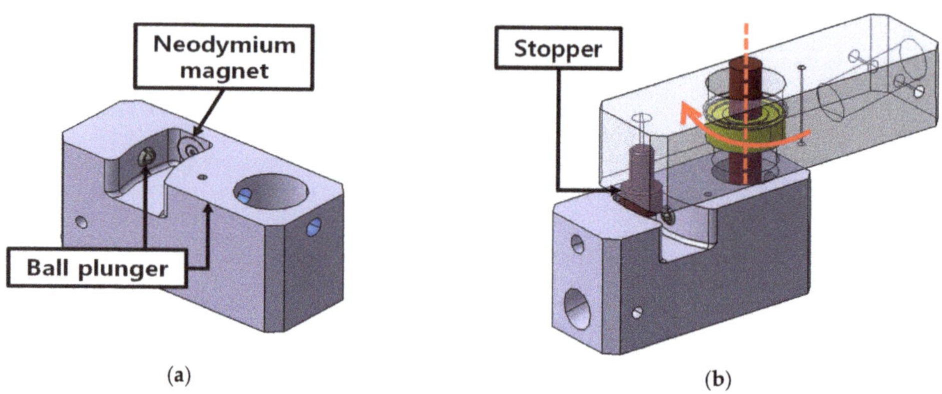

Figure 5. Configuration of latching mechanism: (**a**) Neodymium magnet, ball plungers (**b**) Latching system.

3. Deployment Dynamics

3.1. Modeling of Kinematic Joints and Latching Mechanism

To present the deployment motion, dynamic modeling was conducted with the aid of commercial multibody dynamics (MBD) software, RecurDyn. There were three kinds of joints: revolute joints for one-axis revolution, fixed joints to eliminate rigid-body motion, and on-off joints for latching truss-links. In contrast to the two-dimensional type-2 design, in the actual 3D design, when many truss-links were connected at a rotation joint (see Figure 4a) such as joint F, the over-constrained mechanisms cannot be avoided according to Equation (1). To address this problem, we used the bushing force function of RecurDyn instead of the revolute joint [25]. The bushing force can model any joint with 6 DOF by adjusting the stiffness of six springs independently. In the case of a revolute joint with the bushing force, the translation stiffness was set to be sufficiently high that the translation motion would be constrained; one of the rotational stiffnesses was set to zero for the revolution while the other rotation stiffness was set high. The proposed truss-link mechanism must be latched on each joint, B, E, and H, in a fully deployed state at the end of the deployment process. The on/off joints provided by RecurDyn are used to address these problems, which allows operation when certain conditions are met [26], as shown in Table 6. Using this, we modeled the two links connected to the joint to be fixed without generating additional motion when they are aligned.

Table 6. Expression function for latching mechanism.

Type	Expression
 On/Off joint	Expression: [AZ(1,2) × RTOD > 180°] AZ(1,2): Angle between 1 and 2 (1: body 1, 2: body 2) RTOD: Radian to degree 180°: Latching angle

3.2. Deployment Dynamics Analysis

Deployment behavior was analyzed by RecurDyn. Table 7 shows the mass properties of the two panels, and Figure 6 shows the rotational stiffness of the 90° and 180° torsional spring hinges, which were measured and realized through the spline function of RecurDyn [23]. The initial rotation angles were set to 90° and 180°, respectively, forcing the rotation joints to have driving torques in the stowed configuration.

Table 7. Dimensions and mass properties of inner and outer panels.

	Panel 1	Panel 2
Mass [kg]	4.23	4.28
Moment of inertia [kg·mm²] w.r.t. C.G.	$I_{xx} = 0.81 \times 10^6$ $I_{yy} = 0.36 \times 10^6$ $I_{zz} = 0.45 \times 10^6$ $I_{xy} = 9.18 \times 10^3$ $I_{yz} = -0.52 \times 10^1$ $I_{zx} = 0.99 \times 10^1$	$I_{xx} = 0.81 \times 10^6$ $I_{yy} = 0.36 \times 10^6$ $I_{zz} = 0.45 \times 10^6$ $I_{xy} = 9.52 \times 10^3$ $I_{yz} = -0.21 \times 10^1$ $I_{zx} = 0.45 \times 10^1$
Dimension	800 mm × 800 mm × 20 mm	

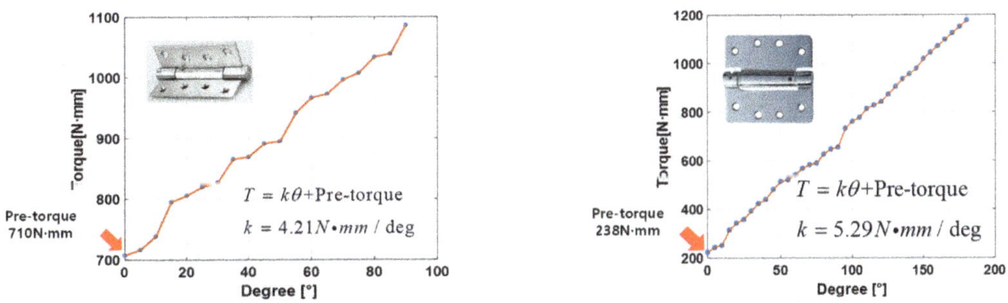

(a) (b)

Figure 6. Configuration of rotational spring hinges and the relationship between torque and rotation angles: (**a**) 90° hinge (Joint O), (**b**) 180° hinge (Joint I).

4. Deployment Test

4.1. Test Configuration and Test Cases

Dummy panels accommodating rotational torsion spring hinges between panels were established to verify the design of the deployment performance. To make a weightless environment for structures, a zero-G device was used [27] by hanging the structure with spring-loaded wires to minimize deflection and friction against gravity on the air bearings (Figure 7). Before the test, alignment of the test fixture was adjusted within 0.01° using a two-axis digital protractor, to place the panel close to horizontal from the ground. As shown in Table 8, three test cases were chosen to determine whether the panel was fully deployed according to changes of the driving torque. The video camera was installed at the ceiling of the test facility. During the tests, we employed a set of red markers on top of the frames to record the deployment angle history of the inner and outer panels using the MATLAB color extraction algorithm (see Figure 8). Finally, the locations of color were used to evaluate the angles of panels using inverse kinematics from the position history. The deployment test showed the dummy panels were fully deployed with the joints of each truss-link latched successfully (see Figure 9).

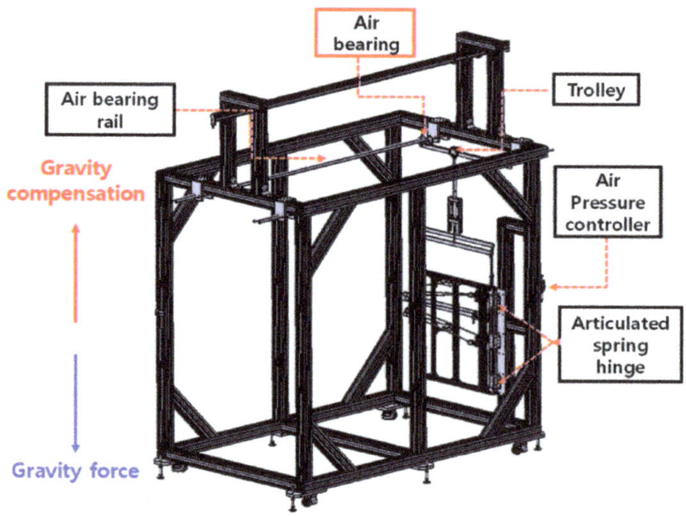

Figure 7. Test configuration of panels with truss-link mechanism [27].

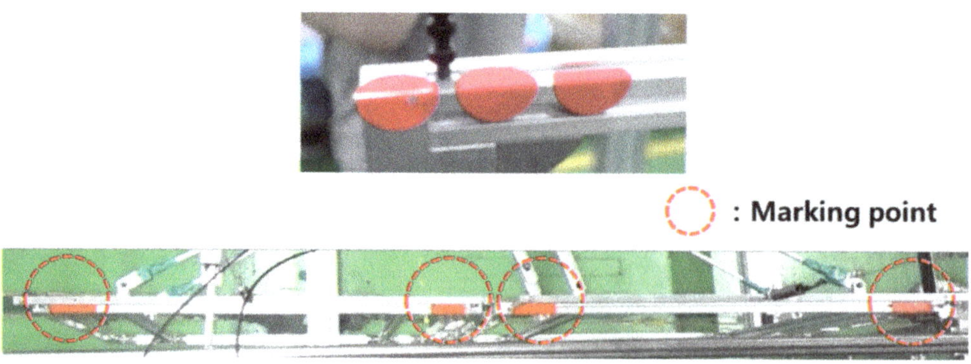

Figure 8. Configuration of red markers on dummy panels.

Table 8. Test cases.

Test-ID	Joint O	Joint I	Truss-Links
Case 1	2.17 N·m (2ea)	2.38 N·m (2ea)	×
Case 2	2.17 N·m (2ea)	2.38 N·m (2ea)	O
Case 3	2.17 N·m (2ea)	3.57 N·m (3ea)	O

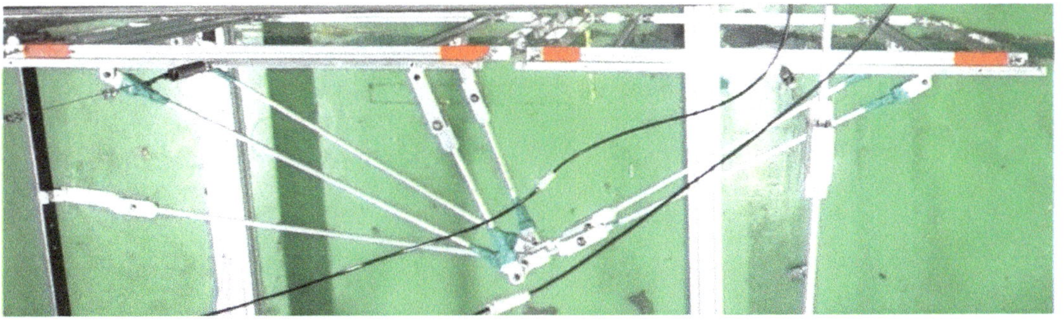

Figure 9. Fully deployed configuration (case 2 and case 3).

4.2. Modeling of Friction and Trajectory Error

For friction compensation, a rotational friction model considering the Coulomb friction torque was employed as given below [27,28]:

$$EFT \approx T_C \cdot \tanh\left(\frac{\omega}{\omega_{coul}}\right) \tag{13}$$

where T_C, ω and ω_{Coul} are the Coulomb friction torque, relative angular velocity, and Coulomb threshold velocity, respectively; ω_{Coul} is a parameter used to alleviate numerical instability caused by Coulomb friction. In this work, 0.01 rad/s was used as the reference value [28]. As seen in Figure 10, EFT1, EFT2, and EFT3 are the equivalent friction torques at joints O, I, and F; ϕ_1 and ϕ_2 indicate the rotation angles of joint O and I; trajectory error was defined and computed using Equation (14) as follows:

$$Trajectory\ error(\phi^a) = \frac{1}{n}\sum_{t=0}^{n}\left(\frac{\phi^e(t_i) - \phi^a(t_i)}{\phi^e(t_i)} \times 100\right) \tag{14}$$

where $\phi^a = \phi^a(EFT1, EFT2, EFT3, t)$ in degrees are the rotation angles according to the time evolution; the initial time was $t_0 = 0$ s and the time step $\Delta t = 0.05$ s; n, a, e denote the number of time data used, analysis, and experiment, respectively. The data were compared for four seconds until all panels were fully deployed in the test. We added the test frames in the MBD model to consider the effect of test fixture. Contact stiffness of 37,000 N/mm$^{1.5}$ and contact damping of 3.7 N·s/mm$^{1.25}$ were applied based on Hertz contact theory [29] between panel to panel and between truss-links to panels to allow for movement within a limited area in case of contact.

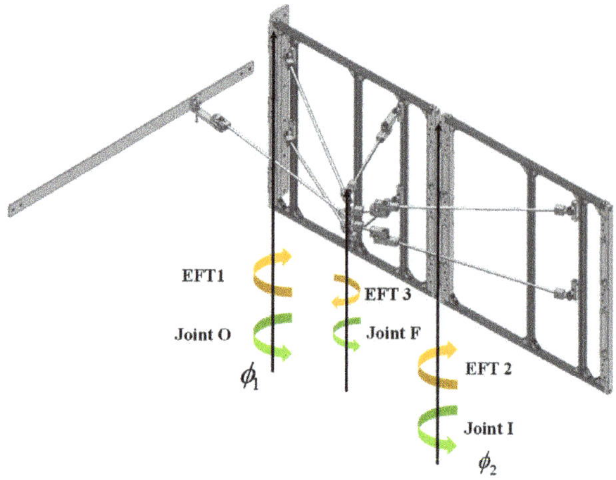

Figure 10. Definition of the equivalent friction torques.

4.3. Response Surface Methodology for Friction Identification

From comparison results between the test and simulation, as shown in Figure 11, it was observed that all simulations deployed earlier than those of the test. This is believed to be caused by a slight misalignment of the test frame, friction at the joints, and air drag from the panels and truss-links during the deployment test. Therefore, we attempted to adjust the deployment simulation model by adding the equivalent friction torques (EFT) at the three revolute joints defined as EFT1, EFT2, and EFT3, including those from all sources of friction, as defined in Figure 10. The EFT values can be easily determined by solving an optimization problem with the response surface method (RSM), making the rotation angle history of the analysis results similar to that of the test. The response surface y was generated with three design parameters, as in Equation (15). It is a polynomial function having ten terms corresponding to unknown coefficients, and the optimal response is found through the response function.

$$y = \beta_0 + \beta_1 x_1 + \beta_2 x_2 + \beta_3 x_3 + \beta_{12} x_1 x_2 + \beta_{13} x_1 x_3 + \beta_{23} x_2 x_3 + \beta_{11} x_1^2 + \beta_{22} x_2^2 + \beta_{33} x_3^2 \quad (15)$$

where x_i, β_i and β_{ij} are the values of the design variables and the coefficients of the polynomials of design variables, respectively. Here, the design variables indicate the equivalent friction torques: EFT1, EFT2, and EFT3. To determine the coefficients of the polynomials, 15 design cases were determined by the central composite design, popularly adopted in the design of experiments (DOE). The response variables were assigned to the trajectory error of the angles history of the inner panel (90°) at joint O and the outer panel (180°) at joint I. After deployment simulation for 15 design cases for test case 2, two output responses were computed and summarized, shown in Table 9. Finally, the coefficients of the polynomials of design variables were determined by regression analysis. To check the suitability of the derived response surface, R^2 (coefficient of determination) was obtained as 0.978, indicating a high correlation between the design variables and response variables. By determining the minimum point of the response surface, it was found that EFT1, EFT2, and EFT3 were 0.54 N·m, 0.472 N·m, and 0.065 N·m, respectively, showing an excellent correlation between the test and analysis for test case 2 with the corresponding trajectory error 1 and 2 as 7.69%, 6.62%, respectively, as presented in Figure 11b. The coefficients of the polynomials were obtained in the Equation (16).

$$\begin{aligned} \boldsymbol{\beta} &= [\beta_0, \beta_1, \beta_2, \beta_3, \beta_{12}, \beta_{13}, \beta_{23}, \beta_{11}, \beta_{22}, \beta_{33}] \\ &= [16.329,\ 3.713,\ -1.703,\ 1.107,\ 0.823,\ -1.290,\ 0.658,\ 2.040,\ 1.529,\ 1.262] \end{aligned} \quad (16)$$

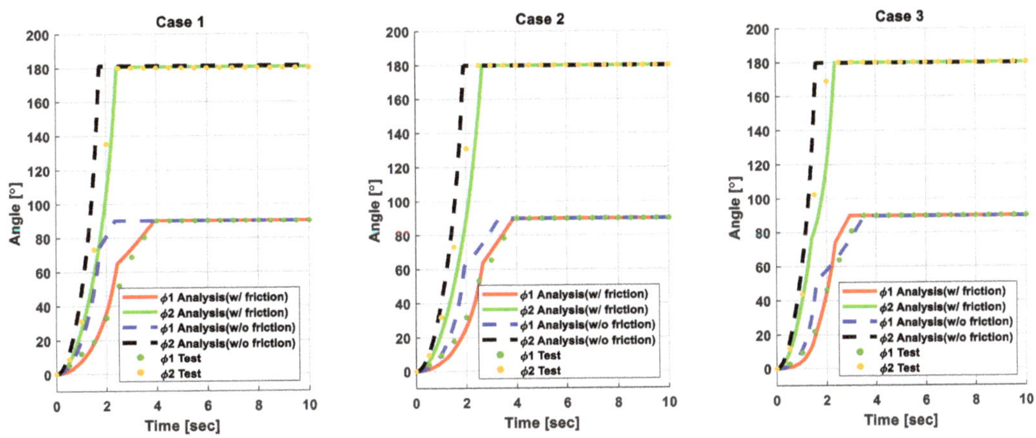

Figure 11. Comparison of the rotation angle history between test and analysis before/after friction compensation: (**a**) case 1, (**b**) case 2, (**c**) case 3.

Table 9. Design cases of central composite design.

No.	Design Variables			Response Variables	
	EFT1	EFT2	EFT3	Trajectory Error 1 (%)	Trajectory Error 2 (%)
1	1.17	0.60	0.10	28.65	20.75
2	0.75	1.10	0.10	15.90	28.03
3	0.75	0.60	0.10	16.32	13.07
4	1.00	0.30	0.15	25.69	15.20
5	0.33	0.60	0.10	15.49	13.32
6	0.75	0.095	0.10	25.35	3.66
7	0.50	0.90	0.05	11.10	22.03
8	1.00	0.30	0.05	23.98	8.73
9	1.00	0.90	0.05	24.64	21.46
10	0.50	0.30	0.05	18.07	6.70
11	0.75	0.60	0.016	19.46	13.29
12	0.50	0.30	0.15	20.60	10.84
13	1.00	0.90	0.15	24.64	25.36
14	0.75	0.60	0.18	20.28	15.88
15	0.50	0.90	0.15	13.71	28.34

Although there remains a slight gap between the analysis and the test after friction compensation, this is explained by all the sources of friction occurring in the test which cannot be considered. In addition, these EFT values were applied in test case 1 and test case 3, which additionally produce a good correlation between the test and analysis, as shown in Figure 11a,c. A detailed comparison is summarized in Figures 12–14.

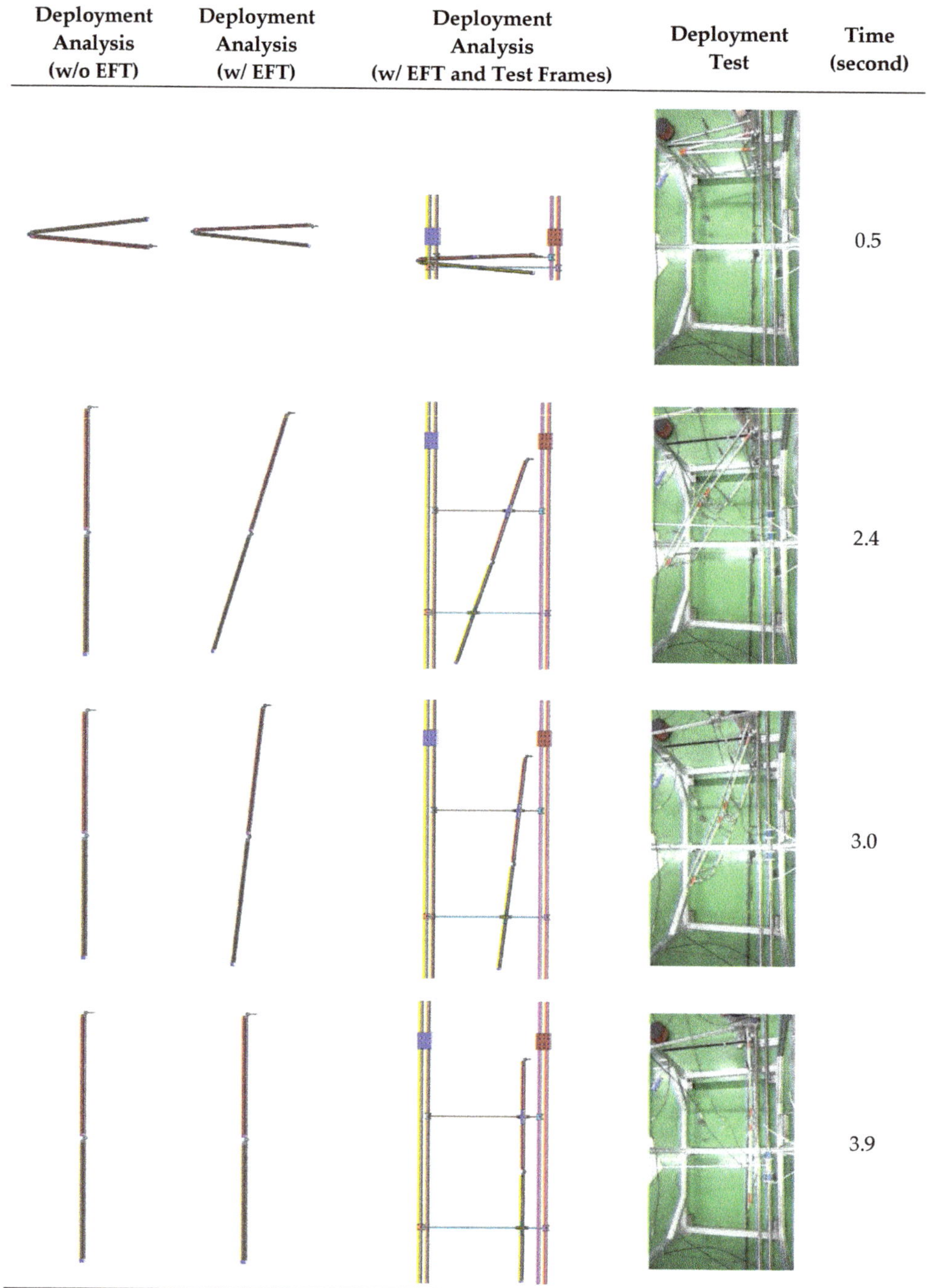

Figure 12. Comparison of the deployed motion history for case 1.

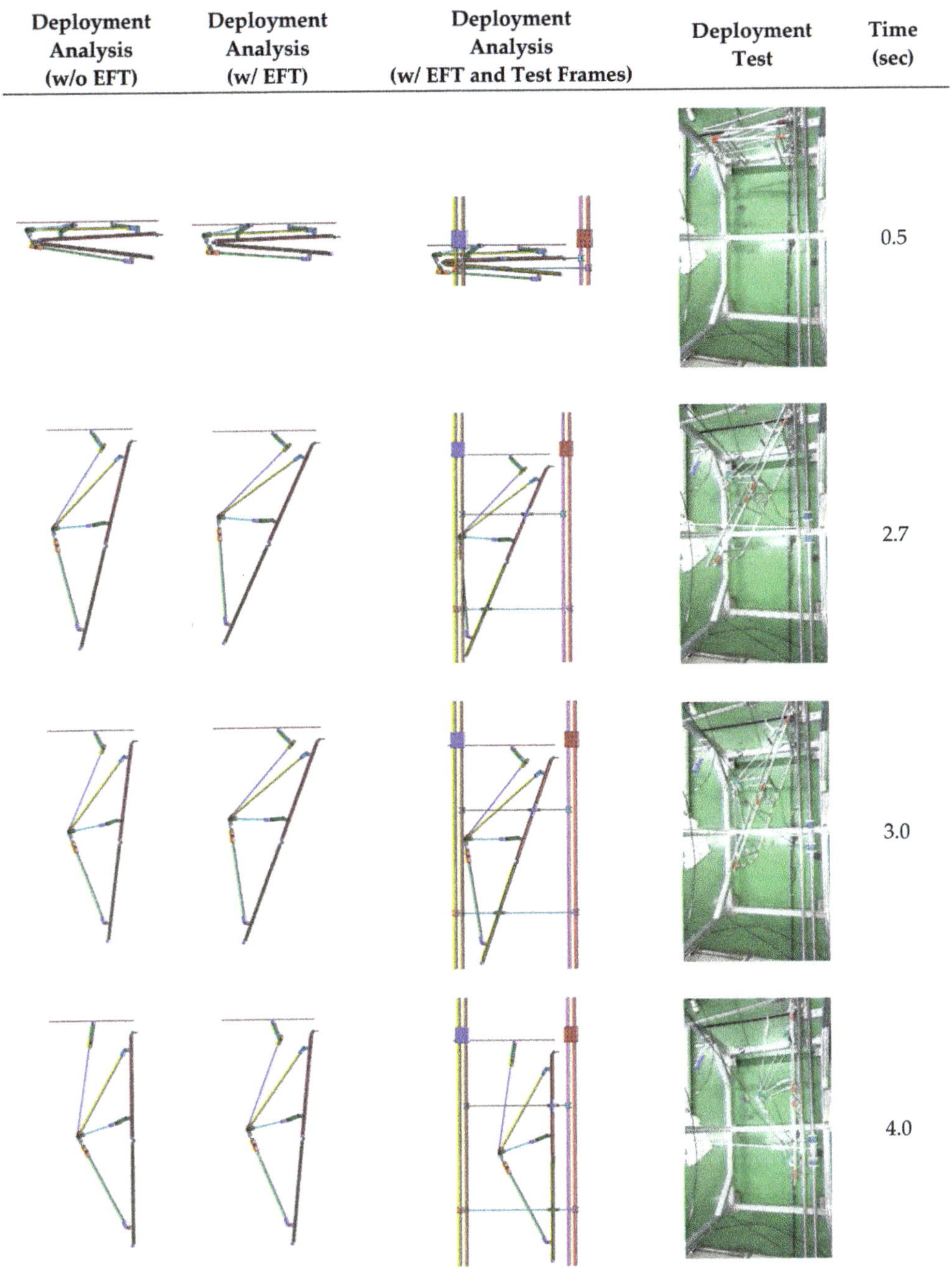

Figure 13. Comparison of the deployed motion history for case 2.

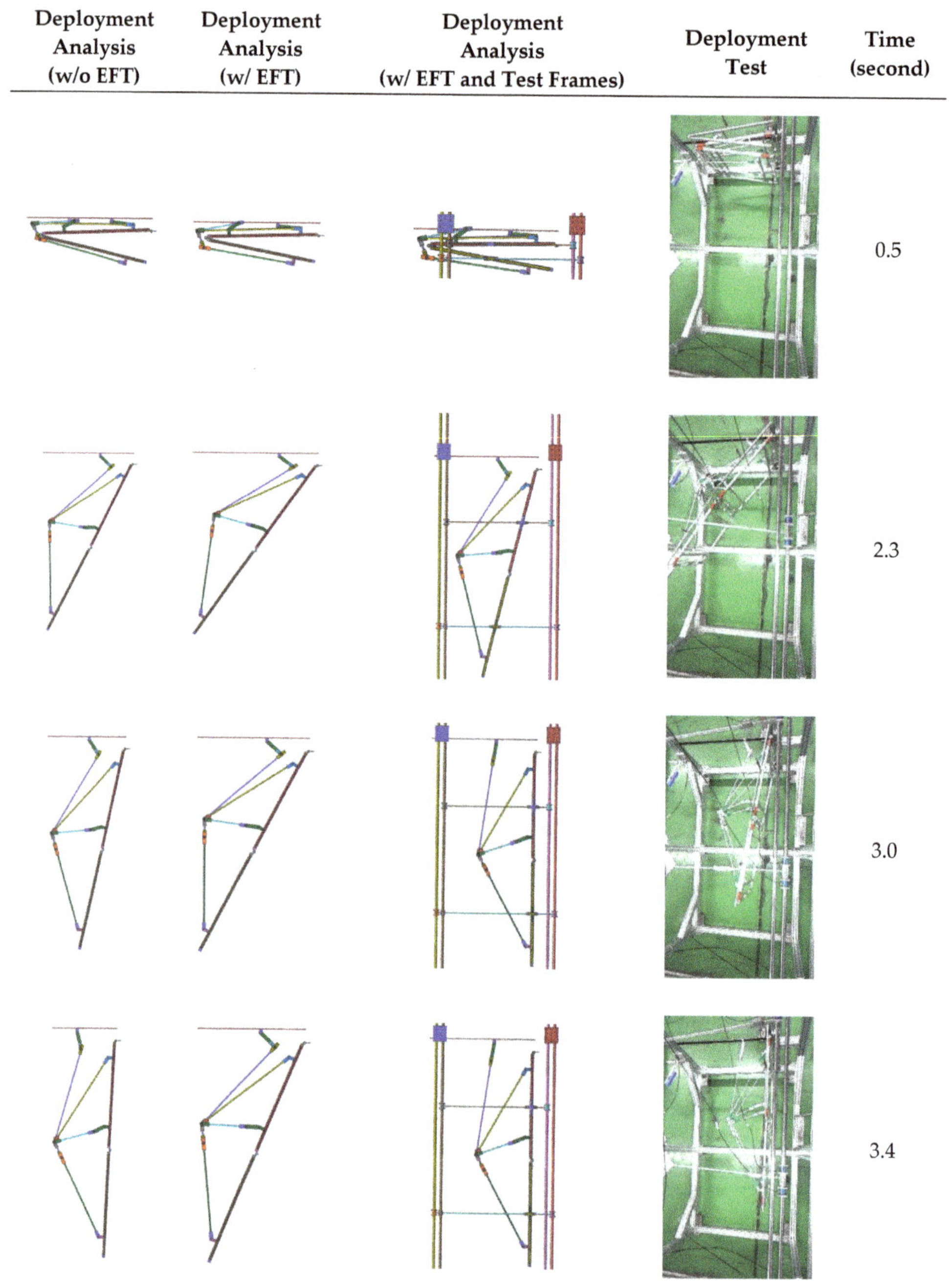

Figure 14. Comparison of the deployed motion history for case 3.

4.4. *Torque Margin Evaluation*

From the deployment test result, for all test cases the deployment of the truss-link mechanism was successful. However, we re-validated the EFTs with torque margin analysis [30]. If the torque margin was negative, the deployment would fail due to inadequate driving torque. If the torque margin was positive, the selected driving torque would be appropriate and completely deployed. To analyze this phenomenon, the torque margin was calculated as

$$E_D^{90^\circ} = \int_0^{\frac{\pi}{2}} T_D d\theta \quad (17)$$

$$E_D^{180^\circ} = \int_0^{\pi} T_D d\theta \quad (18)$$

$$E_R^{90^\circ} = \int_0^{\frac{\pi}{2}} T_R d\theta \quad (19)$$

$$E_R^{180^\circ} = \int_0^{\pi} T_R d\theta \quad (20)$$

$$Torque\ Margin = \left(\frac{E_D}{E_R} - 1\right) \times 100(\%) \quad (21)$$

where T_D, T_R are the driving torque and the resistance torque of Joint O (90°) and Joint I (180°), respectively. Moreover, E_D, E_R are the driving torque energy and the resistance torque energy value of Joint O and Joint I. The torque margin calculation results for Joint O and I are summarized in Tables 10 and 11. If the torque margin value was negative, then the deployment of panels at the corresponding joints would not be achieved. For the test cases, all torque margin values were positive. As seen in Figures 12–14, all the test cases were fully deployed. Through this evaluation, it can be predicted that the worst case cannot be deployed in advance. Therefore, the EFT estimation method developed in this work was verified with excellent reliability.

Table 10. Driving and resistance torque energy according to the test cases.

Test-ID	Driving Torque Energy [N·m]	Resistance Torque Energy [N·m]
Case 1	2.818	0.848
	4.484	1.473
Case 2	2.818	0.848
	4.484	1.476
Case 3	2.818	0.848
	6.726	1.476

Table 11. Torque margin calculation for all test cases.

Test-ID	Joint	Driving Torque [N·m]	Friction Torque [N·m]	Torque Margin [%]	Test (Pass/Fail)	Analysis (Pass/Fail)
Case 1	Joint O	2.17	0.54	232.31	Pass	Pass
	Joint I	2.38	0.47	204.41	Pass	Pass
Case 2	Joint O	2.17	0.54	232.31	Pass	Pass
	Joint I	2.38	0.47	203.80	Pass	Pass
Case 3	Joint O	2.17	0.54	232.31	Pass	Pass
	Joint I	3.57	0.47	355.69	Pass	Pass

5. Conclusions

In this paper, the modeling and validation of a passive truss-link mechanism applicable to a large-scale deployable structure for small satellites were designed to achieve a successful deployment. To design the truss-link mechanism, we referred to the truss-link structure of RADARSAT-2 satellites. A detailed design of the 3D truss-link structure followed based on the 2D configuration defined by the equations consisting of geometric construction methods and bar-groups methods.

Note that in contrast to the many truss-link mechanisms [11–17], we do not accommodate any active driving mechanisms as they lack simplicity, are not highly reliable, and are expensive. As driving mechanisms, only conventional torsion spring hinges with proper latching mechanisms were employed in this work.

To demonstrate the performance of the deployable structure with truss-links, a deployment simulation with the MBD model was conducted with proper joint modeling and rotational spring stiffness. Furthermore, a deployment test was conducted to check the correlation between the test and analysis. The results revealed that the time history of the deployment test was slightly behind that of the deployment analysis. To compensate for this, an inverse identification technique for equivalent friction torque (EFT) was proposed based on the results of the response surface method combined with the central composite design technique. Finally, with the EFTs, the angular rotation history in the deployment analysis was much closer to that of the deployment test with a trajectory error of less than 8%. Therefore, the adequacy of the proposed design of the deployable structure with truss-links was verified. This kind of truss-links mechanism has strong potential to extend a multi-modular deployable structure for small satellites. It will be further discussed in our future works.

Author Contributions: Conceptualization, J.H.L.; methodology, H.-S.C., T.S.J.; software, J.-H.P., H.-S.C.; validation, H.-S.C., J.-H.P.; investigation, D.-Y.K., H.-S.C.; data curation, H.-S.C.; writing—original draft preparation, J.H.L.; writing—review and editing, T.S.J.; supervision, J.H.L. All authors have read and agreed to the published version of the manuscript.

Funding: This research was supported by the National Research Foundation of Korea (NRF) funded by the Ministry of Science and ICT (NRF-2018M1A3A02065478).

Institutional Review Board Statement: Not applicable.

Informed Consent Statement: Not applicable.

Data Availability Statement: Not applicable.

Conflicts of Interest: The authors declare no conflict of interest.

References

1. Kim, S.-Y. SAR Antenna Technology. In Proceedings of the Korea Electromagnetic Engineering Society, Ilsan, Korea, 25 November 2011.
2. Park, T.-Y.; Kim, S.-Y.; Yi, D.-W.; Jung, H.-Y.; Lee, J.-E.; Yun, J.-H.; Oh, H.-U. Thermal Design and Analysis of Unfurlable CFRP Skin-Based Parabolic Reflector for Spaceborne SAR Antenna. *Int. J. Aeronaut. Space Sci.* **2021**, *22*, 433–444. [CrossRef]
3. Kim, D.-Y.; Lim, J.H.; Jang, T.-S.; Cha, W.H.; Lee, S.-J.; Oh, H.-U.; Kim, K.-W. Optimal design of stiffness of torsion spring hinge considering the deployment performance of large scale sar antenna. *J. Aerosp. Syst. Eng.* **2019**, *13*, 78–86.
4. Hoffait, S.; Brüls, O.; Granville, D.; Cugnon, F.; Kerschen, G. Dynamic analysis of the self-locking phenomenon in tape-spring hinges. *Acta Astronaut.* **2010**, *66*, 1125–1132. [CrossRef]
5. Kim, K.-W.; Park, Y. Systematic design of tape spring hinges for solar array by optimization method considering deploy-ment performances. *Aerosp. Sci. Technol.* **2015**, *46*, 124–136. [CrossRef]
6. Lee, Y.; Lee, J.E.; Kim, M.-G.; Jung, S.Y. A Study of a Mobile Launch Platform Effect on Bi-folded Wing Deployment. *Int. J. Aeronaut. Space Sci.* **2021**, *22*, 560–566. [CrossRef]
7. Calassa, M.C.; Kackley, R. Solar Array Deployment Mechanism. In Proceedings of the 29th Aerospace Mechanisms Symposium, Pasadena, CA, USA, 17–19 May 1995.
8. Cha, W.-H.; Seo, J.-K.; Jang, T.-S.; Lee, S.Y.; Chae, J.-S.; Shin, G.-H.; Lee, S.-H.; Ryu, K.-S.; Kim, D.-K. NEXTSat-1 Solar Panel Deployment Device Test. In Proceedings of the Korean Society for Aeronautical and Space Sciences, Jeju-si, Korea, 18–20 November 2015.
9. Kim, K.-W.; Park, Y. Solar array deployment analysis considering path-dependent behavior of a tape spring hinge. *J. Mech. Sci. Technol.* **2015**, *29*, 1921–1929. [CrossRef]
10. Jeong, J.W.; Yoo, Y.I.; Shin, D.K.; Lim, J.H.; Kim, K.W.; Lee, J.J. A novel tape spring hinge mechanism for quasi-static deployment of a satellite deployable using shape memory alloy. *Rev. Sci. Instrum.* **2014**, *85*, 025001. [CrossRef] [PubMed]
11. Lin, F.; Chen, C.; Chen, J.; Chen, M. Dimensional synthesis of antenna-deployable support structure. *Int. J. Aeronaut. Space Sci.* **2020**, *21*, 404–417. [CrossRef]
12. Francis, R.; Graf, G.; Edwards, P.G.; McCaig, M.; McCarthy, C. The ers-1 spacecraft and its payload. *ESA Bull.* **1991**, *65*, 26–48.
13. Ahmed, S.; Parashar, S.; Langham, E.; McNally, J. RADARSAT Mission Requirements and Concept. *Can. J. Remote Sens.* **1993**, *19*, 280–288.
14. Thomas, W.D.R. RADARSAT-2 extendible support structure. *Can. J. Remote Sens.* **2004**, *30*, 282–286. [CrossRef]
15. Thompson, A.A. Overview of the RADARSAT Constellation Mission. *Can. J. Remote Sens.* **2015**, *41*, 401–407. [CrossRef]

16. Kankaku, Y.; Suzuki, S.; Osawa, Y. ALOS-2 mission and development status. In Proceedings of the 2013 IEEE International Geoscience and Remote Sensing Symposium—IGARSS, Melbourne, Australia, 21–26 July 2013.
17. Born, G.H.; Dunne, J.A.; Lame, D.B. Seasat Mission Overview. *Science* **1979**, *204*, 1405–1406. [CrossRef] [PubMed]
18. The European Space Agency. Earth Online. Available online: https://earth.esa.int/eogateway/missions/worldview-4 (accessed on 11 November 2021).
19. The European Space Agency. Sentinel-5p. Available online: https://www.esa.int/Applications/Observing_the_Earth/Copernicus/Sentinel-5P (accessed on 11 November 2021).
20. Wang, Y.; Deng, Z.; Liu, R.; Yang, H.; Guo, H. Topology Structure Synthesis and Analysis of Spatial Pyramid Deployable Truss Structures for Satellite SAR Antenna. *Chin. J. Mech. Eng.* **2014**, *27*, 2014. [CrossRef]
21. Han, B.; Xu, Y.; Yao, J.; Zheng, D.; Guo, L.; Zhao, Y. Type synthesis of deployable mechanisms for ring truss antenna based on constraint-synthesis method. *Chin. J. Aeronaut.* **2020**, *33*, 2445–2460. [CrossRef]
22. Uicker, J.J. *Theory of Machines and Mechanisms*, 5th ed.; OXFORD University Press: New York, NY, USA, 2017; pp. 12–17.
23. Xu, Y.; Lin, Q.; Wang, X.; Li, L.; Cong, Q.; Pan, B. Mechanism Design and Dynamic Analysis of a Large-Scale Spatial Deployable Structure for Space Mission. In Proceedings of the Seventh International Conference on Electronics and Information Engineering, Nanjing, China, 17–18 September 2017.
24. *ASTM B209M, Standard Specification for Aluminum and Aluminum-Alloy Sheet and Plate (Metric)*; ASTM International: West Conshohocken, PA, USA, 21 November 2014.
25. Available online: https://support.functionbay.com/ko/faq/single/79/revolute-joint-modeling-using-bushing-force (accessed on 8 October 2021).
26. FunctionBay, Recurdyn/Solver Theoretical Manual. Available online: https://functionbay.com/documentation/onlinehelp/default.htm#!Documents/introduction.htm (accessed on 8 October 2021).
27. Kim, D.-Y.; Choi, H.-S.; Lim, J.H.; Kim, K.-W.; Jeong, J. Experimental and Numerical Investigation of Solar Panels Deploy-ment with Tape Spring Hinges Having Nonlinear Hysteresis with Friction Compensation. *Appl. Sci.* **2020**, *10*, 7902. [CrossRef]
28. Armstrong, B.; De Wit, C.C. Friction modeling and compensation. *Control. Handb.* **1996**, *77*, 1369–1382.
29. Barber, J.R. *Contact Mechanics*; Springer: Cham, Switzerland, 2018; pp. 29–40.
30. Postma, R.W. Force and Torque Margins for Complex Mechanical Systems. In Proceedings of the 37th Aerospace Mechanisms Symposium, Galveston, TX, USA, 18–21 May 2014; pp. 107–118.

applied sciences

MDPI

Article

Gaussian Process Surrogates for Modeling Uncertainties in a Use Case of Forging Superalloys

Johannes G. Hoffer [1,*], Bernhard C. Geiger [1] and Roman Kern [2,*]

[1] Know-Center GmbH, Research Center for Data-Driven Business & Big Data Analytics, Inffeldgasse 13, 8010 Graz, Austria; geiger@ieee.org

[2] Institute of Interactive Systems and Data Science, Graz University of Technology, Inffeldgasse 16c, 8010 Graz, Austria

* Correspondence: jhoffer@know-center.at (J.G.H.); rkern@tugraz.at (R.K.)

Abstract: The avoidance of scrap and the adherence to tolerances is an important goal in manufacturing. This requires a good engineering understanding of the underlying process. To achieve this, real physical experiments can be conducted. However, they are expensive in time and resources, and can slow down production. A promising way to overcome these drawbacks is process exploration through simulation, where the finite element method (FEM) is a well-established and robust simulation method. While FEM simulation can provide high-resolution results, it requires extensive computing resources to do so. In addition, the simulation design often depends on unknown process properties. To circumvent these drawbacks, we present a Gaussian Process surrogate model approach that accounts for real physical manufacturing process uncertainties and acts as a substitute for expensive FEM simulation, resulting in a fast and robust method that adequately depicts reality. We demonstrate that active learning can be easily applied with our surrogate model to improve computational resources. On top of that, we present a novel optimization method that treats aleatoric and epistemic uncertainties separately, allowing for greater flexibility in solving inverse problems. We evaluate our model using a typical manufacturing use case, the preforming of an Inconel 625 superalloy billet on a forging press.

Keywords: GP regression; FEM; surrogate modeling; multi-objective optimization; hot metal forming; Inconel 625

Citation: Hoffer, J.G.; Geiger, B.C.; Kern, R. Gaussian Process Surrogates for Modeling Uncertainties in a Use Case of Forging Superalloys. *Appl. Sci.* **2022**, *12*, 1089. https://doi.org/10.3390/app12031089

Academic Editors: Jin-Gyun Kim, Jae Hyuk Lim and Peter Persson

Received: 16 December 2021
Accepted: 18 January 2022
Published: 20 January 2022

1. Introduction

Conducting experiments to better understand manufacturing processes is crucial, with real physical experiments being considered the gold standard. However, conducting real physical experiments for each new experimental setting is impractical because of expensive materials, production stoppages and labor hours for monitoring and evaluation. One good alternative is conducting experiments via simulation, where numerical methods–such as Finite Element Method (FEM)–present a well-observed method in the field of structural analysis. However, solving complex problems with FEM is time-consuming and computationally expensive. In order to reduce the computational effort, surrogate modeling may offer a promising solution [1]. Surrogate models are trained in a supervised manner and are designed to learn the function mapping between inputs and outputs. With a sufficient amount of training data with respect to the observed use case, a customized surrogate model is able to substitute for a FEM simulation up to a certain accuracy. When only specific dimensions with a controlled reduction in accuracy of a simulation result are desired, reduced-order surrogate modeling is an already known technique. Thus, reduced-order surrogate modeling aims to substitute the high-resolution simulation domain with some carefully selected dimensions of importance, e.g., selected displacement measures of a deformed part can be predicted by a reduced-order surrogate modeling with low computational effort, instead of performing a computationally intensive FEM simulation that predicts the displacement of each node representing the deformed part.

Meanwhile, Gaussian process regression (GP) has been successfully used as a surrogate model in the past. In literature, GP regression is also called "kriging" after the statistician and mining engineer Danie G. Krige [2]. However, for consistency, we use only the term GP regression or plain GP in this paper. Regarding GP regression, one of the biggest advantages is that it predicts a distribution (described by mean and standard deviation) rather than just a point estimate. The predicted standard deviation can be seen as a quality criterion related to the corresponding predicted mean value. In the following, we will refer to that standard deviation of a prediction as epistemic uncertainty, i.e., how certain the model is with respect to its prediction. Considering real manufacturing processes, another source of uncertainty can be observed with regard to the lack of complete control over all influence parameters. These deviations occurring during repeated process iterations under the same conditions are referred to as aleatoric uncertainty.

Recapitulating, we want to shed light on two types of uncertainties in surrogate modeling: (1) epistemic uncertainty referring to the lack of knowledge in respect to a simulation model and can be minimized by adding additional sources of information (with respect to machine learning models, it is mainly increasing the number of training instances at new locations in the feature space) and (2) aleatoric uncertainty referring to deviations of an observed manufacturing process itself, i.e., aleatoric uncertainty cannot be minimized even if more data is generated. Since epistemic and aleatoric uncertainties describe different properties, it seems natural to treat them separately when making predictions or optimization. However, it should be mentioned that in certain circumstances it may be useful to consider uncertainty as a whole rather than dividing it into aleatoric and epistemic uncertainty. In such cases, heteroskedastic GP regression represents a common approach for optimization with surrogate models [3–5]. In our problem definition, especially in solving inverse problems, we argue that the distinction of epistemic and aleatoric uncertainty shows clear advantages.

There is a wealth of literature on surrogate models, reduced-order surrogate models, and optimization with GP regression. We present in the following the main related works to our research field organized in (1) GP regression and FEM simulations, (2) GP regression trained with pure sensor data and (3) optimization with GP regression.

In the work of Roberts et al. [6], they predict damage development in forged brake discs reinforced with Al-SiC particles using damage maps. In addition to Multilayer Perceptron (MLP), Roberts et al. [6] utilize GP regression as a surrogate model. Loghin and Ismonov [7] predict the stress intensity factors using GP regression trained with FEM results of a classical bolt-nut assembly use case. Ming et al. [8] model an electrical discharge machining process with GP regression. Su et al. [9] utilize GP regression as a surrogate in a structural reliability analysis of a large suspension bridge. In the work of Guo and Hesthaven [10], GP regression is used as a reduced-order model for nonlinear structural analysis in a 1D and 3D use case, where data generation was performed with active learning. Hu et al. [11] use GP regression to estimate residual stresses field of machined parts from two-dimensional numerical simulations. Yue et al. [12] propose two active learning approaches using GP regression for a composite fuselage use case. In the work of Ortali et al. [13] GP regression is used as a reduced-order surrogate model for fluid dynamics use cases. Venkatraman et al. [14] use GP regression as a surrogate model of texture in micro-springs. GP regression can also be used on data with multiple fidelity levels, where Lee et al. [15] investigate GP regression surrogate modeling with uncertain material properties of soft tissues and multi-fidelity data. Brevault et al. [16] provide an overview of multi-fidelity GP regression techniques in the field of aerospace systems. GP regression can also be extended by methods that stack them or use them in a tree model. Civera et al. [17] predict imperfections in pultruded glass fiber reinforced polymers with a treed method of GP regression trained with experimental data and FEM simulation results. Abdelfatah et al. [18] propose a stacked GP regression to integrate different datasets and propagate uncertainties through the stacked model.

GP regression can also be used for calibrating simulations, where Mao et al. [19] use GP regression as a surrogate model in a Bayesian model updating method to calibrate FEM simulation of a long-span suspension bridge.

In addition to the use of FEM data, GP regression also finds application in the use of pure sensor data, which we will discuss in the following. Tapia et al. [20] use a GP regression based surrogate model of a laser powder-bed fusion process to predict melt pool depth. Yu et al. [21] utilize–besides other thriving methods–a GP regression to model the relationship between geological variables and the broken rock zone thickness. Lee [22] uses GP regression trained with experimental data to optimize wire arc additive manufacturing process deposition parameters. Saul et al. [23] propose chained GP regression models based on non-linear latent function combination. Binois et al. [24] provide a heteroskedastic GP regression approach and results of two use cases, namely manufacturing and management of epidemics.

In the course of function maximization with GP regression surrogate models, Dai Nguyen et al. [25] propose a robust optimization approach based on Upper Confidence Bound (UCB) Bayesian Optimization (BO). In another field of optimization, namely solving inverse problems, there is related work found where BO with generalized chi-squared distribution is researched by Huang et al. [26], and Uhrenholt and Jensen [27], where besides standard GP regression Uhrenholt and Jensen [27] utilized warped GP regression from the work of Snelson et al. [28]. An extension of the standard BO can be found in the work of Plock et al. [29], where they combine BO with the Levenberg-Marquardt method. While in maximization and minimization problems aleatoric and epistemic uncertainties can often be treated in the same way, in most cases robust results can be obtained by distinguishing between these two sources of uncertainty [30]. We refer to robust results when mean predictions are associated with low aleatoric uncertainty.

There is already considerable related work in reduced-order surrogate modeling and optimization using GP regression surrogates. However, to the best of our knowledge, we could not identify related work for solving inverse problems in which aleatoric and epistemic uncertainties are treated differently. Optimization approaches for solving inverse problems usually use only epistemic uncertainty. When epistemic and aleatoric uncertainties are taken into account, they are often simply combined, resulting in the potential loss of important information.

To sum up, we identify the following drawbacks:

- Related work shows that mainly epistemic uncertainty is used for prediction or optimization with GP regression.
- In research using aleatoric and epistemic uncertainties, they are not considered separately when solving inverse problems.

As a response, we present the following main contributions of our research to tackle the identified drawbacks of related work:

1. We present a GP based surrogate that models (a) the mean result, (b) the aleatoric and (c) the epistemic uncertainty of a manufacturing process outcome.
2. We utilize aleatoric and epistemic uncertainties in solving inverse problems for robust optimization results.

With the proposed surrogate model and novel multi-objective optimization strategy, we pave the way for surrogate modeling and inverse problem-solving for practical applications that make use of explicit modeling of sources of uncertainties. Our findings are validated on a typical hot metal forming manufacturing process: preforming an Inconel 625 superalloy billet on a forging press.

This paper is structured as follows. In Section 2, we present the proposed surrogate model, providing an introduction to GP regression in Section 2.1 and describe the GP based parts of our surrogate model in Sections 2.2 and 2.3. The data generation of aleatoric uncertainty for our surrogate model approach is presented in Section 2.4. Section 3 deals with optimization, where we outline active learning in Section 3.1 and solving inverse problems in Section 3.2. In Section 4 we present the studied use case, preforming an Inconel 625 superalloy billet on a forging press, where we give insights on the design of the forging aggregate characteristics in Section 4.1 and all information regarding the corresponding FEM simulation in Section 4.2. Section 5 shows the results, which are discussed in Section 6. In Section 7, we present the conclusion of our work and an outlook for the future.

2. GP based Surrogate Model

In this section, we first introduce briefly the general idea behind our surrogate modeling approach. We familiarize in Section 2.1 the reader with the general functionality of GP regression to provide an appropriate foundation for the content that follows. In Sections 2.2 and 2.3 we provide more detailed descriptions of each individual GP of our surrogate model. After describing our surrogate model, we move on to uncertainty propagation analysis with FEM simulation in Section 2.4, where we present the procedure for obtaining aleatoric uncertainties.

GP regression is already well researched for surrogate modeling, replacing expensive target labellers (e.g., numerical simulations, expensive manually labelling, conducting real physical experiments, etc.). One reason is their ability to work with low-dimensional data. Another big advantage of using GP regression is that predictions are made in a probabilistic way, i.e., a prediction is represented by a posterior distribution. Thus, a prediction of GP regression is described by a mean and a covariance. The covariance of a prediction can be used as a metric of prediction confidence, i.e., epistemic uncertainty. We specify that outputs of GP regression describe a distribution with mean m and epistemic uncertainty σ.

The proposed surrogate model consists of two individual GPs and takes manufacturing process-specific parameters x_m, part-specific parameters x_p and aleatoric process uncertainty $\bar{\Sigma}_{al}(Z)$ as input and predicts the mean manufacturing result μ and aleatoric uncertainty σ_{al} of the manufacturing result, see Figures 1b, 2 and 3. A similar simulation approach using FEM is shown in Figure 1a. We define Z as a parameter that describes the manufacturing process characteristics, e.g., velocity profile of a forming tool. Our model assumes that $\bar{\Sigma}_{al}(Z)$ can be efficiently obtained for every x_m. This assumption is justified in our running example, where we focus on the first of two directly successive forging strokes. That means that measurements of the manufacturing process are available (i.e., velocity profile of the forging tool), but measurements in respect to the forged part are not possible due to the short time span between the first and second stroke.

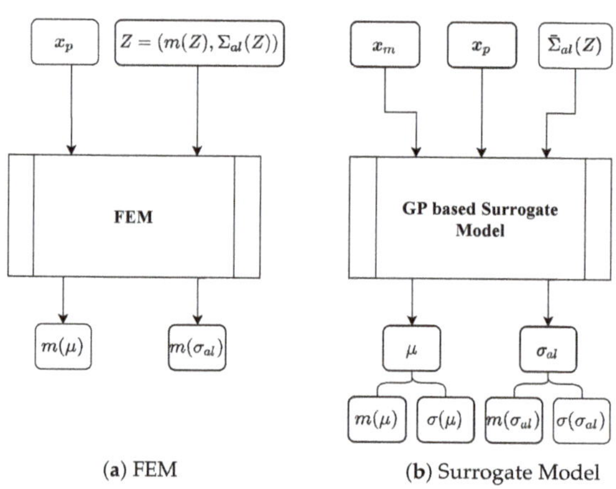

Figure 1. Simulation of Manufacturing Processes with Uncertainties: (**a**) FEM simulation scenario and (**b**) GP based surrogate model with manufacturing process-specific parameters x_m, part-specific parameters x_p and distribution Z that describes a manufacturing process-specific characteristic by

mean $m(Z)$ and aleatoric manufacturing process uncertainty $\Sigma_{al}(Z)$ where $\bar{\Sigma}_{al}(Z)$ is an aggregated form of $\Sigma_{al}(Z)$. Outputs are the mean of the manufacturing process result $m(\mu)$ and mean of the aleatoric uncertainty $m(\sigma_{al})$, each with corresponding epistemic uncertainties $\sigma(\mu)$ and $\sigma(\sigma_{al})$ in the GP based surrogate model.

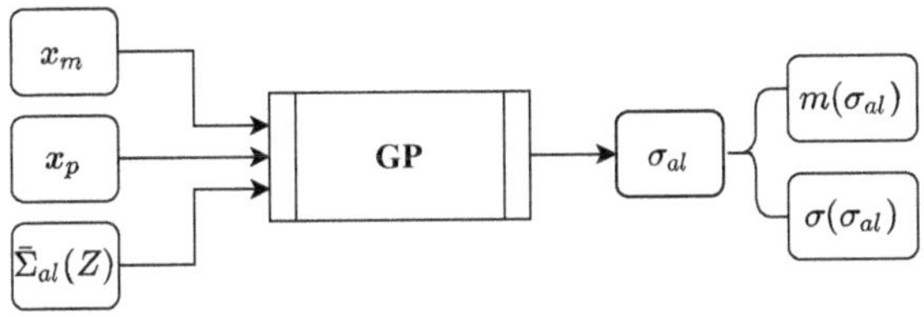

Figure 2. GP takes manufacturing process parameters x_m, part specific parameters x_p and aleatoric manufacturing process uncertainty $\bar{\Sigma}_{al}(Z)$ as input and predicts the mean $m(\sigma_{al})$ and epistemic uncertainty $\sigma(\sigma_{al})$ of the aleatoric uncertainty of the manufacturing process result.

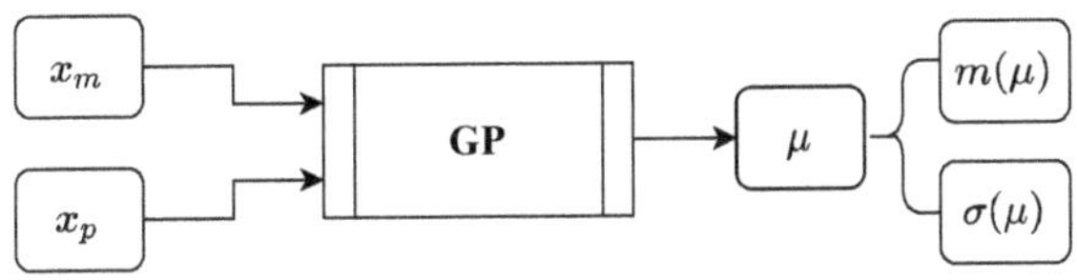

Figure 3. GP takes manufacturing process-specific parameters x_m and part specific parameters x_p as input and predicts the mean $m(\mu)$ and epistemic uncertainty $\sigma(\mu)$ of the manufacturing process result.

2.1. Gaussian Process

A GP is a generalization of the Gaussian distribution. The Gaussian distribution describes random variables or random vectors, while a GP describes functions $f(x)$. In general, a GP is completely specified by its mean function $m(x)$ and covariance function $k(x,x')$, also called kernel. If the function $f(x)$ under consideration is modeled by a GP, we have

$$\mathbb{E}[f(x)] = m(x) \tag{1}$$

$$\mathbb{E}[(f(x)-m(x))(f(x')-m(x'))] = k(x,x') \tag{2}$$

for all x and x'. Where x refers to training and x' to test data. Thus, we can define the GP by

$$f(x) \sim \mathcal{GP}(m(x), k(x,x')). \tag{3}$$

We use the following notation for explanatory purposes only in this section. Matrix $D_{train} = (X,Y)$ contains the training data with input data matrix $X = (x_1,\dots,x_n)$ and output data matrix $Y = (y_1,\dots,y_n)$, and test data matrix $D_{test} = (X',Y')$ contains the test data with $X' = (x'_{n+1},\dots,x'_{n+m})$ as input and $Y' = (y'_{n+1},\dots,y'_{n+m})$ as output. We define that they are jointly Gaussian and have zero mean with consideration of the prior distribution, further, we assume an additive independent identically distributed Gaussian noise with variance σ_n^2 and identity matrix I for noisy observations:

$$\begin{bmatrix} Y \\ Y' \end{bmatrix} \sim \mathcal{N}\left(0, \begin{bmatrix} k(X,X)+\sigma_n^2 I & k(X,X') \\ k(X',X) & k(X',X') \end{bmatrix}\right) \tag{4}$$

The GP predicts the function values Y' at positions X' in a probabilistic way, where, the posterior distribution can be fully described by the mean and the covariance.

$$\begin{aligned} Y'|X',X,Y \sim \; & \mathcal{N}(k(X',X)[k(X,X)+\sigma_n^2 I]^{-1}Y, \\ & k(X',X') - k(X',X)[k(X,X)+\sigma_n^2 I]^{-1}k(X,X')) \end{aligned} \tag{5}$$

Resulting in mean

$$m(Y') = \mathbb{E}[Y'|X,Y,X'] = k(X',X)[k(X,X)+\sigma_n^2 I]^{-1}Y \tag{6}$$

covariance

$$COV(Y') = k(X',X') - k(X',X)[k(X,X)+\sigma_n^2 I]^{-1}k(X,X') \tag{7}$$

and epistemic standard deviation σ

$$\sigma(Y') = \sqrt{\text{diag}(COV(Y'))} \tag{8}$$

where the diagonal of the covariance matrix COV is extracted as a vector and the square root is calculated for each element to determine the epistemic standard deviation σ. It can be observed that the selection or design of the covariance function is the main ingredient when using GP regression. In the following, we describe the two covariance functions we use in our approach: the popular Radial Basis Function (RBF) (also called squared exponential covariance function)

$$k_{\text{RBF}}(x,x') = \exp\left(\frac{||x-x'||^2}{l^2}\right) \tag{9}$$

with characteristic length-scale parameter l and $||\cdot||$ denoting the Euclidean distance and the Matérn covariance function

$$k_{\text{Matérn}}(x,x') = \frac{1}{\Gamma(\nu)2^{\nu-1}}\left(\frac{\sqrt{2\nu}}{l}||x-x'||\right)^{\nu} K_\nu\left(\frac{\sqrt{2\nu}}{l}||x-x'||\right) \tag{10}$$

with gamma function Γ, modified Bessel function K_ν and parameter ν that controls the smoothness of the resulting function. For more information on GP regression and covariance functions, we refer the reader to the book of Williams and Rasmussen [31].

2.2. Aleatoric Uncertainty GP

A GP is used to predict a manufacturing process related aleatoric uncertainty $\sigma_{al} = \sigma_{al}(x_m, x_p, \bar{\Sigma}_{al}(Z))$ of the manufactured part. Aleatoric uncertainty data are generated by uncertainty propagation analysis with FEM simulation. The inputs are the setting parameters from a real physical manufacturing process x_m, properties of the part to be manufactured x_p and aleatoric manufacturing process uncertainty $\bar{\Sigma}_{al}(Z)$ obtained from, e.g., sensor data of the real physical manufacturing process, see Figure 2. Here, Z describes a characteristic of the manufacturing process, e.g., the velocity profile of a forming tool. The output σ_{al} is predicted by a GP regression, such that $\sigma_{al} \sim \mathcal{GP}(m(x_m, x_p, \bar{\Sigma}_{al}(Z)), k((x_m, x_p, \bar{\Sigma}_{al}(Z)), (x_m, x_p, \bar{\Sigma}_{al}(Z))'))$ with mean $m(x_m, x_p, \bar{\Sigma}_{al}(Z))$ and covariance function $k((x_m, x_p, \bar{\Sigma}_{al}(Z)), (x_m, x_p, \bar{\Sigma}_{al}(Z))')$.

Of course, a wide variety of manufacturing process characteristics can be implemented, e.g., rolling speeds, cutting forces, heating times etc. As a running example, we choose as a manufacturing process hot metal forming on a friction screwpress, where x_m contains different input features which control the forging aggregate (clutch pressure between flywheels and rotation speed of the electric motor), x_p describes the part to be forged by different dimensions and part temperature and Z is a resulting velocity profile of the forging tool over time for a given input x_m, where $\bar{\Sigma}_{al}(Z)$ represents aggregated aleatoric deviations in respect to forging velocity. σ_{al} then describes the deviations of the final forged part, i.e., deviations from important final part geometries. All relevant details of our running example can be found in Section 4.

2.3. Mean Result GP

Besides the GP that predicts the aleatoric uncertainty of a manufactured part, a second GP is used to predict the mean result μ of the manufactured part. The inputs for the second GP are the setting parameters from the real physical manufacturing process x_m and properties of the to be manufactured part x_p. The output μ is predicted by the GP regression, such that $\mu \sim \mathcal{GP}(m(x_m, x_p), k((x_m, x_p), (x_m, x_p)'))$. In respect to our running example, μ describes the final forged part by important final part geometries.

2.4. Uncertainty Propagation Analysis

In uncertainty propagation analysis, the effect of uncertainties related to an input on uncertainties of the corresponding output is investigated. In our case, $\Sigma_{al}(Z)$ refers to the aleatoric deviations of a manufacturing process characteristics (i.e., deviations in velocity profile data) due to different input settings. We refer to uncertainties with respect to a manufacturing process output obtained by uncertainty propagation analysis as aleatoric uncertainty σ_{al}.

We vary input values $x^{(j)} = (x_m^{(j)}, x_p^{(j)})$ with $j \in \{1, \ldots, N\}$ where N is the number of different input setting scenarios. For each case of process-specific input parameters $x_m^{(j)}$, we obtain a process-specific characteristic $Z^{(j)}$ that is a distribution with mean $m(Z^{(j)})$ and standard deviation $\Sigma_{al}(Z^{(j)})$. Such distributions occur because, with identical input parameters, process characteristics in reality can show deviations when repeated. We simulate that behavior with a separate GP, thus, a random variable $Z^{(j)}$ is assumed to be Normally distributed, such that $Z^{(j)} = \mathcal{N}(m(Z^{(j)}), \Sigma_{al}(Z^{(j)}))$. From the posterior, we randomly draw M predictions $z^{(i)(j)}$ with $i \in \{1, \ldots, M\}$ (i.e., different curves characterizing the manufacturing process) and with each $z^{(i)(j)}$ and $x_p^{(j)}$ we execute FEM simulations to obtain targets $y^{(i)(j)}$. We collect the individual targets $y^{(i)(j)}$, such that we obtain for each input setting j a distribution with mean $\mu^{(j)}$ and aleatoric standard deviation (i.e., aleatoric uncertainty) $\sigma_{al}^{(j)}$. With that, we are able to describe each target by its distribution.

Thus, we obtain a dataset $D = \left\{D^{(1)}, \ldots, D^{(N)}\right\}$ where each datapoint $D^{(j)} = (X^{(j)}, Y^{(j)})$ can then be separated into input $X^{(j)} = (x_m^{(j)}, x_p^{(j)}, \bar{\Sigma}_{al}(Z^{(j)}))$ and output $Y^{(j)} = (\mu^{(j)}, \sigma_{al}^{(j)})$. Here $\bar{\Sigma}_{al}(Z^{(j)})$ is an aggregated manufacturing process uncertainty obtained from data. We model each output with a GP regression, thus the outputs are described again by a distribution with mean m and epistemic standard deviation σ (i.e., epistemic uncertainty), such that $\mu^{(j)} = \mathcal{N}(m(\mu^{(j)}), \sigma(\mu^{(j)}))$ and $\sigma_{al}^{(j)} = \mathcal{N}(m(\sigma_{al}^{(j)}), \sigma(\sigma_{al}^{(j)}))$.

3. Active Learning and Solving Inverse Problems

For optimization, we evaluate our surrogate model in two different areas: (1) active learning and (2) solving multi-objective inverse problems. We refer to active learning as a method to find the most informative data points in the feature space for the best overall performance of the surrogate model, i.e., predicting the mean result of a manufacturing process μ and corresponding aleatoric uncertainty of the manufacturing result σ_{al}. When solving multi-objective inverse problems, we try to find inputs where the error between a given target vector and a prediction as well as the aleatoric uncertainty is minimal, leading to robust optimization results.

3.1. Active Learning

Active learning is already well researched in terms of optimal use of resources for parameter optimization of a model, i.e., generating training data, see [12,32–34]. The process of generating training data means obtaining labels Y_{train} for an input X_{train}, such that a dataset $D_{train} = (X_{train}, Y_{train})$ can be used to fit or optimize parameters of a model. Labels Y_{train} are obtained by an oracle, where an oracle can be a domain expert, results of real physical experiments or like in our case results of expensive numerical FEM simulations.

In the following, we present the idea behind the researched optimization approach and highlight the applicability of active learning with our proposed surrogate model.

In active learning, a number of n_{AL} datapoints connected to maximum epistemic uncertainty σ_{ep} are queried from a pool of candidates X_{pool} to build a training dataset $D_{train} = (X_{train}, Y_{train})$ that is used for training a surrogate model. Thus, we select ideal training data, i.e., we use a minimum amount of training data such that the overall epistemic uncertainty in respect to making prediction on X_{pool} is minimized. We define in (11) the active learning query strategy with loss function $\mathcal{L}_{AL} = \mathcal{L}_{AL}(\sigma_{ep}) = \sigma_{ep}(x)$ to select a new query datapoint $d_q^{AL} = (x_q^{AL}, y_q^{AL})$ with input x_q^{AL} and output y_q^{AL}.

$$d_q^{AL} = \underset{x\ in\ X_{pool}}{\operatorname{argmax}}\ \sigma_{ep}(x) \tag{11}$$

A query datapoint d_q^{AL} is then moved to the training dataset D_{train}, the surrogate model is fitted and the iterative generation of training data starts again. In respect to our proposed surrogate model, we are able to utilize directly the epistemic uncertainty predictions of the two GPs, i.e., $\sigma(\mu)$ and $\sigma(\sigma_{al})$. Thus, we define $\sigma_{ep}(x) = \sigma(\mu(x)) + \sigma(\sigma_{al}(x))$ and select training data by utilizing (11).

3.2. Inverse Problem

In real physical manufacturing processes, it is commonly required that the result of the manufacturing process lies within a given tolerance range. Therefore, the parameters that control the manufacturing process and the properties of the part must be carefully selected. Moreover, the process of finding inputs to obtain a given target can be formulated as an inverse problem, i.e., finding causal factors for a required effect. In our work, we define that a basic solution of an inverse problem is to find an input x_{inv}, minimizing a distance $d = d(y_{inv}, y_{target})$ between prediction y_{inv} and target vector y_{target}. However, such solutions neglect the existence of process variations, i.e., aleatoric uncertainty. With no consideration of aleatoric uncertainty, the found ideal inputs can lead to quite good results regarding mean values but very high deviations, such that no robustness assertions can be made.

Therefore, we present a novel multi-objective optimization approach in (12) based on BO with a modified UCB acquisition function, where we make a clear separation of uncertainties, such that a loss function $\mathcal{L}_{inv}$, dependent of a distance function d, respective aleatoric σ_{al} and epistemic σ_{ep} uncertainties is minimized.

$$x_{inv} = \underset{X_{pool}}{\operatorname{argmin}}\ \mathcal{L}_{inv}(d, \sigma_{al}, \sigma_{ep}) \tag{12}$$

As a distance function d, we select the absolute error between mean target μ_{target} and mean manufacturing process result $m(\mu)$ as the metric. However, our approach is not limited to a specific distance metric, so any can be used.

$$d = d(\mu_{target}, m(\mu)) = |\mu_{target} - m(\mu)| \tag{13}$$

Utilizing $m(\sigma_{al})$, $\sigma(\sigma_{al})$, $m(\mu)$ and $\sigma(\mu)$ from our proposed surrogate model, we define epistemic uncertainty $\sigma_{ep} = \sigma_{ep}(\sigma(\sigma_{al}), \sigma(\mu)) = \sigma(\sigma_{al}) + \sigma(\mu)$ and construct a loss function $\mathcal{L}_{inv}$ with tuning parameters α and β, where α controls the influence of the aleatoric uncertainty and β controls exploration vs. exploitation, i.e., the influence of the epistemic uncertainty.

$$\begin{aligned} &\mathcal{L}_{inv}(d(\mu_{target}, m(\mu)), m(\sigma_{al}), \sigma_{ep}(\sigma(\sigma_{al}), \sigma(\mu))) = \\ &d(\mu_{target}, m(\mu)) + \alpha \cdot m(\sigma_{al}) - \beta \cdot \sigma_{ep}(\sigma(\sigma_{al}), \sigma(\mu))) \end{aligned} \tag{14}$$

Thus, with our approach, we find inputs that provide robust outputs close to a given target while keeping aleatoric uncertainty low. As a result, we obtain robust optimization outcomes when solving multi-objective inverse problems with our approach. In the work of

Dai Nguyen et al. [25] we found a similar handling of uncertainty in the observation of the acquisition function, however, the authors only focus on maximizing black box functions, while we present an extension that solves multi-objective inverse problems.

4. Case Study on Forging Superalloys

We evaluate the proposed surrogate model and novel optimization method with a classic use case from the field of hot metal forming, preforming an Inconel 625 superalloy billet on an artificially designed forging press. First, we design the forging press characteristics with a parameterized curve and a GP and second, we design the forming process itself in a FEM simulation environment where we provide all the relevant information so that it is possible for researchers to link directly to our work.

4.1. Forging Aggregate Characteristic

We calculate the mean forming velocity values of an artificially designed forging process on the example of a forging screwpress by a self-designed parameterized curve in (15) that models the die velocity v_{die} in mm/s as a function of the process timestep t in *seconds*, clutch pressure x_1 in *bar* and rotation speed of the electric motor x_2 in rpm, such that $v_{die} = v_{die}(x_1, x_2, t)$. Where, x_1 and x_2 are two process-specific setting parameters, i.e. $x_m = (x_1, x_2)$.

$$v_{die}(x_1, x_2, t) = \kappa_1 \cdot x_1 \cdot x_2 \cdot t^2 - \kappa_2 \cdot x_1 \cdot x_2 \cdot t^3 \tag{15}$$

where $\kappa_1 = \frac{5}{3}$ mm^2/kg and $\kappa_2 = \frac{5}{3}$ mm^2/kgs are constants. We utilize a designed forging press specific GP with data generated by using (15) to model the mean and input dependent deviations in respect to the manufacturing process characteristic Z (i.e., Z represents the velocity profile of the forging die v_{die}). Z is defined by a distribution with mean $m(Z)$ and aleatoric standard deviation $\Sigma_{al}(Z)$. With respect to our use case, the forging press specific GP with output $Z^{(j)}$ is at the very beginning of the uncertainty propagation analysis, see Figure 4. The inputs for the forging press GP are $x_m = (x_1, x_2)$ and time increments $t = \{0, \dots, T\}$, where T represents the duration of the manufacturing process. The output of the forging press GP is Z, such that $Z \sim \mathcal{GP}(m(x_m, t), k((x_m, t), (x_m, t)'))$ with mean $m(x_m, t)$ and covariance function $k((x_m, t), (x_m, t)')$. Thus, we obtain for each time increment a distribution describing the velocity at time t. The principle GP design for the forging press can be seen in Figure 5. As covariance function, k we found out that an RBF kernel is appropriate.

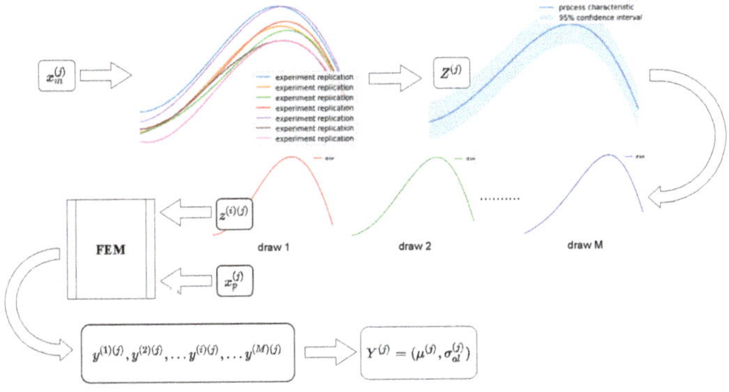

Figure 4. Uncertainty Propagation Analysis: the characteristic $Z^{(j)}$ of the manufacturing process is described by a distribution, since deviations occur when the process is repeated with identical $x_m^{(j)}$. With M draws of $z^{(i)(j)}$ out of the distribution $Z^{(j)}$ as manufacturing process characteristic and to be manufactured part parameters $x_p^{(j)}$, FEM simulations are executed to obtain targets $y^{(i)(j)}$ that describe a distribution $Y^{(j)} = (\mu^{(j)}, \sigma_{al}^{(j)})$ with mean $\mu^{(j)}$ and aleatoric standard deviation, i.e., uncertainty $\sigma_{al}^{(j)}$ for given inputs $x_m^{(j)}$ and $x_p^{(j)}$.

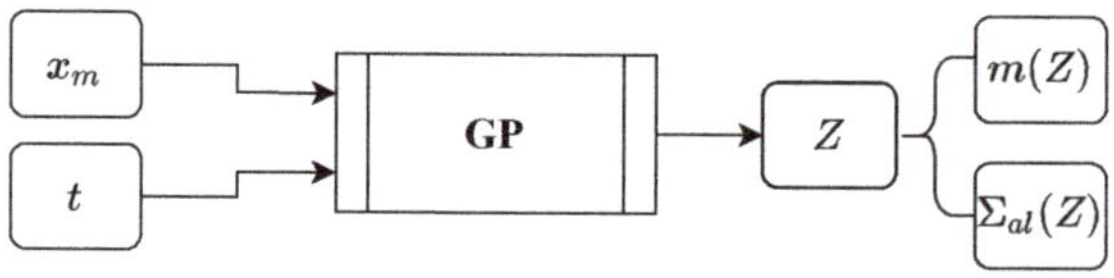

Figure 5. GP takes manufacturing process-specific parameters x_m and manufacturing process time steps t as input and predicts a manufacturing process-specific characteristic (i.e., velocity profile of the forging die) Z with mean $m(Z)$ and uncertainty $\Sigma_{al}(Z)$.

We utilize (15) and different input parameter combinations to generate training data for the forging press GP, see Table 1. In terms of time step size t, we assume that each forging stroke lasts one second, and we model each stroke with a resolution of 100 time steps.

Table 1. Input parameter combinations to generate training data for the forging press GP.

Training Data for Forging Press GP									
x_1	12	12	12	16	16	16	20	20	20
x_2	50	60	70	50	60	70	50	60	70

To obtain different deviations connected to different x_1 and x_2 combinations, we use the underlying inference properties of GP regression and vary inter- and extrapolation tasks in respect to the input values for forging process representation, see Table 2.

Table 2. Input parameter combinations to model forging press characteristics.

Evaluation Data for Forging Press Characteristics																
x_1	10	10	10	10	14	14	14	14	18	18	18	18	22	22	22	22
x_2	45	55	65	75	45	55	65	75	45	55	65	75	45	55	65	75

We define interpolation such that a value is within the training range (i.e., x_1 equals 14 or 18 and x_2 equals 55 or 65) and extrapolation such that a value is out of the training range (i.e., x_1 equals 10 or 22 and x_2 equals 45 or 75).

Exemplary forging press characteristics can be seen in Figure 6, where Figure 6a shows low deviation because x_1 and x_2 are both lie within the range of training data, Figure 6b,c show moderate deviation because one of the process parameters is within and the other is outside the range of the training data and Figure 6d shows the highest deviation because both of the process-parameters lie outside the range of training data. Thus, our forging press GP represents a forging aggregate characteristics with uncertainties dependent on the inputs. In our approach, we intentionally generate deviations depending on input parameters and assume that uncertainty is aleatoric to approximate reality, i.e., we abuse epistemic uncertainty and assume that it is aleatoric. When working with sensor data coming from a real manufacturing process, it is obvious that deviations, i.e., aleatoric process uncertainties Σ_{al}, can be directly measured from data.

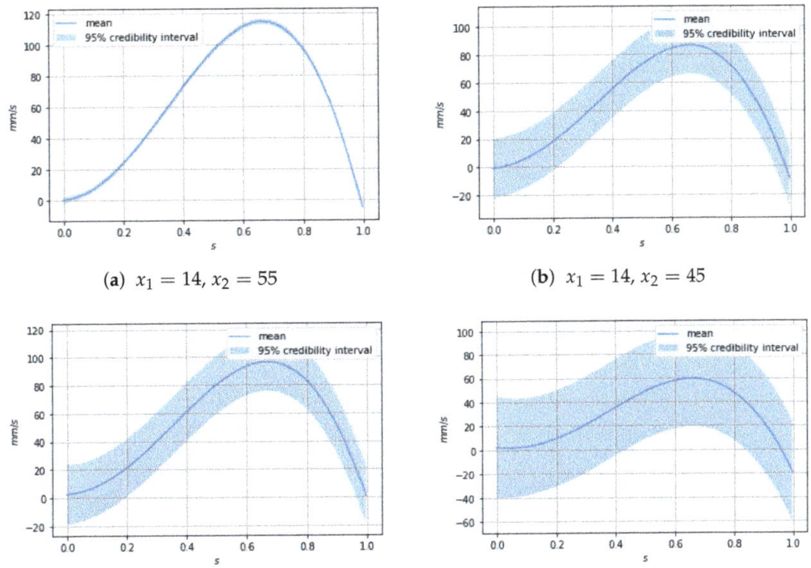

(**a**) $x_1 = 14, x_2 = 55$ (**b**) $x_1 = 14, x_2 = 45$

(**c**) $x_1 = 10, x_2 = 55$ (**d**) $x_1 = 10, x_2 = 45$

Figure 6. Exemplary forging press characteristics $Z^{(j)}$ represented by mean and 95% credibility interval of v_{die} over t with (**a**) low deviation, (**b**,**c**) moderate deviation and (**d**) high deviation.

4.2. FEM Simulation

The considered use case, preforming an Inconel 625 superalloy billet on a forging press machine, is observed by utilizing a corresponding FEM simulation. Manufacturing process related FEM inputs $Z^{(j)}$, i.e., different velocity profiles of the upper die over time, are modeled by the forging press GP. Inputs for the forging press GP are x_1, x_2 and t, such that $Z^{(j)(t)} = Z^{(j)(t)}(x_1^{(j)}, x_2^{(j)}, t)$. All 16 possible combinations for manufacturing process related FEM inputs are shown in Table 2. Billet related inputs $x_p^{(j)}$ that are shared with our proposed surrogate model and FEM simulation are diameter, height and temperature, such that $x_p^{(j)} = (d^{(j)}, h^{(j)}, \theta^{(j)})$. One possible billet configuration is shown in Figure 7 and possible billet parameters for different configurations are shown in Table 3. We define the radius of the rounded edges to be constant 10 mm across all configurations.

Table 3. Key parameters for billet configurations $x_p^{(j)}$, values in mm.

	Configuration Data		
Diameter d	220	240	260
Height h	200	210	220
Temperature θ	900	1000	1100

We observe in total 27 different billets. Connecting manufacturing process related combinations with different billets, we obtain 432 combinations, i.e., $j \in \{1, \ldots, 432\}$. For evaluation of the uncertainty propagation, we randomly draw $z^{(i)(j)}$ with $i \in \{1, \ldots, 20\}$ from each distribution $Z^{(j)}$, i.e., 20 FEM simulations are performed for each input setting. Thus, a total of 8640 FEM simulation results are generated for our experiments. Selected FEM output variables for our surrogate model are the final diameter and height of the preformed billet, such that $y^{(i)(j)} = (d_{final}^{(i)(j)}, h_{final}^{(i)(j)})$ and $Y^{(j)} = (\mu(d_{final}^{(j)}), \mu(h_{final}^{(j)}), \sigma_{al}(d_{final}^{(j)}), \sigma_{al}(h_{final}^{(j)}))$.

In respect to the final diameter d_{final}, we calculate the empiric mean by $\mu(d_{final}^{(j)}) = \frac{1}{20}\sum_{i=1}^{20} d_{final}^{(i)(j)}$ and aleatoric standard deviation by $\sigma_{al}(d_{final}^{(j)})^2 = \frac{1}{20}\sum_{i=1}^{20}(d_{final}^{(i)(j)} - \mu(d_{final}^{(j)}))^2$. The calculations are analogous with respect to h_{final}. Thus, we obtain a dataset with 432 instances described by six input features and four output features. For our running example, input features are clutch pressure, rotation speed, initial billet diameter, initial billet height, initial billet temperature and aggregated manufacturing process uncertainties obtained from data, i.e., the aggregated output of the forging press GP $\bar{\Sigma}_{al}(Z^{(j)}) = \sum_{t=1}^{T} \Sigma_{al}(Z)^{(j)(t)}$. Output features are the mean of the final billet diameter, the mean of the final billet height, the aleatoric uncertainty of the final billet diameter and the aleatoric uncertainty of the final billet height.

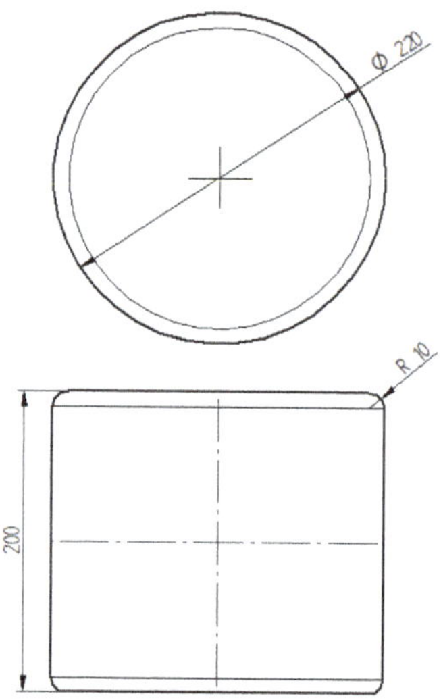

Figure 7. Billet configuration with Diameter $d = 220$ mm, Height $h = 200$ mm and rounded edges with Radius = 10 mm.

The problem is defined as a 2D axisymmetric simulation task to utilize symmetries and make efficient use of computational resources. We utilize isotropic elasto-plastic Inconel 625 material behavior from literature. The Young's modulus is temperature-dependent and the yield stress depends on plastic strain, strain-rate and temperature. We set contact properties to tangential behavior with isotropic directionality and a friction coefficient of 0.3 between the billet and upper and lower forging tool, which means that we assume lubricated hot forging conditions. The lower tool is encastred and the upper tool's boundary conditions are set so that the vertical movement $z^{(i)(j)}$ is drawn from distribution $Z^{(j)}$ and there is no horizontal movement. An exemplary simulation definition can be seen in Figure 8, where (a) shows the initial state of the billet loaded with a randomly drawn screwpress velocity profile $z^{(i)(j)}$ and (b) the end result of the simulation with selected FEM output variables $y^{(i)(j)}$, i.e., the final diameter of 288 mm and the final height of 92.83 mm.

All billets are meshed with an approximate global element size of 7 mm, using 4-node bilinear axisymmetric quadrilateral elements with reduced integration and hourglass control. We obtain our FEM simulation results in the context of general static simulations. Details of the simulation steps are shown in Table 4. Simulation control parameters that are not listed are left at default values.

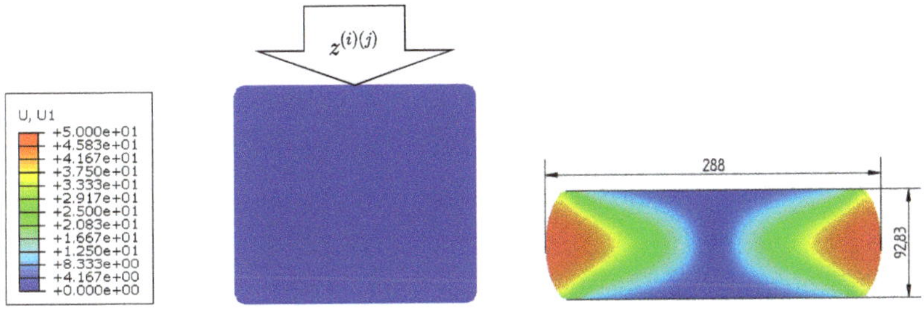

(**a**) initial billet (**b**) preformed billet

Figure 8. Preforming an Inconel 625 superalloy billet: (**a**) initial billet and randomly drawn velocity profile $z^{(i)(j)}$, (**b**) FEM simulation result with graphical presentation of the horizontal displacement $U, U1$ and selected output variables $y^{(i)(j)}$, i.e., final diameter of 288 mm and final height of 92.83 mm.

Table 4. Abaqus FEM simulation control parameters for our use case.

Abaqus FEM Simulation Settings	
Simulation type	Static, General
Time period	1
Nlgeom	On
Max number of increments	1000
Initial increment size	0.001
Min increment size	1×10^{-5}
Max increment size	1
Equation solver method	Direct
Solution technique	Full Newton

5. Results

5.1. GPs

Before utilizing optimization methods, we evaluate each individual GP, see Table 5. The screwpress GP is trained with data that is generated by using inputs from Table 1 with (15) and tested on data generated by using inputs from Table 2 with (15). As covariance function, k we found out that an RBF kernel is appropriate for this GP. The GPs of our proposed surrogate model are both designed with a Matérn kernel with $\nu = 2.5$ and are independently evaluated by 10-fold cross-validation with inputs from Table 2 and $Z^{(j)}$ obtained from the screwpress GP. Outputs are obtained from FEM simulations, see Section 4.2. In each cross-validation step, we split the respective data randomly such that 10 percent are in the test dataset and the remaining 90 percent are used for model training.

Table 5. Evaluation of individual GPs by average R2-Scores over 10 folds.

Individual GP Evaluations					
	Screwpress	Aleatoric Uncertainty		Mean Result	
	$m(Z)$	$m(\sigma_{al,d_{final}})$	$m(\sigma_{al,h_{final}})$	$m(\mu_{d_{final}})$	$m(\mu_{h_{final}})$
R2-Score	0.9923	0.8146	0.8455	0.9586	0.9555

In addition, we calculate mean Pearson kurtosis

$$kurt_{Pearson} = \frac{1}{N}\sum_{j=1}^{N}\frac{1}{M}\sum_{i=1}^{M}\left(\frac{y^{(i)(j)} - \mu^{(j)}}{\sigma_{al}^{(j)}}\right)^4 \quad (16)$$

and mean Fisher-Pearson coefficient of skewness

$$skew_{Fisher-Pearson} = \frac{1}{N}\sum_{j=1}^{N}\frac{1}{M}\sum_{i=1}^{M}\left(\frac{y^{(i)(j)} - \mu^{(j)}}{\sigma_{al}^{(j)}}\right)^3 \tag{17}$$

to describe the distribution shapes obtained from uncertainty propagation analysis, see Table 6.

Table 6. Mean values of Pearson kurtosis and Fisher-Pearson coefficient of skewness calculated from uncertainty propagation analysis results.

Distribution Properties		
	d_{final}	h_{final}
$kurt_{Pearson}$	3.003	2.685
$skew_{Fisher-Pearson}$	0.449	0.015

GP models were implemented with the GPflow library version 2.2.1 and Python 3.8.10. Inferences were run on a machine with 16 GB RAM, 8 CPUs and Intel(R) i7-8565 2.0 GHz processor. We utilized a L-BFGS-B algorithm to train the models. Training our surrogate model on all available data took an average of 1.36 s based on 10 measurements. For one prediction our model needs on average 0.046 s. A FEM simulation lasted on average 149.78 s.

5.2. Active Learning

We evaluate our proposed surrogate model by using active learning and compare it with an approach based on random training data selection. Evaluation is based on 10-fold cross-validation. In each cross-validation step, models are initially trained on two randomly selected datapoints out of the pool dataset containing 432 instances. Evaluation metrics are R2-Score and mean-squared-error (MSE) and are computed on a 20 percent hold-out test set that is randomly generated in each cross-validation step. Results for the mean of reduced-order predictions and corresponding aleatoric uncertainties regarding final diameter and height are shown respectively in Figures 9 and 10, where solid lines depict the mean R2-Score values and shaded areas are obtained by adding and subtracting one standard deviation. Mean values and standard deviations are calculated from the 10 cross-validation results.

5.3. Solving Inverse Problem

We evaluate our proposed multi-objective optimization strategy by solving inverse problems, i.e., we try to find input settings that lead to an output that is as near as possible to an initially defined target vector. In addition to minimize distances between a target vector y_{target} and random mean vector $m(\mu^{(j)})$, we try to achieve results that also keep mean aleatoric uncertainty $m(\sigma_{al}^{(j)})$ low. We utilize a 10-fold cross-validation, where in each cross-validation step the target vector is randomly drawn from the pool dataset and the best found prediction after drawing 50 datapoints out of the pool dataset is used for evaluation. This means that for each method, a dataset of 50 datapoints is generated, and each best prediction is found by evaluating the respective acquisition function on the corresponding generated dataset. We compare our approach with two other baselines, namely:

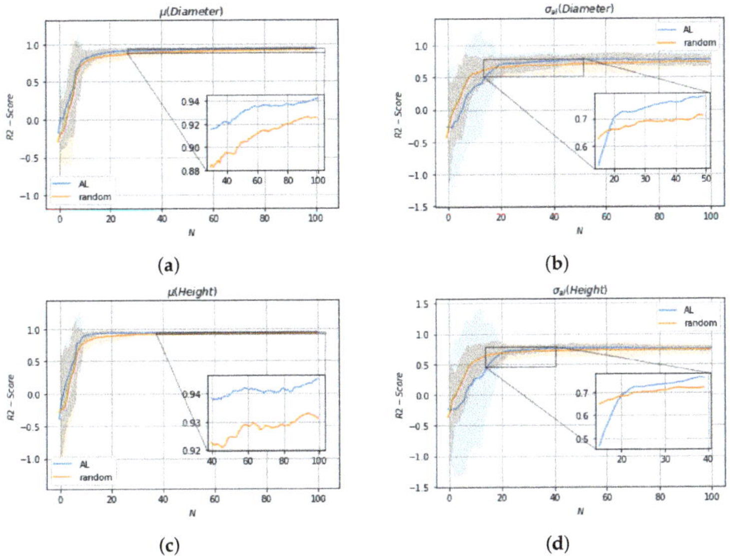

Figure 9. R2-Scores of 10-fold cross-validation over number of drawn training data N. In each cross-validation step, models are initially trained on two randomly selected datapoints drawn from the pool dataset. Solid lines depict the mean R2-Score values and shaded areas the upper and lower confidence bounds obtained by adding and subtracting the standard deviations, calculated from the obtained results. (**a**) $m(\mu_{Diameter})$; (**b**) $m(\sigma_{al,Diameter})$; (**c**) $m(\mu_{Height})$; (**d**) $m(\sigma_{al,Height})$.

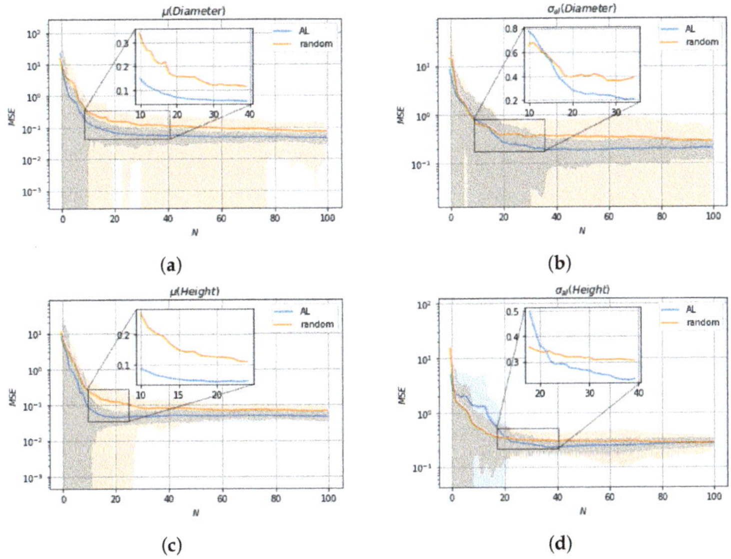

Figure 10. MSEs of 10-fold cross-validation over number of drawn training data N. In each cross-validation step, models are initially trained on two randomly selected datapoints drawn from the pool dataset. Solid lines depict the mean R2-Score values and shaded areas the upper and lower confidence bounds obtained by adding and subtracting the standard deviations, calculated from the 10 obtained results. (**a**) $m(\mu_{Diameter})$; (**b**) $m(\sigma_{al,Diameter})$; (**c**) $m(\mu_{Height})$; (**d**) $m(\sigma_{al,Height})$.

1. Combined (This baseline can be considered as an approximation to the use of heteroskedastic GP in UCB BO.): no distinction of uncertainties in UCB based BO, i.e., simply adding aleatoric and epistemic uncertainty with loss:

$$\mathcal{L}_{combined}(d(\mu_{target}, m(\mu)), m(\sigma_{al}), \sigma_{ep}(\sigma(\sigma_{al}), \sigma(\mu))) = d(\mu_{target}, m(\mu)) - [\alpha \cdot m(\sigma_{al}) + \beta \cdot \sigma_{ep}(\sigma(\sigma_{al}), \sigma(\mu)))] \quad (18)$$

2. Epistemic: neglecting aleatoric uncertainty in UCB based BO with loss:

$$\mathcal{L}_{epistemic}(d(\mu_{target}, m(\mu)), m(\sigma_{al}), \sigma_{ep}(\sigma(\sigma_{al}), \sigma(\mu))) = d(\mu_{target}, m(\mu)) - \beta \cdot \sigma_{ep}(\sigma(\sigma_{al}), \sigma(\mu))). \quad (19)$$

Figure 11 shows representative plots of optimization results for one random target vector (i.e., one cross-validation step) over 50 draws of x_{inv}, where solid lines depict squared errors and dotted lines show mean aleatoric uncertainty $m(\sigma_{al})$. Figures 12 and 13 show different distributions of optimization results obtained by 10-fold cross-validation in respect to squared errors and mean aleatoric uncertainty $m(\sigma_{al})$. Distributions are visualized by kernel density estimation.

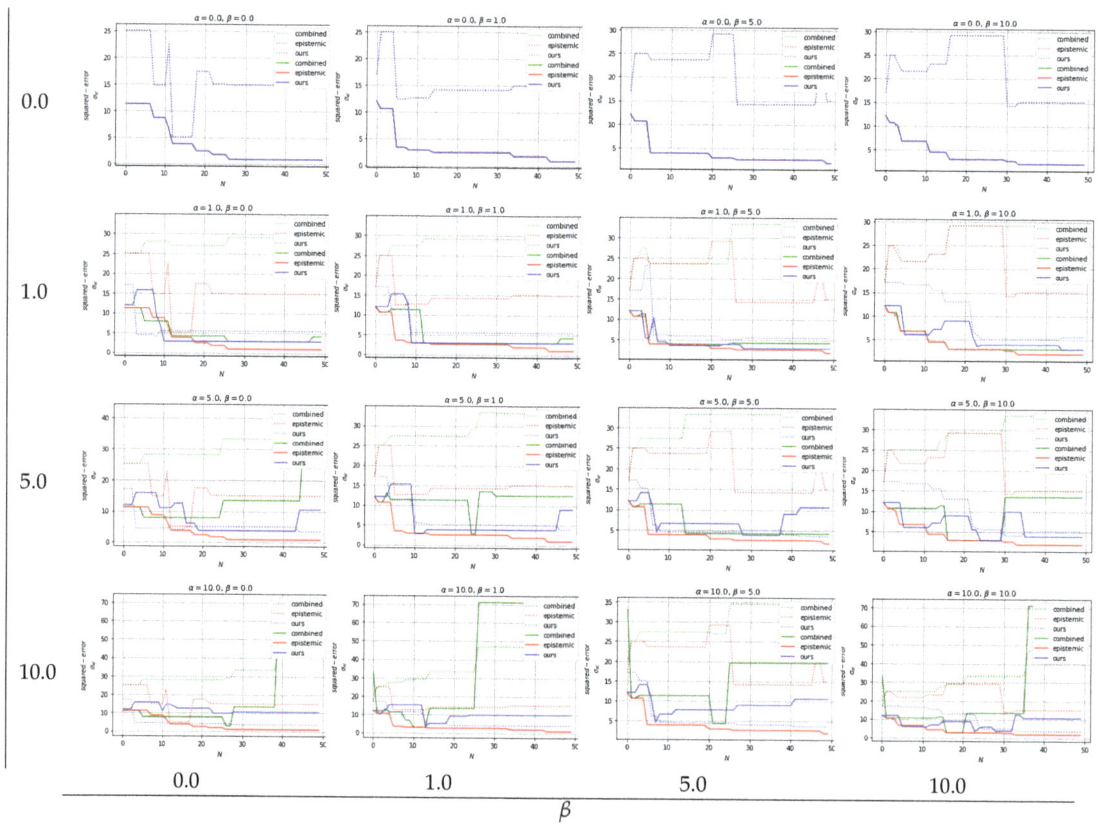

Figure 11. Representative plots of multi-objective optimization results for different hyperparameter settings α and β over number of optimization steps N. Solid lines depict squared error values, and dotted lines represent corresponding mean aleatoric uncertainty $m(\sigma_{al})$. The plots for $\alpha = 0$ show only blue lines, because the results of the different methods are the same and the lines are on top of each other.

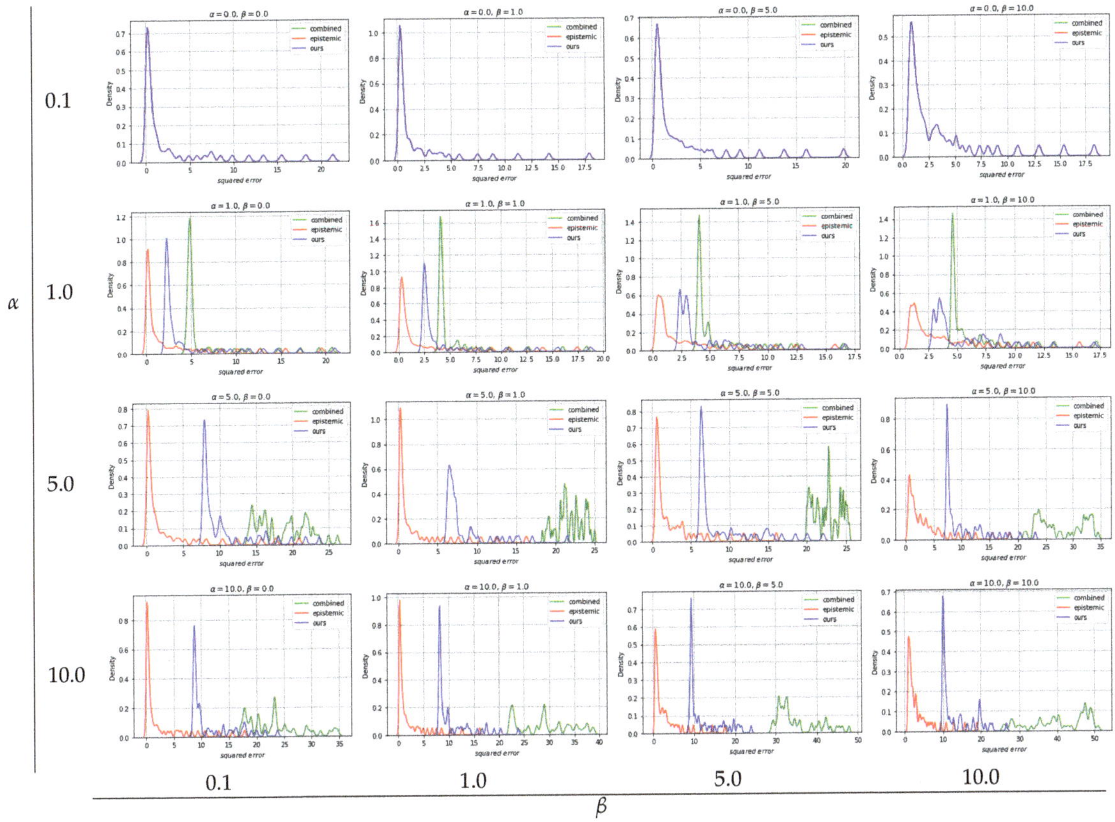

Figure 12. Kernel density estimate plots of squared errors for different hyperparameter settings α and β, distributions are obtained by 10-fold cross-validation, where in each fold a target vector is randomly selected. The plots for $\alpha = 0$ show only blue lines, because the results of the different methods are the same and the lines are on top of each other.

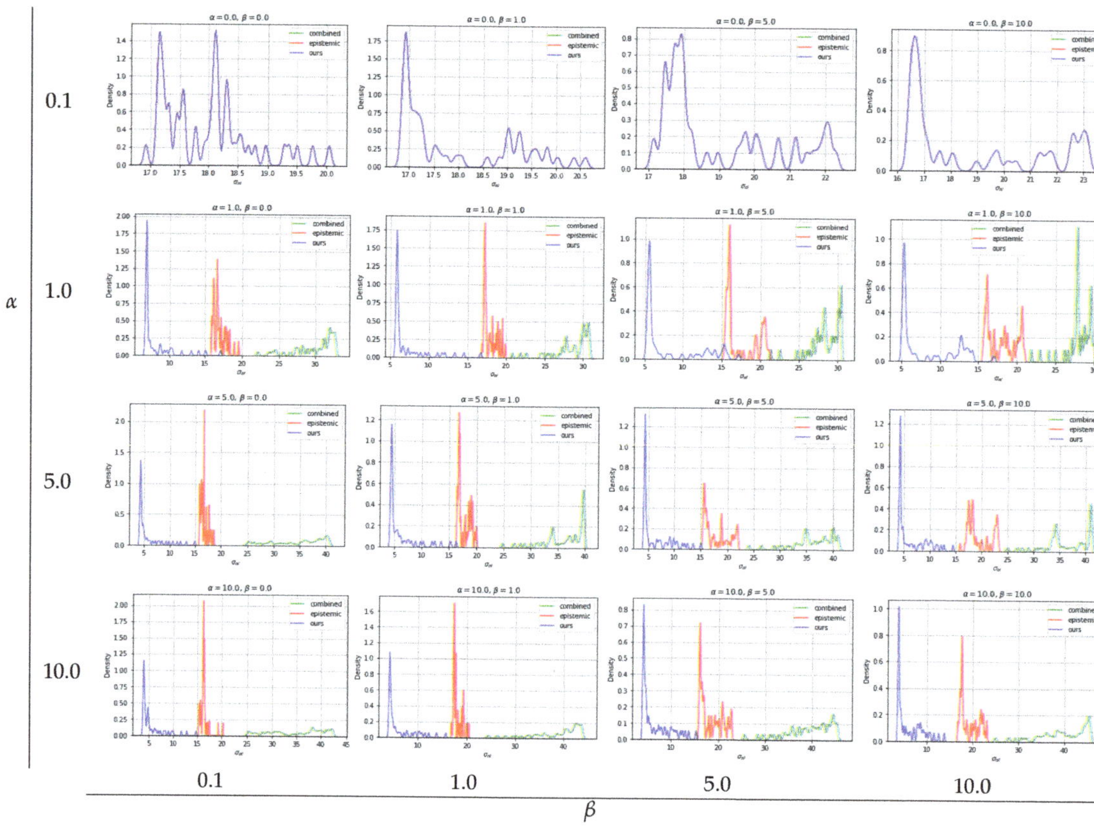

Figure 13. Kernel density estimate plots of mean aleatoric uncertainties $m(\sigma_{al})$ for different hyperparameter settings α and β, distributions are obtained by 10-fold cross-validation, where in each fold a target vector is randomly selected. The plots for $\alpha = 0$ show only blue lines, because the results of the different methods are the same and the lines are on top of each other.

6. Discussion

Evaluation of the individual GPs with 10-fold cross-validation shows promising R2-Scores (lowest: 0.8146, mean: 0.89355, highest: 0.9586), i.e., hyperparameters appear to be appropriate for further evaluations. Observation of generated manufacturing process uncertainties, i.e., $\bar{\Sigma}_{al}(Z^{(j)})$ shows a diverse data landscape, thus, we assume that further uncertainty propagation analysis is meaningful.

We observe the distributions obtained from uncertainty propagation analysis by calculating Pearson kurtosis and Fisher-Pearson coefficient of skewness (A Pearson kurtosis of 3.0 and Fisher-Pearson coefficient of skewness of 0.0 describe a normal distribution). Regarding kurtosis, results shows that distributions are near to Normal distributions, where the distribution of h_{final} is slightly platykurtic ($kurt_{Pearson}(h_{final}) = 2.685 < 3.0$), i.e., it is less peaked than a Normal distribution and the distribution of d_{final} is little leptokurtic ($kurt_{Pearson}(d_{final}) = 3.003 > 3.0$), i.e., the distribution is more peaked compared to a Normal distribution. In terms of skewness, the distribution of d_{final} is more skewed compared to h_{final}, however, both values are less than 0.5 so that approximate symmetry can be assumed.

We evaluate the impact of data selection for model training using two metrics, R2-Score and MSE, with a 10-fold cross-validation comparing active learning with random sample selection. With respect to mean values μ, active learning shows overall an improvement compared to random sample selection. In terms of aleatoric uncertainties σ_{al}, random sample selection is superior to active learning up to the selection of about 20 samples, but after that active learning shows superior performance compared to random sample selection. The initially worse performance of active learning with respect to σ_{al} is due to a trade-off in the active learning cost function between $\sigma(\mu)$ and $\sigma(\sigma_{al})$ with the influence of $\sigma(\mu)$ dominating. A possible solution for this would be the introduction of appropriate tuning parameters that regulate the influence of the respective epistemic uncertainties $\sigma(\mu)$ and $\sigma(\sigma_{al})$. Moreover, it should be noted that random sample selection shows only better performance at a stage where the tuning of parameters is far from complete, so the better performance is not applicable in practice.

With regard to solving inverse problems, we compare our novel robust UCB based BO multi-objective optimization algorithm with two baselines: (1) combined: no distinction of uncertainties in UCB based BO and (2) epistemic: neglecting aleatoric uncertainty in UCB based BO. We show that over different values of tuning parameters α and β there are clear tendencies of the different approaches. By disabling the influence of aleatoric uncertainty ($\alpha = 0$), all three approaches show the same results as expected: low squared errors and neglected aleatoric uncertainty. For all approaches, slight differences can be seen over different β values while $\alpha = 0$, regulating the exploration vs. exploitation trade-off.

Due to the fundamentals of the epistemic approach, there are no differences in the optimization result when α values are changed for constant β values, see Figure 11. Differences in kernel density estimate plots over varying α values are from random target vector selection. Overall, the epistemic approach yields the best optimization results in terms of squared errors, see Figures 11 and 12, however, as expected, aleatoric uncertainty is ignored and thus high, see Figure 13. The combined approach, where aleatoric and epistemic uncertainties are simply added and handled as quasi-epistemic, shows the overall worst results. At low α values, the squared errors are acceptable, but the aleatoric uncertainty is high due to inappropriate handling of information, see Figures 11–13. To arrive at our approach, once aleatoric uncertainty is considered, i.e., $\alpha > 0.0$ results for the inverse problem show low squared errors and low aleatoric uncertainty which we recognize as robust results. Moreover, by increasing α one can see that our approach leads to results where lowering aleatoric uncertainty σ_{al} is more preferred than lowering squared errors, see Figure 11 $\alpha = 1.0$ and $\alpha = 10.0$. Kernel density estimate plots generated from 10-fold cross-validation results confirm those findings, where clear tendencies of optimization results in respect to tuning parameters α and β can be seen. While an approach considering only epistemic uncertainties delivers overall best results in respect to squared errors, aleatoric uncertainties are out of scope, thus, optimization results lead to less robust outcomes. An approach considering aleatoric and epistemic uncertainties combined by summing them up shows overall worst results and can not compete with the remaining. Our approach, where aleatoric and epistemic uncertainties are considered to deliver different information, depicts that overall good results are achieved with respect to squared errors while keeping aleatoric uncertainty low, thus robust solutions for solving multi-objective inverse problems are provided.

Moreover, our model is directly applicable in an industrial framework where the forging press characteristics are represented by measured sensor data of the aggregate (e.g., velocity over time, forging force over time, forging force over the forming path, etc.), which can be used in an appropriately designed FEM simulation for uncertainty propagation analysis and, moreover, for surrogate model training.

7. Conclusions

In this work, we present a GP based reduced-order surrogate model approach with a novel multi-objective target vector optimization strategy to obtain more robust optimization results by concerning aleatoric and epistemic uncertainties. Evaluation on a classic hot metal forming use case, preforming an Inconel 625 forging billet on a self-designed forging press, depicts the advantages of our approach compared to baselines. Our major findings include that our surrogate model produces fast results (over 3000 times faster) compared to FEM simulation, with a calculated loss of accuracy and information. Moreover, active learning can be used directly with our model to make optimal use of computational resources, and solving inverse problems leads to robust optimization results, i.e., finding results close to a defined objective while keeping aleatoric uncertainty low. With our work, we pave one promising way for faster and more realistic simulation and optimization methods.

In future work, we will evaluate our GP based surrogate model and multi-objective optimization strategy on manufacturing process use cases concerning other domains, with real sensor data describing the characteristics of a manufacturing process. Additionally, we will research other Bayesian machine learning and deep learning models as components instead of GP in our surrogate model approach. Moreover, we will experiment with further active learning approaches.

Author Contributions: Conceptualization, J.G.H. and B.C.G.; methodology, J.G.H. and B.C.G.; software, J.G.H.; validation, J.G.H., B.C.G. and R.K.; formal analysis, J.G.H.; investigation, J.G.H., B.C.G. and R.K.; resources, J.G.H. ; data curation, J.G.H.; writing—original draft preparation, J.G.H.; writing—review and editing, J.G.H., B.C.G. and R.K.; visualization, J.G.H.; supervision, B.C.G. and R.K.; project administration, J.G.H., B.C.G. and R.K.; funding acquisition, J.G.H. All authors have read and agreed to the published version of the manuscript.

Funding: This research was funded by Österreichische Forschungsförderungsgesellschaft (FFG) Grant No. 881039 and Open Access Funding by the Graz University of Technology.

Institutional Review Board Statement: Not applicable.

Informed Consent Statement: Not applicable.

Data Availability Statement: Not applicable.

Acknowledgments: The project BrAIN–Brownfield Artificial Intelligence Network for Forging of High Quality Aerospace Components (FFG Grant No. 881039) is funded in the framework of the program 'TAKE OFF', which is a research and technology program of the Austrian Federal Ministry of Transport, Innovation and Technology. The Know-Center is funded within the Austrian COMET Program—Competence Centers for Excellent Technologies—under the auspices of the Austrian Federal Ministry of Transport, Innovation and Technology, the Austrian Federal Ministry of Economy, Family and Youth and by the State of Styria. COMET is managed by the Austrian Research Promotion Agency FFG. We would like to thank our colleagues at voestalpine Böhler Aerospace GmbH for the fruitful discussions.

Conflicts of Interest: The authors declare no conflict of interest.

References

1. Yang, X.S.; Koziel, S.; Leifsson, L. Computational optimization, modelling and simulation: Recent trends and challenges. *Procedia Comput. Sci.* **2013**, *18*, 855–860. [CrossRef]
2. Krige, D.G. A statistical approach to some basic mine valuation problems on the Witwatersrand. *J. S. Afr. Inst. Min. Metall.* **1951**, *52*, 119–139.
3. Makarova, A.; Usmanova, I.; Bogunovic, I.; Krause, A. Risk-averse Heteroscedastic Bayesian Optimization. In Proceedings of the Thirty-Fifth Conference on Neural Information Processing Systems, Virtual, 6 December 2021.
4. Binois, M.; Gramacy, R.B. hetGP: Heteroskedastic Gaussian process modeling and sequential design in R. *J. Stat. Softw.* **2021**, *98*, 1–44. [CrossRef]
5. Tran, T.; Stough, J.V.; Zhang, X.; Haggerty, C.M. Bayesian Optimization of 2D Echocardiography Segmentation. In Proceedings of the 2021 IEEE 18th International Symposium on Biomedical Imaging (ISBI), Nice, France, 13–16 April 2021; pp. 1007–1011.
6. Roberts, S.; Kusiak, J.; Liu, Y.; Forcellese, A.; Withers, P. Prediction of damage evolution in forged aluminium metal matrix composites using a neural network approach. *J. Mater. Process. Technol.* **1998**, *80*, 507–512. [CrossRef]
7. Loghin, A.; Ismonov, S. Augmenting generic fatigue crack growth models using 3D finite element simulations and Gaussian process modeling. In *Pressure Vessels and Piping Conference*; American Society of Mechanical Engineers: New York, NY, USA, 2019; Volume 58936, p. V002T02A004.
8. Ming, W.; Zhang, G.; Li, H.; Guo, J.; Zhang, Z.; Huang, Y.; Chen, Z. A hybrid process model for EDM based on finite-element method and Gaussian process regression. *Int. J. Adv. Manuf. Technol.* **2014**, *74*, 1197–1211. [CrossRef]

9. Su, G.; Peng, L.; Hu, L. A Gaussian process-based dynamic surrogate model for complex engineering structural reliability analysis. *Struct. Saf.* **2017**, *68*, 97–109. [CrossRef]
10. Guo, M.; Hesthaven, J.S. Reduced order modeling for nonlinear structural analysis using Gaussian process regression. *Comput. Methods Appl. Mech. Eng.* **2018**, *341*, 807–826. [CrossRef]
11. Hu, X.; Li, Y.; Zhao, Z.; Liu, C.; Salonitis, K. Residual stresses field estimation based on deformation force data using Gaussian Process Latent Variable Model. *Procedia Manuf.* **2021**, *54*, 279–283. [CrossRef]
12. Yue, X.; Wen, Y.; Hunt, J.H.; Shi, J. Active learning for gaussian process considering uncertainties with application to shape control of composite fuselage. *IEEE Trans. Autom. Sci. Eng.* **2020**, *18*, 36–46. [CrossRef]
13. Ortali, G.; Demo, N.; Rozza, G. Gaussian process approach within a data-driven POD framework for fluid dynamics engineering problems. *arXiv* **2020**, arXiv:2012.01989.
14. Venkatraman, A.; de Oca Zapiain, D.M.; Lim, H.; Kalidindi, S.R. Texture-sensitive prediction of micro-spring performance using Gaussian process models calibrated to finite element simulations. *Mater. Des.* **2021**, *197*, 109198. [CrossRef]
15. Lee, T.; Bilionis, I.; Tepole, A.B. Propagation of uncertainty in the mechanical and biological response of growing tissues using multi-fidelity Gaussian process regression. *Comput. Methods Appl. Mech. Eng.* **2020**, *359*, 112724. [CrossRef] [PubMed]
16. Brevault, L.; Balesdent, M.; Hebbal, A. Overview of Gaussian process based multi-fidelity techniques with variable relationship between fidelities, application to aerospace systems. *Aerosp. Sci. Technol.* **2020**, *107*, 106339. [CrossRef]
17. Civera, M.; Boscato, G.; Fragonara, L.Z. Treed gaussian process for manufacturing imperfection identification of pultruded GFRP thin-walled profile. *Compos. Struct.* **2020**, *254*, 112882. [CrossRef]
18. Abdelfatah, K.; Bao, J.; Terejanu, G. Environmental Modeling Framework using Stacked Gaussian Processes. *arXiv* **2016**, arXiv:1612.02897.
19. Mao, J.; Wang, H.; Li, J. Bayesian finite element model updating of a long-span suspension bridge utilizing hybrid Monte Carlo simulation and kriging predictor. *KSCE J. Civ. Eng.* **2020**, *24*, 569–579. [CrossRef]
20. Tapia, G.; Khairallah, S.; Matthews, M.; King, W.E.; Elwany, A. Gaussian process-based surrogate modeling framework for process planning in laser powder-bed fusion additive manufacturing of 316L stainless steel. *Int. J. Adv. Manuf. Technol.* **2018**, *94*, 3591–3603. [CrossRef]
21. Yu, Z.; Shi, X.; Zhou, J.; Huang, R.; Gou, Y. Advanced prediction of roadway broken rock zone based on a novel hybrid soft computing model using Gaussian process and particle swarm optimization. *Appl. Sci.* **2020**, *10*, 6031. [CrossRef]
22. Lee, S.H. Optimization of cold metal transfer-based wire arc additive manufacturing processes using gaussian process regression. *Metals* **2020**, *10*, 461. [CrossRef]
23. Saul, A.D.; Hensman, J.; Vehtari, A.; Lawrence, N.D. Chained gaussian processes. In Proceedings of the Artificial Intelligence and Statistics, PMLR, Cadiz, Spain, 9–11 May 2016; pp. 1431–1440.
24. Binois, M.; Gramacy, R.B.; Ludkovski, M. Practical heteroscedastic gaussian process modeling for large simulation experiments. *J. Comput. Graph. Stat.* **2018**, *27*, 808–821. [CrossRef]
25. Dai Nguyen, T.; Gupta, S.; Rana, S.; Venkatesh, S. Stable bayesian optimization. In *Pacific-Asia Conference on Knowledge Discovery and Data Mining*; Springer: Cham, Switzerland, 2017; pp. 578–591.
26. Huang, C.; Ren, Y.; McGuinness, E.K.; Losego, M.D.; Lively, R.P.; Joseph, V.R. Bayesian optimization of functional output in inverse problems. *Optim. Eng.* **2021**, *22*, 2553–2574. [CrossRef]
27. Uhrenholt, A.K.; Jensen, B.S. Efficient Bayesian optimization for target vector estimation. In Proceedings of the 22nd International Conference on Artificial Intelligence and Statistics, PMLR, Naha, Japan, 16–18 April 2019; pp. 2661–2670.
28. Snelson, E.; Rasmussen, C.E.; Ghahramani, Z. Warped gaussian processes. *Adv. Neural Inf. Process. Syst.* **2004**, *16*, 337–344.
29. Plock, M.; Burger, S.; Schneider, P.I. Recent advances in Bayesian optimization with applications to parameter reconstruction in optical nano-metrology. *Proc. SPIE* **2021**, *11783*, 117830J.
30. Hüllermeier, E.; Waegeman, W. Aleatoric and epistemic uncertainty in machine learning: An introduction to concepts and methods. *Mach. Learn.* **2021**, *110*, 457–506. [CrossRef]
31. Williams, C.K.; Rasmussen, C.E. *Gaussian Processes for Machine Learning*; MIT Press: Cambridge, MA, USA, 2006; Volume 2.
32. Burbidge, R.; Rowland, J.J.; King, R.D. Active learning for regression based on query by committee. In Proceedings of the International Conference on Intelligent Data Engineering and Automated Learning, Birmingham, UK, 16–19 December 2007; Springer: Berlin/Heidelberg, Germany, 2007; pp. 209–218.
33. Wu, D.; Lin, C.T.; Huang, J. Active learning for regression using greedy sampling. *Inf. Sci.* **2019**, *474*, 90–105. [CrossRef]
34. Meka, R.; Alaeddini, A.; Oyama, S.; Langer, K. An active learning methodology for efficient estimation of expensive noisy black-box functions using Gaussian process regression. *IEEE Access* **2020**, *8*, 111460–111474. [CrossRef]

applied sciences

MDPI

Article

Modeling and Analysis of FBV Movements for Automotive Driveshafts in the PPR Region

Mihai Bugaru * and Ovidiu Vasile

Department of Mechanics, University Politehnica of Bucharest, 060042 București, Romania; ovidiu_vasile2002@yahoo.co.uk
* Correspondence: skmbugaru@yahoo.com; Tel.: +40-749-222333

Featured Application: The asymptotic method approach (AMA) is used to compute the amplitudes of forced bending vibrating (FBV) movements versus the excitation frequency in the region of principal parametric resonance (PPR) for multibody mechanical systems. An FBV movements model was designed for an automotive driveshaft to realize a powerful design tool for FBV movement control of the automotive transmission.

Abstract: This research's goal is to model and analyze the forced bending vibrating (FBV) movements for the elements of an automotive driveshaft using a perturbation technique, the asymptotic method approach (AMA), in the region of principal parametric resonance (PPR). The PPR region was chosen because the principal parametric resonance region is one of the essential resonance regions. The model of FBV movements for the automotive driveshaft (AD) considers the aspects of the following phenomena: geometric nonuniformity of the AD elements and shock excitation due to the road. To overcome the equations for the FBV movements of the AD elements, all inertia characteristics were reduced to the longitudinal ax of the midshaft using the variation of the geometric moments of inertia with the concurrent axis and Stener's theorem. The midshaft of the AD was considered a Timoshenko simply supported beam with a concentrated mass at both ends and springs and dampers for linear and rotational movements at both ends. To determine the equations describing the FBV movements of the AD elements, Hamilton's principle was used. After establishing the equations of motion for each AD element coupled with the specific boundary conditions, the amplitude and the phase angle were computed for stationary and nonstationary motion in the PPR region using the first order of the AMA, and the dynamic instability frontiers were determined based on the same equations. The dynamic behavior of the AD was investigated concerning the variation of the damping ratio and the variation of the parametric excitation coefficient. The AMA coupled with the model of FBV movements for the AD exhibits the future research directions for analyzing FBV movements for the AD in the regions of superharmonic resonances, subharmonic resonances, combination resonances, internal resonances, and simultaneous resonances. Additionally, the AMA can predict the endurance of the AD and design control of car damping systems.

Keywords: multibody dynamics; automotive driveshafts; forced bending vibrations; asymptotic method; principal parametric resonance; dynamic stability frontiers

Citation: Bugaru, M.; Vasile, O. Modeling and Analysis of FBV Movements for Automotive Driveshafts in the PPR Region. *Appl. Sci.* **2022**, *12*, 3237. https://doi.org/10.3390/app12073237

Academic Editors: Jin-Gyun Kim, Jae Hyuk Lim and Peter Persson

Received: 9 March 2022
Accepted: 21 March 2022
Published: 22 March 2022

1. Introduction

The present work presents a perturbation approach, specifically the asymptotic method approach (AMA) [1], to investigate the principal parametric resonance (PPR) for the forced bending vibrating (FBV) movements of the elements of an automotive driveshaft (AD), with the AMA being a powerful tool for the investigation of vibrations induced by shocks, as mentioned by Webber in the literature [2]. To analyze the PPR region, a model for FBV movements of the AD elements was designed, with a specific multibody dynamic structure considering the following aspects: the geometric nonuniformity of the AD elements and the shock excitation induced by the road geometry. The excitations of the FBV movements of the AD elements are due to the shock excitations acting on the automotive wheel, being an excitation induced by the road geometry [3]. The geometric nonuniformities of the AD elements are undoubtedly the leading cause of nonlinear parametric vibrations of the AD in the range of 0.1–15 kHz, as established by experimental data in the literature [4] (pp. 98–123). The first researchers who considered the specific geometry of the AD elements were Mazzei and Scott. They first analyzed the nonlinear dynamic behavior of the AD elements in the PPR region [5]. The lateral vibration of the AD was analyzed by Browne and Palazzolo for superharmonic resonance in [6]. In [7], Xia et al., studied the bending vibrations of a shaft for 4WD (four-wheel drive) drivetrains to design the control methods at low frequencies. Analysis of the parametric bending vibrations for a cardan shaft was performed by Alugongo, and the data are presented in [8]. The NVH phenomena related to the vehicle driveline architecture were investigated by Wellmann et al. in [9] within different frequency ranges. At the same time, in [10], the same authors realized the driveline integration process.

In [11], Yang et al., performed dynamic analysis and vibration tests on the carbon-fiber-reinforced plastics drive-line transmission for machine tools. Yao studied the vibration control of driveshafts considering the Sommerfeld effect on multiple linked shafts, and the data are presented in [12]. Jadhav et al. [13] conducted an extended analysis of the vibrations for a driveshaft with a crack using experimental modal analysis and finite element analysis (FEA) for isotropic materials such as structural steel. The increased interest in achieving a high standard of comfort in the automotive industry led to developing research on dynamic vibration absorbers for the bending vibration of the vehicle propeller shaft [14]. In the same area, Wu et al. investigated the use of photonic crystals in the vibration dampers of the AD [15]. In recent years, composite materials gained advancements in industrial utilization. Therefore, analysis of their dynamic behavior was needed, so Prakash and Sinha estimated the deflection and the natural frequencies of an AD made of such materials using the FEA [16]. The authors of [17] analyzed an AD carrying torque from the principal car differential to the rear wheel differential using the FEA for lateral and torsional vibration, considering the conventional and composite materials. Alam et al. in [18] performed a detailed investigation of the vibration characteristics of the composite AD using the finite element method (FEM). This paper aims to investigate the dynamic behavior of the AD elements for the FBV movements in the PPR region. This investigation supposes the design of a model for FBV movements for an AD's elements, the computation of the stationary and nonstationary amplitude spectrum in the PPR region, and the analysis of the dynamic instability frontiers in the same area. The novelty for this investigation is the design of the model of FBV movements of the AD elements and the use of a perturbation approach (AMA) to compute the amplitude, the phase angle, and the dynamic instability frontiers for both stationary and nonstationary motions. The designed model for FBV movements for an AD's elements contains the phenomena of geometric inertia nonuniformity of the AD and the shock excitation of the AD due to the road geometry. An AD is a multibody mechanical structure mechanism that allows transmitting a moment from the gearbox to the wheel, as shown in Figure 1. The principal components of such an AD, designed for heavy-duty SUVs, are presented in Figure 2 and consist of the bowl (a) fixed with the steering wheel and the midshaft (b) that links both the bowl and the tulip (c).

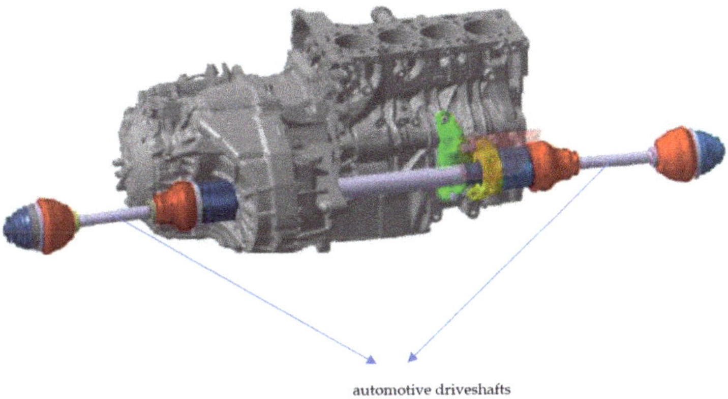

Figure 1. The ADs are mounted in the gearbox.

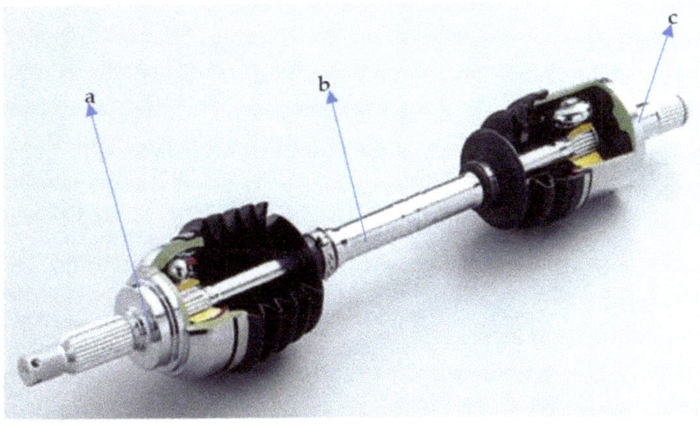

Figure 2. The details of an AD.

The elements of the tulip–tripod joint is presented in Figure 3, where the elements include the midshaft (a), on which the tripod (b) is mounted through the splines and linked with the tulip's bell (c), and the tulip ax (d), which is fixed in the gearbox through the splines.

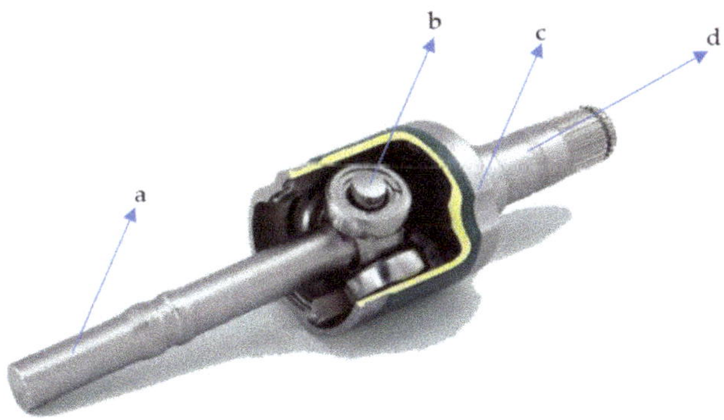

Figure 3. Details of tulip-tripod joint (tripod fixed by splines on the midshaft).

The elements of the bowl–inner race joint can be seen in detail in Figure 4, showing the midshaft (a) on which is mounted the inner race (b) is fixed through splines, linked with the bowl's bell (c), and the bowl ax (d) that is fixed to the steering wheel through the splines.

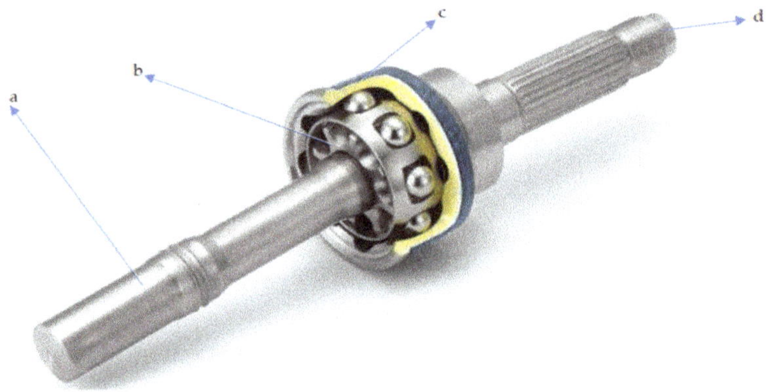

Figure 4. Details of the bowl-inner race (inner race fixed by splines on the midshaft) joint.

2. Computation of the Inertial Characteristics of the AD Elements for the FBV Movements

To use the AMA, it is necessary to compute the equations of the FBV movements for the AD elements, so it must be established from the beginning the movements of each AD element. Therefore, it was considered that the tulip and the bowl had deflections and rotations specific to the rigid body movements. In contrast, the midshaft had continuous deflections and rotations typical to a continuous elastic beam. These assumptions agreed with the technical reality because the rigidity of the tulip and the bowl are more significant compared with the midshaft's rigidity. Each AD element has its referential system, presented in Figure 5, namely the cartesian systems $X_1Y_1Z_1, X_2Y_2Z_2, X_3Y_3Z_3$.

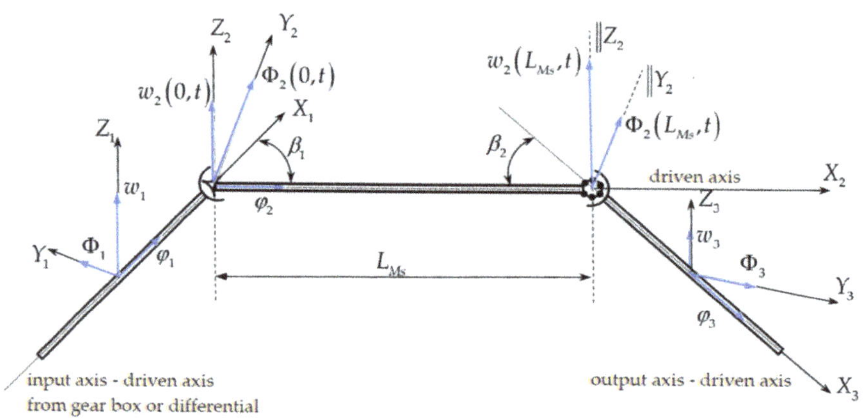

Figure 5. Schematic representation of an automotive driveshaft.

The planes $(X_1Y_1), (X_2Y_2), (X_3Y_3)$ are coplanar planes, so the axes Z_1, Z_2, Z_3 are parallel, and the movements considered for each AD element for the FBV movements are as follows (see Figure 5):

- The rigid torsion rotation angles $\varphi_1, \varphi_2, \varphi_3$ concerning the axes X_1, X_2, X_3 for the tulip, midshaft, and bowl;
- The rigid deflections in bending w_1, w_3 concerning the axes Z_1, Z_3 for the tulip and bowl;
- The rigid twisting angles in bending Φ_1, Φ_3 concerning the axes Y_1, Y_3 for the tulip and bowl;
- The rigid rotation angle β_1 of the tulip axis X_1 concerning the midshaft axis X_2;
- The rigid rotation angle β_2 of the bowl axis X_3 concerning the midshaft axis X_2;
- The elastic deflection in bending $w_2(x,t)$ concerning the midshaft axis Z_2;
- The elastic twisting angle in bending $\Phi_2(x,t)$ concerning the midshaft axis Y_2.

To compute the equations of the FBV movements of the AD elements using Hamilton's principle, it was necessary to reduce the mass and geometric moments of inertia of the tulip and bowl to the cartesian system of reference of the midshaft $X_2Y_2Z_2$ using the variation of the geometric moments of inertia concerning the concurrent axis and parallel axis (Steiner's theorem). The geometric tulip's design (see Figure 3) is composed of the tulip's bell (TB) and tulip ax (TA), while the geometric bowl's design (see Figure 4) is composed of the bowl's bell (BB) and bowl ax (BA). The geometric and mass moments of inertia for the tulip $J_{Y_2T}, J_{Z_2T}, I_{Y_2T}, I_{Z_2T}$ reduced to the Y_2, Z_2 axes of the midshaft are given by the following equations:

$$J_{Y_2T} = J_{Y_2TB} + J_{Y_2TA}, \tag{1}$$

$$J_{Z_2T} = J_{Z_2TB} + J_{Z_2TA}, \tag{2}$$

$$I_{Y_2T} = J_{Y_2TB}\rho L_{TB} + J_{Y_2TA}\rho L_{TA}, \tag{3}$$

$$I_{Z_2T} = J_{Z_2TB}\rho L_{TB} + J_{Z_2TA}\rho L_{TA}, \tag{4}$$

$$J_{Y_2TB} = 0.5(J_{1T} + J_{2T})\left[1 + \sin^2\beta_1 + \chi_{nT}\cos(2\varphi_1)\cos^2\beta_1\right] + \frac{S_T L_{TB}^2}{12}\cos^2\beta_1 + S_T(d_{CT})^2, \tag{5}$$

$$\chi_{nT} = \frac{J_{1T} - J_{2T}}{J_{1T} + J_{2T}}, \tag{6}$$

$$J_{Y_2TA} = \frac{\pi d_{TA}^4}{64}\left(1 + \sin^2\beta_1\right) + \frac{\pi d_{TA}^2}{4}\frac{L_{TA}^2}{12}\cos^2\beta_1 + \frac{\pi d_{TA}^2}{4}(L_{TB} + 0.5L_{TA})^2, \tag{7}$$

$$J_{Z_2TB} = 0.5(J_{1T} + J_{2T})[1 - \chi_{nT}\cos(2\varphi_1)] + \frac{S_T L_{TB}^2}{12} + S_T(d_{CT})^2, \tag{8}$$

$$J_{Z_2TA} = \frac{\pi d_{TA}^4}{64} + \frac{\pi d_{TA}^2}{4}\frac{L_{TA}^2}{12} + \frac{\pi d_{TA}^2}{4}(L_{TB} + 0.5L_{TA})^2, \tag{9}$$

where J_{1T}, J_{2T} are the principal geometric moments of inertia for the tulip's bell, $J_{Y_2TB}, J_{Y_2TA}, J_{Z_2TB}, J_{Z_2TA}$ are the geometric moments of inertia for the tulip's elements reduced to the axes Y_2, Z_2, ρ is the mass density of the AD elements, d_{CT} is the distance between the center mass of the tulip and the tripod's center mass, S_T is the surface of the cross-section for the tulip's bell, χ_{nT} is the nonuniformity of the geometric moments of inertia for the tulip (see Figure 3), L_{TB} is the length of the tulip's bell, L_{TA} is the length of the tulip ax, and d_{TA} is the diameter of the tulip ax. The geometric and mass moments of inertia for the bowl $J_{Y_2B}, J_{Z_2B}, I_{Y_2B}, I_{Z_2B}$ reduced to the Y_2, Z_2 axes of the midshaft are given by the following equations:

$$J_{Y_2B} = J_{Y_2BB} + J_{Y_2BA}, \tag{10}$$

$$J_{Z_2B} = J_{Z_2BB} + J_{Z_2BA}, \tag{11}$$

$$I_{Y_2B} = J_{Y_2BB}\rho L_{BB} + J_{Y_2BA}\rho L_{BA}, \tag{12}$$

$$I_{Z_2B} = J_{Z_2BB}\rho L_{BB} + J_{Z_2BA}\rho L_{BA}, \tag{13}$$

$$J_{Y_2BB} = 0.5(J_{1B} + J_{2B})\left[1 + \sin^2\beta_2 + \chi_{nB}\cos(2\varphi_3)\cos^2\beta_2\right] + \frac{S_B L_{BB}^2}{12}\cos^2\beta_2 + S_B(d_{CB})^2, \quad (14)$$

$$\chi_{nB} = \frac{J_{1B} - J_{2B}}{J_{1B} + J_{2B}}, \quad (15)$$

$$J_{Y_2BA} = \frac{\pi d_{BA}^4}{64}\left(1 + \sin^2\beta_2\right) + \frac{\pi d_{BA}^2}{4}\frac{L_{BA}^2}{12}\cos^2\beta_2 + \frac{\pi d_{BA}^2}{4}(L_{BB} + 0.5L_{BA})^2, \quad (16)$$

$$J_{Z_2BB} = 0.5(J_{1B} + J_{2B})[1 - \chi_{nB}\cos(2\varphi_3)] + \frac{S_B L_{BB}^2}{12} + S_B(d_{CB})^2, \quad (17)$$

$$J_{Z_2BA} = \frac{\pi d_{BA}^4}{64} + \frac{\pi d_{BA}^2}{4}\frac{L_{BA}^2}{12} + \frac{\pi d_{BA}^2}{4}(L_{BB} + 0.5L_{BA})^2, \quad (18)$$

where J_{1B}, J_{2B} are the principal geometric moments of inertia for the bowl's bell, $J_{Y_2BB}, J_{Y_2BA}, J_{Z_2BB}, J_{Z_2BA}$ are the geometric moments of inertia for the bowl's elements reduced to the axes Y_2, Z_2, d_{CB} is the distance between the center mass of the bowl and the inner race's center mass, S_B is the surface of the cross-section for the bowl's bell, χ_{nB} is the nonuniformity of the geometric moments of inertia for the bowl (see Figure 4), L_{BB} is the length of the bowl's bell, L_{BA} is the length of the bowl ax, and d_{BA} is the diameter of the bowl ax. By analyzing Equations (1)–(18), it can be remarked that the geometric and mass moments of inertia of the tulip and the bowl, reduced to the referential of the midshaft, are functions of the angles $\varphi_1, \varphi_3, \Phi_1, \Phi_3, \beta_1, \beta_2$, meaning that they depend on the dynamic behavior of the tulip and the bowl, with this aspect being a novelty in designing dynamic models for the AD elements' movements.

3. The Dynamic Model for FBV Movements of an AD's Elements

The dynamic model for the FBV (DMFFBV) movements of an AD's elements is presented in Figures 6 and 7.

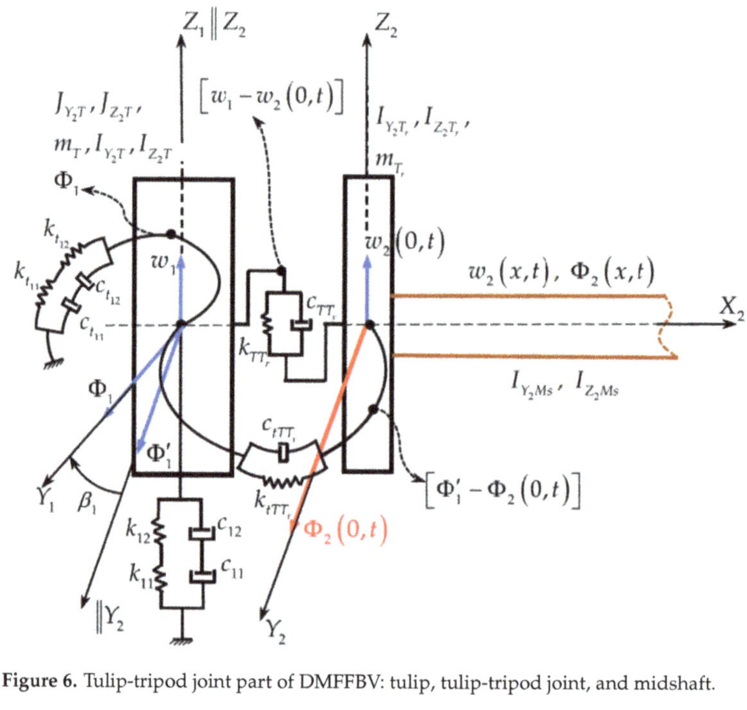

Figure 6. Tulip-tripod joint part of DMFFBV: tulip, tulip-tripod joint, and midshaft.

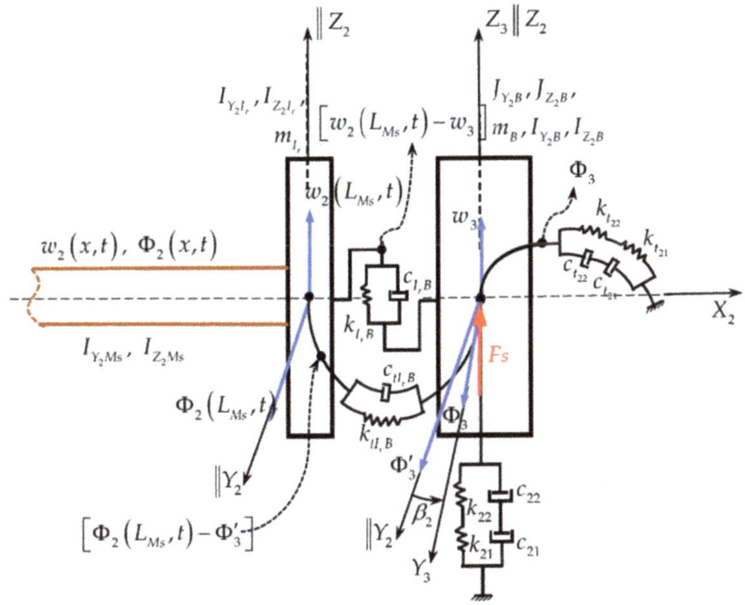

Figure 7. Bowl-inner race joint part of DMFFBV: bowl, bowl-inner race joint, and midshaft.

By surveying the literature already presented in the introduction, it can be remarked that there has not been detailed research performed for the dynamic behavior of each AD element. The dynamic model enhanced by the present work allows the investigation of the dynamic behavior of each component of the AD, the internal mechanisms of excitation transmissions through the AD elements, as well as the causes of producing pitting and micro-cracks in the AD elements, as revealed by the experimental data in the literature [4] (pp. 88–94). The DMFFBV movements, presented in Figures 6 and 7, have three elements: a tulip–midshaft–bowl linked through two joints, the tulip-tripod joint (mounted on the midshaft (see Figure 3)), and the bowl–inner race joint (mounted at the other edge of the midshaft (see Figure 4)). These have the following dynamic characteristics:

1. The tulip has a stiffness k_1, given by the serial springs k_{11}, k_{12} (see Figure 6), and damping c_1, given by the serial dampers c_{11}, c_{12}, for the bending vibration rigid movement of the tulip regarding the axis Z_2 and a stiffness k_{t1}, given by the serial springs $k_{t_{11}}, k_{t_{12}}$, as well as damping c_{t1}, given by the serial dampers $c_{t_{11}}, c_{t_{12}}$, for the angular bending vibration rigid movement of the tulip regarding the axis Y_2, given by the following relations:

$$k_{11} = \frac{3EJ_{Z_2TA}}{L_{TA}^3}, k_{12} = \frac{3EJ_{Z_2TB}}{L_{TB}^3}, k_1 = \frac{k_{11}k_{12}}{k_{11}+k_{12}}, c_1 = \frac{2\Delta}{\sqrt{4\pi^2+\Delta^2}}\sqrt{k_1 m_T}, \quad (19)$$

$$k_{t1} = \frac{GJ_{Y_2TA}}{L_{TA}}, k_{t_{12}} = \frac{GJ_{Y_2TB}}{L_{TB}}, k_{t1} = \frac{k_{t_{11}}k_{t_{12}}}{k_{t_{11}}+k_{t_{12}}}, c_{t1} = \frac{2\Delta}{\sqrt{4\pi^2+\Delta^2}}\sqrt{k_{t1} I_{Y_2} T}, \quad (20)$$

where E is Young's modulus, G is the shearing modulus, m_T is the tulip's mass, and Δ is the logarithmic decrement of the free bending vibrations of the tulip (Δ = 0.001–0.01) [19,20].

2. The uniform midshaft (see Figures 2–7) in continuous FBV movement is assimilated with a uniform Timoshenko beam simply supported at both ends by elastic supports (the tulip-tripod and inner race-bowl joints are elastic supports for the midshaft), having at x = 0 a tripod (see Figure 3) fixed on the midshaft through splines and elastically linked in the tulip-tripod joint with the tulip (see Figure 6) and on the left-hand side at $x = L_{Ms}$ an inner race (see Figure 4) fixed on the midshaft through splines and elastically linked in the

bowl-inner race joint with the bowl (see Figure 7), with the inertial characteristics given by the relations below:

$$J_{Y_2Tr} = 0.5(J_{1Tr} + J_{2Tr})[1 + \chi_{nTr}\cos(2\varphi_2)], J_{Z_2Tr} = 0.5(J_{1Tr} + J_{2Tr})[1 - \chi_{nTr}\cos(2\varphi_2)], \quad (21)$$

$$J_{Y_2Ir} = 0.5(J_{1Ir} + J_{2Ir})[1 + \chi_{nIr}\cos(2\varphi_2)], J_{Z_2Ir} = 0.5(J_{1Ir} + J_{2Ir})[1 - \chi_{nIr}\cos(2\varphi_2)], \quad (22)$$

where J_{1Tr}, J_{2Tr} are the principal geometric moments of inertia for the tripod, J_{1Ir}, J_{2Ir} are the principal geometric moments of inertia for the inner race, χ_{nTr} and χ_{nIr} are the geometric nonuniformities of the tripod and inner race, and $J_{Y_2Tr}, J_{Z_2Tr}, J_{Y_2Ir}, J_{Z_2Ir}$ are the geometric moments of inertia of the tripod and inner race concerning the axes Y_2, Z_2;

3. The bowl has a stiffness k_2, given by the serial springs k_{21}, k_{22} (see Figure 7), and damping c_2, given by the serial dampers c_{21}, c_{22}, for the bending vibration rigid movement of the bowl regarding the axis Z_2 and a stiffness k_{t2}, given by the serial springs $k_{t_{21}}, k_{t_{22}}$, as well as damping c_{t2}, given by the serial dampers $c_{t_{21}}, c_{t_{22}}$, for the angular bending vibration rigid movement of the bowl regarding the axis Y_2, given by the following relations:

$$k_{21} = \frac{3EJ_{Z_2BA}}{L_{BA}^3}, k_{22} = \frac{3EJ_{Z_2BB}}{L_{BB}^3}, k_2 = \frac{k_{21}k_{22}}{k_{21} + k_{22}}, c_2 = \frac{2\Delta}{\sqrt{4\pi^2 + \Delta^2}}\sqrt{k_2 m_B}, \quad (23)$$

$$k_{t_{21}} = \frac{GJ_{Y_2BA}}{L_{BA}}, k_{t_{22}} = \frac{GJ_{Y_2BB}}{L_{BB}}, k_{t2} = \frac{k_{t_{21}}k_{t_{22}}}{k_{t_{21}} + k_{t_{22}}}, c_{t2} = \frac{2\Delta}{\sqrt{4\pi^2 + \Delta^2}}\sqrt{k_{t2} I_{Y_2}B}, \quad (24)$$

where m_B is the bowl's mass.

4. The tulip–tripod joint in FBV movement (see Figures 3 and 6) realizes the elastic link between the tulip and the midshaft through the stiffness k_{TTr} and the damping c_{TTr} for the bending vibrating movements concerning the Z_2 axis and the angular stiffness k_{tTTr} and angular damping c_{tTTr} for the bending vibrating movements concerning Y_2 axis.

5. The bowl–inner race joint in FBV movement (see Figures 4 and 7) realizes the link between the bowl and the midshaft through the stiffness k_{IrB} and the damping c_{IrB} for the bending vibrating movements concerning the Z_2 axis and the angular stiffness k_{tIrB} and the angular damping c_{tIrB} for the bending vibrating movements concerning the Y_2 axis. The wheel induces excitations as a moderate impulsive shock force F_s acting in the Z_2 direction, and the excitation load can be expressed as

$$F_S = \overline{F}_S\left[1 + q_3 t^{q_1} e^{-q_2 t}\right], \quad (25)$$

where $\overline{F}_S$ is the amplitude of the shock on the bowl's longitudinal axis X_3 transmitted from the wheel axis and $q_i, i = \overline{1,3}$, are experimental constants, depending on the type of shock applied at the wheel by the road excitation [4] (pp. 142–172). All the computations were performed considering that the tulip was a cantilever beam fixed in the gearbox with simple elastic supports in the tulip–tripod joint. The bowl was a cantilever beam fixed in the steering wheel by simple elastic supports in the bowl-inner race joint. As mentioned in the literature [4] (pp. 142–172), the shock excitation loads produce huge automotive stress solicitations in the car suspension system and the rim-tire system. These two systems can absorb 90% of the shock energy. Therefore, only 10% of the shock acts on the AD elements as a variation of the quantity of movement during a very short time, estimated at 0.001 s. Because of this phenomenon, the shock is distributed to each AD element, inducing for the bowl stress in the region of the splines, for the midshaft the initial velocity of the "ends", and for the tulip stress in the region of the splines (see Figure 2). Due to this mechanism of excitation, it will be considered that the stress in the tulip's ax and bowl's ax imposes the geometric capability dimensions of the geometric elements d_{TA}, d_{BA}. At the same time, the midshaft dynamic behavior will be an excitation element, like the internal tuned damper (ITD) mentioned in the literature [6], for the tulip and the bowl. The novelty of this DMFFBV of an AD's elements, as can be noted from Equations (19)–(25), consists of linking the rigid movements of the bowl and the tulip with the continuous movement of the midshaft. In addition, another novelty is that the dynamic characteristics concerning stiffness and damping are expressed for the bending displacements and the bending rotation displacements for each AD element, and the serial stiffness and damping of the bowl and tulip are considered, keeping in mind that each of these elements has two distinctive parts: the ax and "bell". In the meantime, the stiffness and damping for the tulip-tripod and bowl-inner race joints were considered for the relative bending displacements and the relative bending rotation displacements of the movements between the bell of the tulip and the tripod, as well as for the inner race and the bell of the bowl.

4. The Equations of FBV Movements of AD Elements Induced by Shock Road Excitations

For the DMFFBV movements of AD elements presented in Figures 6 and 7, using the Hamilton's principle [21] (pp. 265–269) yields

$$\delta \int_{P_1 P_2} L(q_1, \ldots, q_n, \dot{q}_1, \ldots, \dot{q}_n, t)dt = 0, \tag{26}$$

where Lagrange's function $L(q_1, \ldots, q_n, \dot{q}_1, \ldots, \dot{q}_n, t)$ depends on the generalized coordinates $q_1, \ldots, q_n$ and the generalized velocities $\dot{q}_1, \ldots, \dot{q}_n$, while P_1, P_2 are two points in the spatial configurations $\overrightarrow{\left(\dot{q}_1, \ldots, \dot{q}_n\right)} = \vec{\Xi}(q_1, \ldots, q_n)$. Lagrange's function is given by the following equation:

$$L = T + \prod, \tag{27}$$

where the potential energy $\prod$ for the DMFFBV movements of AD elements (see Figures 6 and 7) is given by the following generalized equation [22] (pp. 371–376) [23] (pp. 734–739):

$$\begin{gathered}\prod = \frac{1}{2}\int_0^{L_{Ms}}\left[EJ_{Y_2Ms}\left(\frac{\partial\Phi_2}{\partial x}\right)^2 + kAG\left(\frac{\partial w_2}{\partial x} - \Phi_2\right)^2\right]^2 dx + \frac{1}{2}\left[c_1\left(\dot{w}_1\right)^2 + c_2\left(\dot{w}_3\right)^2 + c_{t1}\left(\dot{\Phi}_1\right)^2 + c_{t2}\left(\dot{\Phi}_3\right)^2\right] + \\ \frac{1}{2}\left[c_{TTr}\left(\dot{w}_1 - \frac{\partial w_2}{\partial t}(0,t)\right)^2 + c_{tTTr}\left(\dot{\Phi}_1\cos\beta_1 - \frac{\partial\Phi_2}{\partial t}(0,t)\right)^2 + c_{IrB}\left(\frac{\partial w_2}{\partial t}(L_{Ms},t) - \dot{w}_3\right)^2 + c_{tIrB}\left(\frac{\partial\Phi_2}{\partial t}(L_{Ms},t) - \dot{\Phi}_3\cos\beta_2\right)^2\right] + \\ + \frac{1}{2}\left[k_1 w_1^2 + k_2 w_3^2 + k_{t1}\Phi_1^2 + k_{t2}\Phi_3^2\right] + \\ + \frac{1}{2}\left[k_{TTr}(w_1 - w_2(0,t))^2 + k_{tTTr}(\Phi_1\cos\beta_1 - \Phi_2(0,t))^2 + k_{IrB}(w_2(L_{Ms},t) - w_3)^2 + k_{tIrB}(\Phi_2(L_{Ms},t) - \Phi_2\cos\beta_2)^2\right],\end{gathered} \tag{28}$$

where A is the cross-section area of the midshaft, $w_2(x,t)$ is the bending deflection (including the shear deformation) of the midshaft concerning the Z_2 axis, Φ_2 (x, t) is the rotation of the cross-section of the midshaft, and concerning the Y_2 axis, due only to the pure bending deflection, k is the shear correction factor, which in the literature [24] is in the range of 0.64–0.846, L_{Ms} is the length of the midshaft, and J_{Y_2Ms} is the geometric moment of inertia of the midshaft concerning the Y_2 direction given by the following equations:

$$J_{Y_2Ms} = \frac{\pi d_{Ms}^4}{64}, \tag{29}$$

$$J_{Y_2Ms} = \frac{\pi\left(d_{eMs}^4 - d_{iMs}^4\right)}{64}, \tag{30}$$

where the midshaft is considered to have a circular or tubular uniform cross-section with a diameter d_{Ms} for the circular cross-section or the diameters d_{eMs}, d_{iMs} for the tubular cross-section. In Equation (28), the generalized Rayleigh's dissipation function [23] (p. 611) specific to the mathematical formulations of the Euler–Lagrange generalized approach was added due to the presence of dampers in the DMFFBV movements, given by the following equation:

$$\Lambda = \frac{1}{2}\left[c_1\left(\dot{w}_1\right)^2 + c_2\left(\dot{w}_3\right)^2 + c_{t1}\left(\dot{\Phi}_1\right)^2 + c_{t2}\left(\dot{\Phi}_3\right)^2 + c_{TTr}\left(\dot{w}_1 - \frac{\partial w_2}{\partial t}(0,t)\right)^2 + c_{tTTr}\left(\dot{\Phi}_1\cos\beta_1 - \frac{\partial \Phi_2}{\partial t}(0,t)\right)^2\right] + \\ +\frac{1}{2}\left[c_{IrB}\left(\frac{\partial w_2}{\partial t}(L_{Ms},t) - \dot{w}_3\right)^2 + c_{tIrB}\left(\frac{\partial \Phi_2}{\partial t}(L_{Ms},t) - \dot{\Phi}_3\cos\beta_2\right)^2\right]. \quad (31)$$

The kinetic energy of the DMFFBV movements (see Figures 6 and 7) for an AD's elements is given by the following generalized equation [22] (p. 374) [23] (p. 721):

$$T = \frac{1}{2}m_T\dot{w}_1^2 + \frac{1}{2}m_{Tr}\left(\frac{\partial w_2}{\partial t}(0,t)\right)^2 + \frac{1}{2}m_B\dot{w}_3^2 + \frac{1}{2}m_{Ir}\left(\frac{\partial w_2}{\partial t}(L_{Ms},t)\right)^2 + \frac{1}{2}I_{Y_2T}\dot{\Phi}_1^2 + \\ +\frac{1}{2}I_{Y_2Tr}\left(\frac{\partial \Phi_2}{\partial t}(0,t)\right)^2 + \frac{1}{2}I_{Y_2B}\dot{\Phi}_3^2 + \frac{1}{2}I_{Y_2Ir}\left(\frac{\partial \Phi_2}{\partial t}(L_{Ms},t)\right)^2 + \int_0^{L_{Ms}}\frac{1}{2}\left[\rho A\left(\frac{\partial w_2}{\partial t}\right)^2 + I_{Y_2Ms}\left(\frac{\partial \Phi_2}{\partial t}\right)^2\right]dx, \quad (32)$$

where m_{Tr} is the mass of the tripod and m_{Ir} is the mass of the inner race (see Figures 6 and 7). Several mathematical manipulations that include integration by parts of the nonlinear system of equations with partial derivatives of the second degree in the unknowns $w_1(t)$, $\Phi_1(t)$, $w_2(x,t)$, $\Phi_2(x,t)$, and $w_3(t)$, $\Phi_3(t)$, yield

$$m_T\ddot{w}_1 + c_1\dot{w}_1 + c_{TTr}\left(\dot{w}_1 - \frac{\partial w_2}{\partial t}(0,t)\right) + k_1 w_1 + k_{TTr}(w_1 - w_2(0,t)) = 0, \quad (33)$$

$$I_{Y_2T}\ddot{\Phi}_1 + c_{t1}\dot{\Phi}_1 + c_{tTTr}\left(\dot{\Phi}_1\cos\beta_1 - \frac{\partial \Phi_2}{\partial t}(0,t)\right) + k_{t1}\Phi_1 + k_{tTTr}(\Phi_1\cos\beta_1 - \Phi_2(0,t)) = 0, \quad (34)$$

$$\rho A\frac{\partial^2 w_2}{\partial t^2} - kAG\left[\frac{\partial^2 w_2}{\partial x^2} - \frac{\partial \Phi_2}{\partial x}\right] = 0, \quad (35)$$

$$I_{Y_2Ms}\frac{\partial^2 \Phi_2}{\partial t^2} - EJ_{Y_2Ms}\frac{\partial^2 \Phi_2}{\partial x^2} - kAG\left(\frac{\partial w_2}{\partial x} - \Phi_2\right) = 0, \quad (36)$$

$$m_B\ddot{w}_3 + c_2\dot{w}_3 + c_{IrB}\left(\dot{w}_3 - \frac{\partial w_2}{\partial t}(L_{Ms},t)\right) + k_2 w_3 + k_{IrB}(w_3 - w_2(L_{Ms},t)) = 0, \quad (37)$$

$$I_{Y_2B}\ddot{\Phi}_3 + c_{t1}\dot{\Phi}_1 + c_{tIrB}\left(\dot{\Phi}_3\cos\beta_2 - \frac{\partial \Phi_2}{\partial t}(L_{Ms},t)\right) + k_{t2}\Phi_3 + k_{tIrB}(\Phi_3\cos\beta_2 - \Phi_2(L_{Ms},t)) = 0, \quad (38)$$

where the boundary conditions are

$$w_2(0,t) = 0, \frac{\partial \Phi_2}{\partial x}(0,t) = 0, \text{ at } x = 0, \quad (39)$$

$$w_2(L_{Ms},t) = 0, \frac{\partial \Phi_2}{\partial x}(L_{Ms},t) = 0, \text{ at } x = L_{Ms}, \quad (40)$$

$$c_{TTr}\dot{w}_1 + k_{TTr}w_1 - kAG\left(\frac{\partial w_2}{\partial x}(0,t) - \Phi_2(0,t)\right) = 0, \text{ at } x = 0, \quad (41)$$

$$c_{IrB}\dot{w}_3 + k_{IrB}w_3 - kAG\left(\frac{\partial w_2}{\partial x}(L_{Ms},t) - \Phi_2(L_{Ms},t)\right) = 0, \text{ at } x = L_{Ms}, \quad (42)$$

$$I_{Y_2Tr}\frac{\partial^2 \Phi_2}{\partial t^2}(0,t) - c_{tTTr}\left(\dot{\Phi}_1\cos\beta_1 - \frac{\partial \Phi_2}{\partial t}(0,t)\right) - k_{tTTr}(\Phi_1\cos\beta_1 - \Phi_2(0,t)) = 0, \text{ at } x = 0, \quad (43)$$

$$I_{Y_2Ir}\frac{\partial^2 \Phi_2}{\partial t^2}(L_{Ms},t) - c_{tIrB}\left(\dot{\Phi}_3\cos\beta_2 - \frac{\partial \Phi_2}{\partial t}(L_{Ms},t)\right) - k_{tTTr}(\Phi_3\cos\beta_2 - \Phi_2(L_{Ms},t)) = 0, \text{ at } x = L_{Ms}, \quad (44)$$

Taking into account that the tulip was fixed in the gearbox (see Figure 6 on the left side) and had simple elastic supports in the tulip-tripod joint (see Figure 6), the mid-shaft was a uniform Timoshenko beam simply supported by elastic at both ends (see Figures 6 and 7) in both the tulip-tripod (left side) and inner race-bowl joints (right side), while the bowl was fixed (see Figure 7, right side) in the steering wheel and had simple elastic support in the inner race-bowl joint (see Figure 7). The system is given by the Equations (33)–(38) together with the boundary conditions from Equations (39)–(44) to represent the nonlinear dynamic behavior of the AD elements in FBV induced by shock force through the wheel by road excitations. By analyzing Equations (35) and (36), it can be remarked that they represent a system of equations with partial derivatives for the bending-shearing vibrations of a uniform shaft that considers the effects of rotary inertia and shear deformation, with the midshaft being a Timoshenko beam. The boundary conditions are given by Equations (39)–(44) and link the bending-shearing vibrations of the midshaft with the tulip and the bowl through the tulip-tripod and bowl-inner race joints, inducing in the solutions of the system of Equations (33)–(38) the next phenomena: the joints of the driveshaft are quasi-isometric [25–27] (p. 78), with the effect of geometric nonuniformity of the inertia characteristics of the joints that vary with the rigid angle of rotation for each element of the driveshaft in the directions X_1, X_2, X_3, the effects of the bending deflection and bending-twisting stiffness for the tulip and the bowl, the effects of the bending deflection and bending-twisting damping for each joint of the driveshaft, the rotary inertia effect in bending, and the shearing effect for the midshaft. The novelty of the Hamilton's Principle approach was that it was used for the first time to compute the equations of motion for each AD element and the determination of the boundary conditions, as is noted in Equations (26)–(44).

5. The Analytical Solutions

The starting point to solve the system differential equations of the FBV movements (SDEOFBVM) (Equations (33)–(38)) for the AD elements (ADEs) was to analyze the vibration mechanism of the midshaft as a Timoshenko beam simply supported at the ends (see Figures 8 and 9). For the midshaft element of the AD, it was considered that $f(x,t) = 0$.

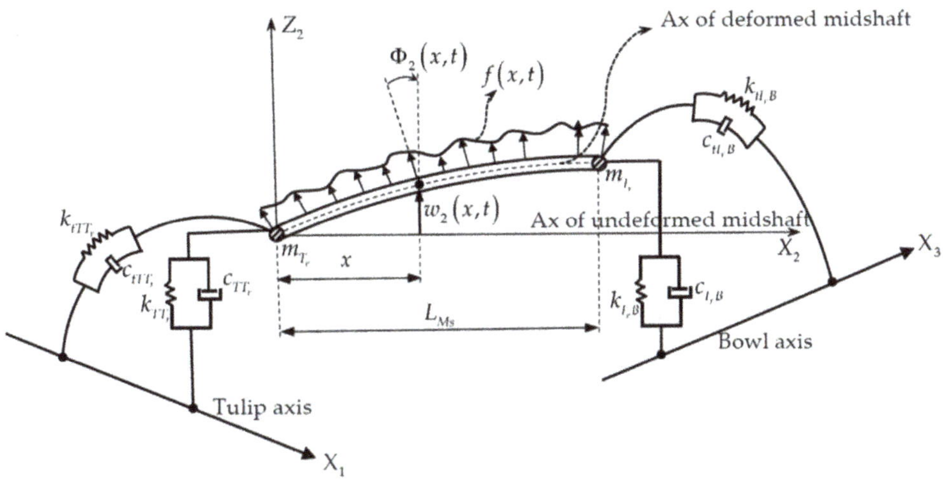

Figure 8. The part of the DMFTDV for the midshaft.

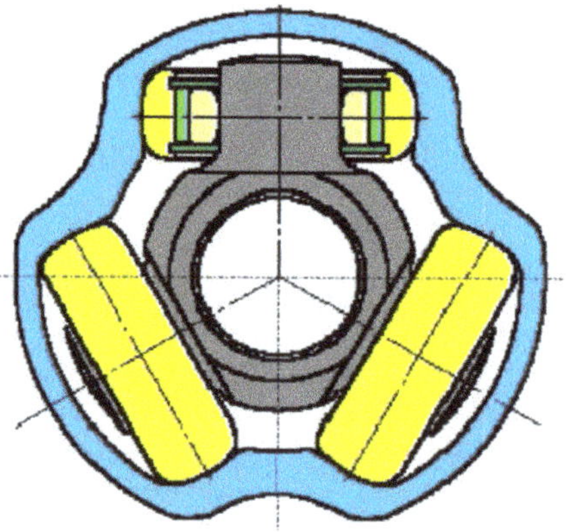

Figure 9. Cross-section of the tulip-tripod joint.

In analyzing the tulip-tripod-midshaft and bowl-inner race-midshaft joints, it became evident that the midshaft was simply supported at both ends where the concentrated mass of the tripod and the inner race m_{Tr}, m_{Ir} were placed, being a uniform shaft linked in bending-shearing vibration with the tulip for x = 0 and with the bowl for $x = L_{Ms}$. Therefore, the general solutions of Equations (35) and (36) that satisfy the boundary conditions of Equations (39) and (40) expressed in normalized bending deflection are [23] (pp. 326–328)

$$w_2(\bar{x},t) = \sin(n\pi\bar{x})[C_{1n}\cosh(\omega_{n1}t) + C_{2n}\sinh(\omega_{n1}t) + C_{3n}\cosh(\omega_{n2}t) + C_{4n}\sinh(\omega_{n2}t)], \bar{x} = \frac{x}{L_{Ms}}, n = 1,2,3,\ldots \quad (45)$$

$$\Phi_2(\bar{x},t) = n\pi\cos(n\pi\bar{x})\left[\left(1+\frac{\rho\omega_{n1}^2}{kG\left(\frac{n\pi}{L_{Ms}}\right)^2}\right)[C_{1n}\cosh(\omega_{n1}t) + C_{2n}\sinh(\omega_{n1}t)] + \left(1+\frac{\rho\omega_{n2}^2}{kG\left(\frac{n\pi}{L_{Ms}}\right)^2}\right)[C_{3n}\cosh(\omega_{n2}t) + C_{4n}\sinh(\omega_{n2}t)]\right], \bar{x} = \frac{x}{L_{Ms}}, n = 1,2,3,\ldots \quad (46)$$

where ω_{n1}, ω_{n2} are the natural frequencies of the Timoshenko beam simply supported at both ends [23] (p. 378), the smaller value corresponds to the bending deformation mode, and the bigger value corresponds to the shear deformation mode, given by the following equation:

$$\omega_{n1,2} = \sqrt{\frac{1+\frac{n^2\pi^2 J_{Y2Ms}}{AL_{Ms}^2}[1+E/kG] \mp \sqrt{\left[1+\frac{n^2\pi^2 J_{Y2Ms}}{AL_{Ms}^2}[1+E/kG]\right]^2 - 4\frac{EJ_{Y2Ms}^2}{kA^2G}\frac{n^4\pi^4}{L_{Ms}^4}}}{2\frac{\rho J_{Y2Ms}}{kAG}}}, n = 1,2,3,\ldots \quad (47)$$

Additionally, the constants $C_{1n}, C_{2n}, C_{3n}, C_{4n}$ are determined by the initial conditions at t = 0. In our case, the constants are given by the equation [4] (pp. 142–172)

$$C_{1n} = 0, C_{2n} \simeq 0.1\frac{\overline{F}_S\Delta t_s}{\omega_{n1}L_{Ms}M}, C_{3n} = 0, C_{4n} = 0, \quad (48)$$

where Δt_s is the time interval for shock excitations due to the road geometry and M is the car's mass. Injecting the solutions given by Equations (45) and (46) in the boundary conditions of Equations (41) and (42) yields the following boundary conditions for the tulip and the bowl in bending:

$$c_{TTr}\dot{w}_1 + k_{TTr}w_1 = -\frac{\rho AL_{Ms}^2}{n\pi}\omega_{n1}^2 C_{2n}\sinh(\omega_{n1}t), n = 1,2,3,\ldots \quad (49)$$

$$c_{IrB}\dot{w}_3 + k_{IrB}w_3 = (-1)^n\frac{\rho AL_{Ms}^2}{n\pi}\omega_{n1}^2 C_{2n}\sinh(\omega_{n1}t), n = 1,2,3,\ldots \quad (50)$$

Equations (49) and (50) can now be injected in Equations (33) and (37) for the DMFFBV, thus yielding DMFFBV equations for the tulip and bowl:

$$m_T\ddot{w}_1 + c_1\dot{w}_1 + +k_1 w_1 = -\frac{\rho A L_{Ms}^2}{n\pi}\omega_{n1}^2 C_{2n}\sinh(\omega_{n1}t), n = 1,2,3,\ldots \tag{51}$$

$$m_B\ddot{w}_3 + c_2\dot{w}_3 + +k_2 w_3 = (-1)^n\frac{\rho A L_{Ms}^2}{n\pi}\omega_{n1}^2 C_{2n}\sinh(\omega_{n1}t), n = 1,2,3,\ldots \tag{52}$$

Taking into account that Browne and Palazzolo in [6] mentioned that excitation terms, as presented on the right side of Equations (51) and (52) could be expressed as cubic internal tuned dampers (ITDs), the forms of Equations (51) and (52) can be modified by dividing Equations (51) and (52) into the mass of the tulip m_T and the mass of the bowl m_B, respectively, expressing the stiffness and damping coefficients as functions of the geometrical characteristics given by Equations (19) and (23) coupled with the expressions of inertial characteristics provided by Equations (8), (9), (17), and (18). Equations (51) and (52) become the normalized differential equations in the time functions $w_1(t), w_3(t)$:

$$\ddot{w}_1 + 2\xi\Omega_1\sqrt{\frac{1-C_1\cos(2\varphi_1)}{1-C_2\cos(2\varphi_1)}}\dot{w}_1 + \Omega_1^2\frac{1-C_1\cos(2\varphi_1)}{1-C_2\cos(2\varphi_1)}w_1 = -\Gamma_0 w_1 - \Gamma_1 w_1^3 - \Gamma_2 w_1^5, \tag{53}$$

$$\ddot{w}_3 + 2\xi\Omega_3\sqrt{\frac{1-C_3\cos(2\varphi_3)}{1-C_4\cos(2\varphi_3)}}\dot{w}_3 + \Omega_3^2\frac{1-C_3\cos(2\varphi_3)}{1-C_4\cos(2\varphi_3)}w_3 = -\Gamma_5 w_3 - \Gamma_3 w_3^3 - \Gamma_4 w_3^5, \tag{54}$$

where the coefficients C_1, C_2, C_3, C_4 are presented in Appendix A, and the damping ratio ξ is given by the following equation:

$$\zeta = \frac{\Delta}{\sqrt{4\pi^2 + \Delta^2}}, \tag{55}$$

The natural frequency of the tulip in bending Ω_1 and the natural frequency of the bowl in bending Ω_3 are given by the following equations:

$$\Omega_1 = \sqrt{\frac{3E}{m_T L_{TA}^3}\frac{b_1(1+a_1)}{1+\frac{b_1}{J_{Z_2TA}}(1+a_1)}}, a_1 = \frac{0.5(J_{1T}+J_{2T})}{S_T\left[\frac{L_{TA}^2}{12}+d_{CT}^2\right]}, b_1 = S_T\left[\frac{L_{TA}^2}{12}+d_{CT}^2\right], \tag{56}$$

$$\Omega_3 = \sqrt{\frac{3E}{m_B L_{BA}^3}\frac{b_2(1+a_2)}{1+\frac{b_2}{J_{Z_2BA}}(1+a_2)}}, a_2 = \frac{0.5(J_{1B}+J_{2B})}{S_B\left[\frac{L_{BA}^2}{12}+d_{CB}^2\right]}, b_2 = S_B\left[\frac{L_{BA}^2}{12}+d_{CB}^2\right]. \tag{57}$$

The coefficients of the cubic and quintic terms expressed in Equations (53) and (54) can be mathematically explained by development in a Taylor infinite series induced by the excitation function $\sinh(x)$ (see Equations (53) and (54)) expressed approximately as

$$\sinh(x) \approx x + \frac{x^3}{3!} + \frac{x^5}{5!}. \tag{58}$$

In the PPR region, the excitation terms $\cos(2\varphi_1(t))$, $\sinh(\omega_{n1}t)$, and $\cos(2\varphi_3(t))$ must satisfy the Equations (59) and (60), as mentioned in [28] (p. 425), since this resonance is one of the most important, as mentioned in [5,6,28]:

$$\eta_1 \simeq 2\Omega_1 \simeq \frac{d\theta_1}{dt} = \frac{2d\varphi_1}{dt} \simeq \omega_{n1}, \tag{59}$$

$$\eta_3 \simeq 2\Omega_3 \simeq \frac{d\theta_3}{dt} = \frac{2d\varphi_3}{dt} \simeq \omega_{n1}, \tag{60}$$

where η_1 is the excitation frequency in the PPR region for the tulip, η_3 is the excitation frequency in the PPR region for the bowl, $\frac{d\varphi_1}{dt}$ is the excitation induced in the FBV movement (Equation (53)) by the rigid twisting angle of the tulip concerning the X_1 axis, and $\frac{d\varphi_3}{dt}$ is the excitation induced in the FBV movement (Equation (54)) by the rigid twisting angle of the bowl concerning the X_3 axis, as given by the following equation:

$$\varphi_3(t) = \varphi_1(t) + \frac{R_{TTr}}{2L_{Ms}}\tan\beta_1\tan^2\frac{\beta_1}{2}\cos(3\varphi_1) + \frac{R_{IrB}}{2L_{Ms}}\tan\beta_2\tan^2\frac{\beta_2}{2}\cos\left[3\left(\varphi_1(t) + \frac{R_{TTr}}{2L_{Ms}}\tan\beta_1\tan^2\frac{\beta_1}{2}\cos(3\varphi_1)\right)\right], \quad (61)$$

This considers the nonuniformity of the kinematic isometry of the AD [25,26]. In Equation (61), the terms R_{TTr}, R_{IrB} represent the radii of the tulip-tripod and inner race-bowl joints. Considering the conditions imposed by Equations (59) and (60) together with Equation (58) and Equations (48)–(50), the coefficients of the cubic and quintic excitation terms on the right-hand side of Equations (53) and (54) are

$$\Gamma_1 = \frac{0.2\rho A}{\pi 3!}\frac{\overline{F}_S\Delta t_s}{Mm_T}\Omega_1, \Gamma_2 = \frac{0.2\rho A}{\pi 5!}\frac{\overline{F}_S\Delta t_s}{Mm_T}\Omega_1, \frac{0.2\rho A}{\pi}\frac{\overline{F}_S\Delta t_s}{\Omega_1 Mm_T} \ll 1 \Rightarrow \Gamma_0 \approx 0, \quad (62)$$

$$\Gamma_3 = \frac{0.2\rho A}{\pi 3!}\frac{\overline{F}_S\Delta t_s}{Mm_B}\Omega_3, \Gamma_4 = \frac{0.2\rho A}{\pi 5!}\frac{\overline{F}_S\Delta t_s}{Mm_B}\Omega_3, \frac{0.2\rho A}{\pi}\frac{\overline{F}_S\Delta t_s}{\Omega_3 Mm_B} \ll 1 \Rightarrow \Gamma_5 \approx 0. \quad (63)$$

By analyzing Equations (53) and (54), they are a generalized form of Mathieu–Hill nonlinear equations coupled through Equation (61), describing the FBV movements of the tulip and the bowl induced by shock excitations due to the road geometry. These equations contain the following phenomena: the non-uniformity of geometric and kinematic isometry of the AD, the non-uniformity of the inertial characteristics for the tulip and the bowl and forced shock excitations due to road excitations. By analyzing the initial system of Equations (33)–(38) with the boundary conditions of Equations (39)–(44), it can be seen that Equations (34) and (38), together with the boundary conditions of Equations (43) and (44), were not used in the mathematical procedure solution because these equations represented the nonlinear dynamic behavior of the FBV angular shearing movements of the AD (the AD's elements of the tulip, midshaft, and bowl), and the present article is dealing with the nonlinear dynamic behavior of the FBV deflection movements of the ADE, namely the bowl, midshaft, and tulip. The FBV movements of the midshaft have the solution given by Equation (45) coupled with Equations (47) and (48). In contrast, the shearing FBV movements of the midshaft have the solution given by Equation (46) coupled with Equations (47) and (48), both representing the solution of Equations (35) and (36) from the SDEOFBVM (Equations (33)–(38)), with the boundary conditions of Equations (39) and (40) from the general system of Equations (39)–(44) that imposed the boundary conditions. Additionally, it is noted that for the nonlinear dynamic behavior of the AD, the dynamic behavior of the midshaft represents an excitation-like internal tuned damper (ITD) used by Browne and Palazzolo in the superharmonic nonlinear lateral vibrations (forced bending vibrations) of an AD [6]. This aspect represents a novelty in the investigation of the dynamic behavior of the ADE, because it allows for the computation of the equations of motion for the tulip and the bowl in a mathematical form adequate for performing a perturbation approach, as can be seen from Equations (53) and (54). Another aspect of this novelty is the capability of computing the natural free frequency in bending for the tulip and the bowl (see Equations (56) and (57)) and to express the excitation frequency in the PPR region, as is noted from Equations (59) and (60). In the meantime, a supplementary aspect of the novelty is the estimation of the shock excitation of the road geometry as cubic and quintic excitation terms in the equations of motion, as is shown by Equations (56) and (57) and the relations of Equations (62) and (63). For the first time, the effect of the nonuniformity of the kinematic isometry of the AD was considered, being expressed by Equation (61), which links the equations of motion for the FBVM regarding the tulip and the bowl.

6. Analysis of the PPR Region for the Tulip and Bowl

As already mentioned above, one of the most important resonant cases of an AD is the PPR [5,6], [28] (p. 425), and for this paper, the authors decided to investigate the dynamic behavior of FBV movement in the PPR region for the tulip and the bowl based on Equations (53) and (54). Similar experimental data for this case study are presented in the literature by Steinwede [4] (pp. 69–144). The study considered a tulip having geometric characteristics such as the geometric moment of inertia and nonuniformity of the geometric moments of inertia, as presented in Table 1 [29], for the tulip AD element. By comparing these presented geometric characteristics with those considered by Steinwede [4] (p. 111), it can be concluded that there is agreement. Using AUTOCAD software, J_{1T}, J_{2T} were computed based on the direct geometric characteristics $L_{TB}, L_{TA}, L_{Ms}, d_{TA}, d_{CT}, S_T$ and the general geometry of the tulip.

Table 1. Geometry characteristics of the tulip [29].

L_{TB} (m)	L_{TA} (m)	L_{Ms} (m)	d_{TA} (m)	d_{CT} (m)	S_T (m^2)	$0.5(J_{1T}+J_{2T})$ (m^4)	χ_{nT}
0.095	0.065	0.470	0.027	0.049	0.0019	9.1531×10^{-7}	0.25

It can be remarked that the nonuniformity of the inertial geometric characteristic χ_{nT} in Table 1 had another value than what was considered in literature because of the geometric characteristic in the cross-section of the tulip presented in Figure 9.

Table 2 presents the physical properties of the material of the tulip and the amplitude and duration of the shock, considering that the material was steel and iron cast. By comparing these presented material properties with those considered by Steinwede [4] (p. 112), it can be concluded that they were in very close agreement.

Table 2. Material properties of the tulip and shock amplitude.

ρ (kg/m^3)	G (Gpa)	E (Gpa)	ζ	$\bar{F}_s$[MN] Δt_s [ms]
7850	77.3	200	0.0016–0.0318	0.05 1–10

To compute the amplitude of the FBV movements of the ADE tulip in the PPR region using a perturbation approach, several methods could be used: the method of harmonic balance [29] (p. 66), the asymptotic method [1] (pp. 299–393), or the method of multiple scales [28] (pp. 424–427). The method of harmonic balance is very efficient, as mentioned in the literature [30] (p. 63), [31], but it allows the computation of the amplitude versus the frequency excitation only for stationary (steady state) motion without the possibility of computing the phase angle [30] (pp. 63–68). Therefore, this method is limited to the steady state FBV. The multiple scales method is challenging to use due to the conditions of zeroing the secular term and the additional necessary study of the convergence of the detuning parameter. Considering these aspects, the authors chose to use the AMA in the first-order approximation. This would allow the investigation of the amplitude and phase angle versus the excitation frequency for the FBV movement of the tulip in the PPR region for both the stationary and nonstationary cases. To compute the solution of Equation (53), it was assumed that the slowing time was $\tau = \varepsilon t$, where ε is a small positive parameter [1] (p. 299). To introduce the slowing time, Equation (53) needed to be transformed to be used in the AMA. The coefficients of the second and third term of Equation (53) on the left side can be expressed as

$$\begin{aligned} &\tfrac{1-C_1\cos(2\varphi_1)}{1-C_2\cos(2\varphi_1)} \approx 1-2\mu\cos(2\varphi_1) = 1-2\mu\cos(\theta_1), \\ &\sqrt{\tfrac{1-C_1\cos(2\varphi_1)}{1-C_2\cos(2\varphi_1)}} \approx (1-2\mu\cos(2\varphi_1))^{-\frac{1}{2}} \approx 1+\mu\cos(2\varphi_1) = 1+\mu\cos(\theta_1), \\ &\mu = \tfrac{1}{2}(C_2-C_1), \end{aligned} \tag{64}$$

where in Equation (64), the condition given by Equation (59) was considered. With the relations in Equation (64), Equation (53) becomes

$$\ddot{w}_1 + \Omega_1^2 w_1 = -2\xi\Omega_1(1+\mu\cos(\theta_1))\dot{w}_1 + 2\mu\Omega_1^2\cos(\theta_1)w_1 - \Gamma_1 w_1^3 - \Gamma_2 w_1^5. \tag{65}$$

The assumption that the damping ratio ξ, the excitation coefficient μ, and the coefficients of cubic and quintic nonlinearity Γ_1, Γ_2 are small is incorporated in the analysis by representing these quantities in the following form:

$$\xi = \varepsilon\xi, \mu = \varepsilon\mu, \Gamma_1 = \varepsilon\Gamma_1, \Gamma_2 = \varepsilon\Gamma_2, \tag{66}$$

where ε is the same small positive parameter used to obtain the slowing time. It is also assumed that the excitation frequency η_1 and the excitation parameter μ vary slowly with time, such that

$$\frac{d\theta_1}{dt} = \eta_1(\tau), \mu = \mu(\tau). \tag{67}$$

Equation (65) becomes the following after neglecting the terms in ε^2:

$$\ddot{w}_1 + \Omega_1^2 w_1 = \varepsilon\left[-2\xi\Omega_1\dot{w}_1 + 2\mu\Omega_1^2\cos(\theta_1)w_1 - \Gamma_1 w_1^3 - \Gamma_2 w_1^5\right]. \tag{68}$$

Regarding Equation (68), the right-hand side represents a perturbation of the mathematical form $H(w_1, \theta_1)$, being a periodic function in θ_1 with period 2π, while the left-hand side of the equation is a linear oscillator. By considering all these physical considerations and confining our attention to the investigation of the PPR region, a solution for Equation (68) is sought after in the following form to the first-order approximation in ε:

$$w_1 = W\cos\left(\frac{1}{2}\theta_1 + \Psi\right), \tag{69}$$

where W and Ψ are functions of time defined by the system of differential equations:

$$\begin{cases} \frac{dW}{dt} = \varepsilon A_1(\tau, W, \psi) \\ \frac{d\psi}{dt} = \Omega_1 - \frac{1}{2}\eta_1(\tau) + \varepsilon B_1(\tau, W, \psi) \end{cases}. \tag{70}$$

Differentiating the right-hand side of Equation (70) and expanding the results in powers of ε yields

$$\begin{aligned} \ddot{W} &= \tfrac{d^2W}{dt^2} = \varepsilon\tfrac{\partial A_1}{\partial\psi}\left(\Omega_1 - \tfrac{1}{2}\eta_1\right) + \varepsilon^2...., \\ \ddot{\psi} &= \tfrac{d^2\psi}{dt^2} = \varepsilon\left[-\tfrac{1}{2}\tfrac{\partial\eta_1}{\partial\tau} + \tfrac{\partial B_1}{\partial\psi}\left(\Omega_1 - \tfrac{1}{2}\eta_1\right)\right] + \varepsilon^2.... \end{aligned} \tag{71}$$

When differentiating the right-hand side of Equation (69) while considering the systems of Equations (70) and (71) and retaining only the first-order terms, we obtained the following:

$$\dot{w}_1 = -W\Omega_1\sin\left(\frac{1}{2}\theta_1 + \psi\right) + \varepsilon\left[A_1\cos\left(\frac{1}{2}\theta_1 + \psi\right) - WB_1\sin\left(\frac{1}{2}\theta_1 + \psi\right)\right], \tag{72}$$

$$\ddot{w}_1 = -W\Omega_1 \cos\left(\frac{1}{2}\theta_1 + \psi\right) + \varepsilon\left[\frac{\partial A_1}{\partial \psi}\left(\Omega_1 - \frac{1}{2}\eta_1\right) - 2W\Omega_1 B_1\right]\cos\left(\frac{1}{2}\theta_1 + \psi\right) - \varepsilon\left[W\frac{\partial B_1}{\partial \psi}\left(\Omega_1 - \frac{1}{2}\eta_1\right) + 2\Omega_1 A_1\right]\sin\left(\frac{1}{2}\theta_1 + \psi\right). \quad (73)$$

Using Equations (72) and (73) in the basic form of Equation (68), equating the terms of the form $\varepsilon\cos\left(\frac{1}{2}\theta_1 + \psi\right), \varepsilon\sin\left(\frac{1}{2}\theta_1 + \psi\right)$ from the left-hand side of Equation (68) with the same terms from the right-hand side of the same equation and neglecting the overtones (the terms having $\cos\left(\frac{n}{2}\theta_1 + \psi\right), \sin\left(\frac{n}{2}\theta_1 + \psi\right), n = 2, 3, 4, 5$) yields the two following coupled first-order differential equations for the unknown functions A_1 and B_1:

$$\frac{\partial A_1}{\partial \psi}\left(\Omega_1 - \frac{1}{2}\eta_1\right) - 2W\Omega_1 B_1 = \mu W\Omega_1^2\cos(2\psi) - \frac{3}{4}\Gamma_1 W^3 - \frac{5}{8}\Gamma_2 W^5, \quad (74)$$

$$W\frac{\partial B_1}{\partial \psi}\left(\Omega_1 - \frac{1}{2}\eta_1\right) + 2\Omega_1 A_1 = -\mu W\Omega_1^2\sin(2\psi) - 2\xi\Omega_1^2 W, \quad (75)$$

This system of Equations (74) and (75) has the following solutions:

$$\begin{cases} A_1 = -\frac{\mu\Omega_1^2 W}{\eta_1}\sin 2\psi - \xi\Omega_1 W, \\ B_1 = -\frac{\mu\Omega_1^2}{\eta_1}\cos 2\psi + \frac{3}{8}\frac{\Gamma_1 W^2}{\Omega_1} + \frac{5}{16}\frac{\Gamma_2 W^4}{\Omega_1} \end{cases}. \quad (76)$$

By introducing the solutions of Equation (76) into the system of Equation (70) and transforming all the system parameters back to their real-time values, the solution of Equation (65), representing the FBV movement of the tulip in the PPR ($\eta_1 \simeq 2\Omega_1$) is given by Equation (69) using the first-order approximation of the AMA, where the amplitude W and the phase angle ψ are given by integrating over time the first-order differential system:

$$\begin{cases} \frac{dW}{dt} = -\frac{\mu\Omega_1^2 W}{\eta_1}\sin 2\psi - \xi\Omega_1 W \\ \frac{d\psi}{dt} = \Omega_1 - \frac{1}{2}\eta_1 - \frac{\mu\Omega_1^2}{\eta_1}\cos 2\psi + \frac{3}{8}\frac{\Gamma_1 W^2}{\Omega_1} + \frac{5}{16}\frac{\Gamma_2 W^4}{\Omega_1} \end{cases}. \quad (77)$$

6.1. Steady State Forced Bending Vibration of the Tulip

For the stationary FBV movement for the tulip of the AD, it would be set to zero $\frac{dW}{dt}$ and $\frac{d\psi}{dt}$ in Equation (77), and expressed from these equations, $\sin 2\psi$ and $\cos 2\psi$ yields the trigonometric equation $\sin^2(2\psi) + \cos^2(2\psi) = 1$, expressing in terms of the system in Equation (77) the bi-quartic equation in the unknown W the amplitude versus the excitation frequency:

$$\lambda_5 W^8 + \lambda_6 W^6 + \lambda_7 W^4 + \lambda_8 W^2 + \lambda_9 = 0, \quad (78)$$

where the coefficients λ_5–λ_9 are given in Appendix A. The solutions to Equation (78) represent the stationary amplitude of the forced bending vibration of the tulip in the PPR region. The stationary phase angle of the FBV movement for the tulip in the PPR region is given from the same system (Equation (77)) set to zero by expressing the tangent of the phase angle, yielding

$$\psi = \frac{1}{2}\mathrm{arctg}\left[-\frac{\xi_1}{1 - \frac{\eta_1}{2\Omega_1} + \frac{3}{8}\frac{\Gamma_1 W^2}{\Omega_1^2} + \frac{5}{16}\frac{\Gamma_2 W^4}{\Omega_1^2}}\right]. \quad (79)$$

6.2. Investigation of Dynamic Instability

The investigation of the dynamic instability of the FBV movement for the tulip represents the computation of the boundaries of the principal parametric region of instability. The base width of the stationary response is the only region in which vibrations may typically initiate. Setting to zero the amplitude in Equation (78) yields the following equation:

$$\left(1-\frac{\eta_1}{2\Omega_1}\right)^2+\xi^2-\frac{1}{4}\mu^2\left(\frac{2\Omega_1}{\eta_1}\right)^2=0, \tag{80}$$

Alternatively, it may yield

$$\left(\frac{\eta_1}{2\Omega_1}\right)^4-2\left(\frac{\eta_1}{2\Omega_1}\right)^3+\left(1+\xi^2\right)\left(\frac{\eta_1}{2\Omega_1}\right)^2-\frac{\mu^2}{4}=0. \tag{81}$$

Equation (81) gives the boundaries of the principal parametric region of instability in the space $\left(\frac{\eta_1}{2\Omega_1}, \xi, \mu\right)$. This is the investigation for the stationary FBV movement of the tulip. For the nonstationary FBV movement of the tulip, the investigation of the dynamic instability consisted of the analysis of the spectral graphs of the nonstationary amplitude and the first derivative of the nonstationary amplitude concerning time, given by the integration of the system of Equation (77). The graphed representation in the configuration space of the "speed" nonstationary amplitude versus the nonstationary amplitude $\left(\frac{dW}{dt}, W\right)$ is evidence of the transition of FBV movement for the tulip through the chaotic movement in the PPR region, and the results are presented in the next paragraph.

6.3. Nonstationary FBVM of the Tulip and Bowl during Transition through the PPR

The resonant regimes of the principal parametric vibration for the forced bending vibration of the tulip were determined. It may be of interest to examine the nature of the nonstationary vibrations performed by the system during a transition of the excitation frequency η_1 through the resonant regime. For this purpose, it was necessary to integrate the system of Equation (77) governing the nonstationary amplitude and the phase angle. This system of differential equations cannot be integrated into the closed form, and therefore, it was obvious to use numerical integration. It may be pointed that numerical integration of first-order equations governing the stationary amplitude and phase angle versus frequency excitation is a much simpler process than the integration of the original system of Equation (77), which is nonlinear since it must evaluate the envelope of an oscillatory function and not the function itself. In the present nonstationary analysis, the sweep of the excitation frequency is taken to be logarithmic and is expressed as

$$\eta_1(t)=\eta_{10}\sigma^{t(1+m)}, \sigma\geq 1, \tag{82}$$

where η_{10} is the initial frequency at t = 0 for the tulip and m is the sweep rate (in octaves/min). The differential system of Equation (77) is integrated numerically using a fourth-to-fifth order Runge–Kutta algorithm from the MATLAB software with a variable integration step of the prediction-correction type. The initial values of the variables W, ψ, and η_1 were chosen as a stationary state starting point. The investigation of the nonstationary forced bending vibration of the tulip was confined to the PPR region, and the results are presented in the next paragraph. For the bowl, the analysis of the PPR region was based on Equation (54) for FBV movement (FBVM), with the excitation frequency η_3 given by Equation (60), the natural frequency Ω_3 established by Equation (57), the cubic and quintic terms Γ_3, Γ_4 given by Equation (63), and the excitation coefficient being $\mu=\frac{1}{2}(C_4-C_3)$, while the terms C_3, C_4 are presented in Appendix A. For the investigation of the nonstationary FBVM of the bowl, the system of differential equations is from Equation (77), using the excitation frequency η_3, the natural frequency Ω_3, and the cubic and quintic terms Γ_3, Γ_4, while the rate of sweep m is established as

$$\eta_3(t)=\eta_{30}\sigma^{t(1+m)}, \sigma\geq 1, \tag{83}$$

where η_{30} is the initial frequency at t = 0 for the bowl. Table 3 presents the geometry and inertial characteristics of the bowl. Using AUTOCAD software, J_{1B}, J_{2B} were computed based on the direct geometric characteristics $L_{BB}, L_{BA}, L_{Ms}, d_{BA}, d_{CB}, S_B$ and the general geometry of the tulip. The material's physical properties are the same as those presented in Table 2.

Table 3. Geometry characteristics of the bowl.

L_{BB} (m)	L_{BA} (m)	L_{Ms} (m)	d_{BA} (m)	d_{CB} (m)	S_T (m^2)	$0.5(J_{1B}+J_{2B})$ (m^4)	χ_{nB}
0.053	0.086	0.470	0.044	0.067	0.0020	10.560×10^{-7}	0.1

Figure 10 presents the axonometric cross-section of the bowl.

Figure 10. Axonometric cross-section of the bowl-balls-inner race joint.

When comparing these presented geometry characteristics (see Table 3) with those considered by Steinwede [4] (p. 112), it can be concluded that they agree. The novelty for investigating the PPR region of an AD is that the AMA was used for such a multibody technical system for the first time. The AMA has no limitations on its use because it allows the development of analytical solutions for the amplitude, phase angle, and dynamic instability frontiers for stationary and nonstationary motion, as can be seen from Equations (65)–(83). Additionally, the AMA allows for significant precision for the computations because it can provide several levels of approximation. In our case, it was used for first-order approximation. For the first time, the amplitude, phase angle, and dynamic instability frontiers for the stationary motion of the FBV of a tulip and bowl in the PPR region were computed, as can be seen by analyzing Equations (78), (79) and (81). In the meantime, the amplitude spectrum, "speed" amplitude spectrum, and representation in the phase space $\left(W, \frac{dW}{dt}\right)$ of the "speed" amplitude versus amplitude for the nonstationary motion in transition through the PPR region of the tulip and the bowl was determined for the first-time using Equations (77), (82) and (83).

7. Results and Discussion

The first results obtained were represented by the natural free frequencies in bending and shearing of the midshaft, given by Equation (56):

$$\nu_{n_{1,2}} = \frac{\omega_{n1,2}}{2\pi}, n_{1,2} = 1, 2, 3, \ldots\ldots. \quad (84)$$

Based on Equations (56) and (83), the natural free frequencies of the midshaft in bending-shearing is an infinite series covering a wide range of frequencies, and thus by comparing the numerical results obtained with the mentioned equations and the experimental data and numerical data (using FEA-ANSYS) presented in the literature [13] and analyzing the data presented in Table 4, agreement for the bending natural free frequencies of the midshaft of the AD was found.

Table 4. Comparison of the experimental natural frequencies (ENFs) of the driveshaft and numerical simulations using the finite element analysis (FEA) program Ansys of the natural frequencies (NSNFs) in [13] with those obtained for the midshaft based on Equations (57) and (83).

ENF (Hz)	NSNF (Hz)	ν_{n1} (Hz)	ν_{n2} (Hz)
683.59	686.77	683.56	1343.3
-	748.01	748.1	1470.1
1406.25	1457.8	1406.2	2763.4
1464.84	1466.6	1464.9	2878.7
-	2207.4	2207.2	4337.3
	2235.6	2235.5	4393

As can be seen from Table 4, the agreement of the results was for the free bending natural frequency, while the second free natural frequency represents the free shearing natural frequency that, as can be noticed, was like the principal parametric resonance compared with the bending. This confirms that the designed DMFFBVM of the AD was much more reliable than the FEA-ANSYS and modal analysis presented in the literature [13]. As mentioned in [13], the dynamic forced vibration of the AD was inducing cracks in the AD's structure, a fact also revealed by Steinwede [4]. Additionally, it was mentioned in [13] that experimental data were found for the midshaft, and this aspect revealed the fact that the midshaft had such a nonlinear dynamic behavior that represented the excitation element for the other components of the AD, namely the tulip and the bowl, playing the same role as the internal tuned damper (ITD) mentioned in the literature [8] for the superharmonic nonlinear lateral vibrations (bending vibrations) of the AD. The fundamental natural free bending vibrations of the tulip given by Equation (56) was $\nu_1 = 519.19$ Hz, considering the geometrical characteristic and the physical properties presented in Tables 1 and 2 for the AD designed for a heavy duty SUV, which conformed with the experiments of Steinwede that imposed an investigation of the forced torsional or bending vibrations of a driveshaft in the range of 0.5–15 kHz [6] (p. 119). Unfortunately, no published experiments investigated the natural free frequency in bending only for the global tulip, midshaft, or global bowl. For the steady state FBV of the tulip and the bowl, specific MATLAB software was developed to compute the amplitude and phase angle in the region of principal parametric resonance based on Equations (78) and (79). Figures 11 and 12 present the variation of the normalized stationary amplitude for the FBV in the PPR region for the tulip of an AD, with this being around 1038.38 Hz.

When analyzing the normalized stationary parametric amplitude presented in the graphs in Figures 11 and 12 for ζ = 0.0016–0.0318 and χ_{nT} = 0.25, it can be found that for the cases, we had a manifestation of a "soft spring" with one branch for $\eta_1 \leq 2\Omega_1$ that indicated the presence of interaction between the principal parametric resonance and the primary resonance, while for $\eta_1 \geq 2\Omega_1$, two branches of a "hard spring" existed, which suggests the manifestation of interaction with the combination resonance or with the internal resonance for the tulip [28] (pp. 132–160). This aspect agrees with the experimental data in the literature [4] (pp. 130–144). It can also be remarked that the increase in the damping ratio in the range of 0.0016–0.0318 increased the minimum normalized stationary amplitude from 0.13 to 0.42 for one branch, and for the second branch, the increase in the minimum normalized stationary amplitude was from 0.32 to 0.425. This can be explained by the existence of dynamic instability in the region of principal parametric resonance, as mentioned in the literature [28] (pp. 132–160). Figures 13 and 14 present the variation of the stationary phase angle for the FBV in the PPR region for the tulip of an AD.

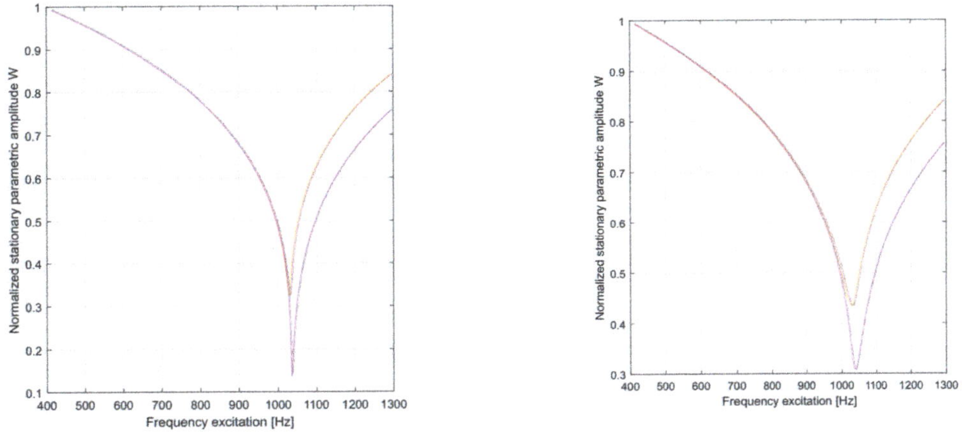

(a) (b)

Figure 11. Normalized stationary parametric amplitude W of the tulip for χ_{nT} = 0.25, where $\mu = 0.623 \times 10^{-4}$. (**a**) ζ = 0.0016. (**b**) ζ = 0.0116.

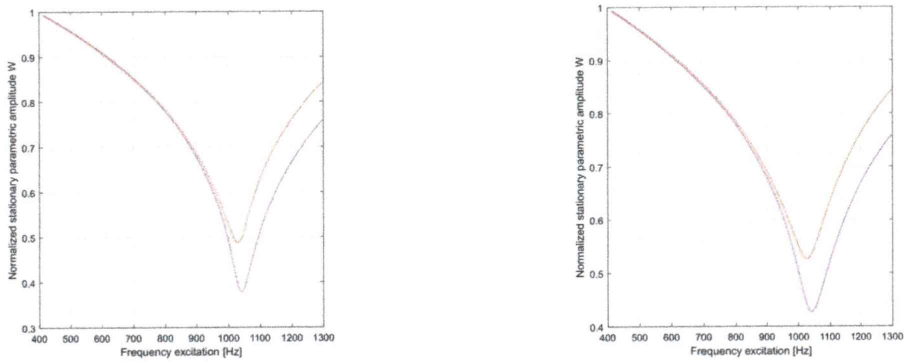

(a) (b)

Figure 12. Normalized stationary parametric amplitude W of the tulip for χ_{nT} = 0.25, where $\mu = 0.623 \times 10^{-4}$. (**a**) ζ = 0.0216. (**b**) ζ = 0.0318.

When analyzing the stationary parametric phase angle presented in the graphs in Figures 13 and 14 for ζ = 0.0016–0.0318 and χ_{nT} = 0.25, it can be seen that for the cases, we had a manifestation of a "soft spring" with one branch for $\eta_1 \leq 2\Omega_1$, while for $\eta_1 \geq 2\Omega_1$, two branches of a "hard spring" exist, but with the increase in the damping ratio passing through the principal parametric resonance for $\eta_1 > 2\Omega_1$, and in the region of a "hard" spring, the two branches are approaching one against the other. In addition, the manifestation of increasing the value of the minimum of the second branch was sensitive to the increase in the value of the damping ratio in the range mentioned above (0.0016–0.0318). Through experiments, Steinwede demonstrated that the nonlinear parametric dynamic behavior of automotive driveshafts is like that of geared systems [4] (p. 117), [32] (pp. 132–160), and therefore, we observed similar pitting phenomena as well as micro-cracks inside the tulip and the bowl of the CVJ tulip-tripod and bowl-inner race joints [6] (pp. 88–94), which was also mentioned in the literature [13]. These aspects will "conduct" dynamic behavior through a chaotic dynamic that has, as a practical effect, an accelerating effect on pitting phenomena and cracking, as mentioned by Steinwede [4] (pp. 88–94) or, in the worst case, results in the manifestation of cracks followed by failure (breaking) of the tulip [4] (p. 89). Figures 15 and 16 present the variation of the normalized stationary amplitude for the FBV in the PPR region for the bowl of an AD (around 6306.6 Hz). The fundamental natural frequency in bending for the bowl (3153.3 Hz) was computed with Equation (57). As can be seen, the natural frequency for the bowl was much bigger than the natural frequency of the tulip, and the PPR region for the bowl was in the PPR region for the forced torsional vibrations of the tulip, as mentioned in the literature [29].

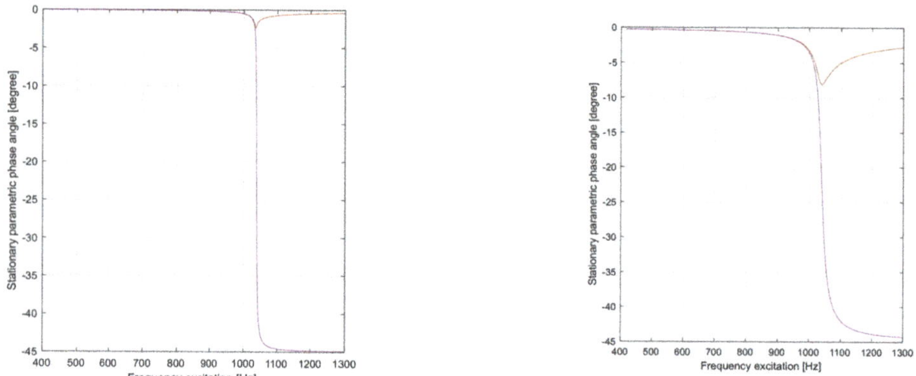

(a) (b)

Figure 13. Stationary parametric phase angle Ψ of the tulip for χ_{nT} = 0.25, where μ = 0.623 × 10^{-4}. (**a**) ζ = 0.0016. (**b**) ζ = 0.0116.

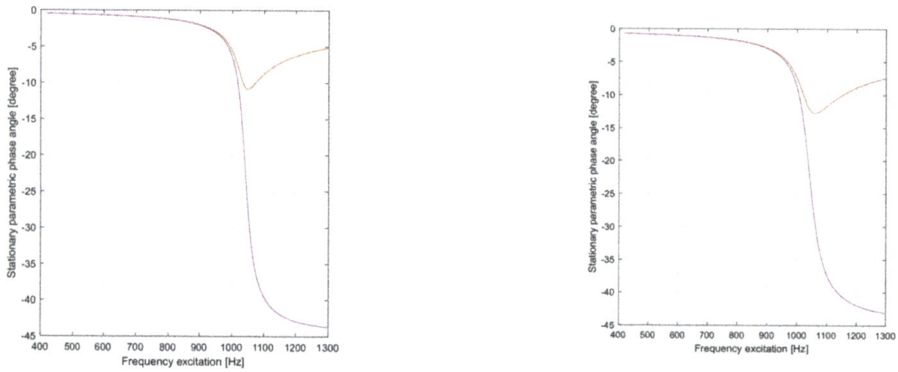

(a) (b)

Figure 14. Stationary parametric phase angle Ψ of the tulip for χ_{nT} = 0.25, where μ = 0.623 × 10^{-4}. (**a**) ζ = 0.0216. (**b**) ζ = 0.0318.

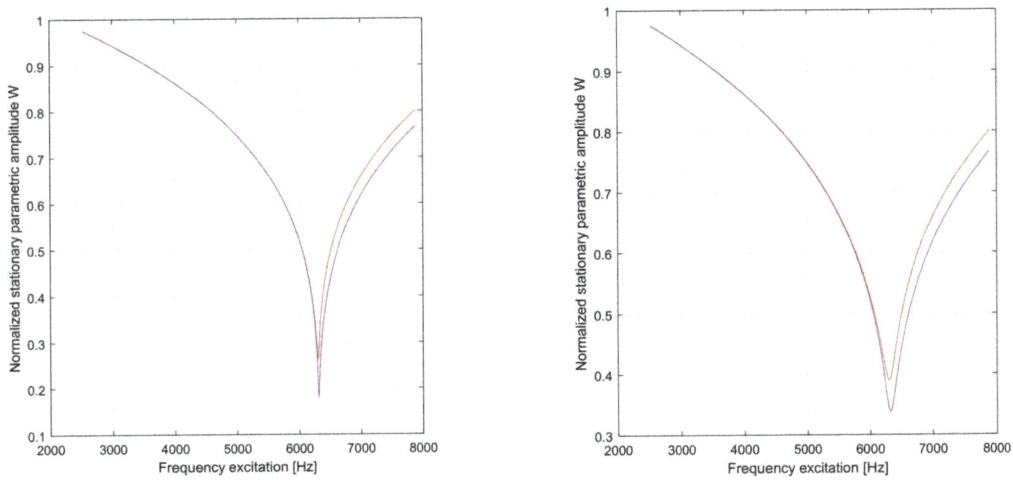

(**a**) (**b**)

Figure 15. Normalized stationary parametric amplitude W of the bowl for χ_{nB} = 0.10, where $\mu = 0.754 \times 10^{-4}$. (**a**) ζ = 0.00016 (**b**) ζ = 0.00016.

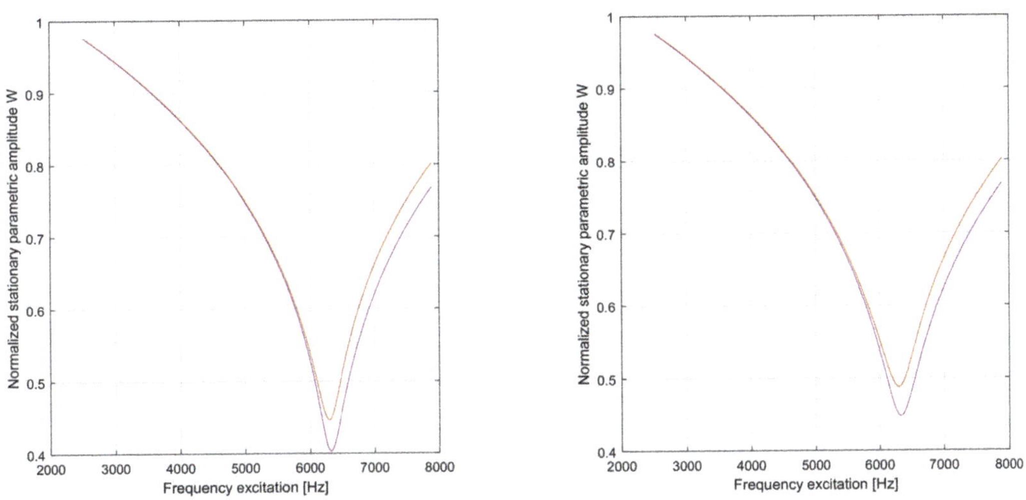

(**a**) (**b**)

Figure 16. Normalized stationary parametric amplitude W of the bowl for χ_{nB} = 0.10, where $\mu = 0.754 \times 10^{-4}$. (**a**) ζ = 0.0216 (**b**) ζ = 0.0318.

When analyzing the normalized stationary parametric amplitude of the bowl presented in the graphs in Figures 15 and 16 for ζ = 0.0016–0.0318 and χ_{nB} = 0.10, the same manifestation in the PPR region for the bowl as for the tulip can be seen but in a different range frequency. The manifestation of a "soft spring" with one branch for $\eta_3 \leq 2\Omega_3$ can be noted, indicating the presence of interaction between the principal parametric resonance and the primary resonance of the bowl, while for $\eta_3 \geq 2\Omega_3$, two branches of a "hard spring" existed, which indicates the manifestation of interaction with the combination resonance or with the internal resonance for the bowl [28] (pp. 132–160). This aspect agrees with the experimental data in the literature [4] (pp. 130–144). It can also be remarked that the increase in the damping ratio in the range of 0.0016–0.0318 increased the minimum normalized stationary amplitude of the bowl from 0.18 to 0.45 for one branch, and for the second branch, the increase in the minimum normalized stationary amplitude was from 0.29 to 0.49. This can be explained by the existence of dynamic instability in the region of principal parametric resonance, as mentioned in the literature [28] (pp. 132–160). Figures 17 and 18 present the variation of the stationary phase angle for the FBV in the PPR region for the AD bowl.

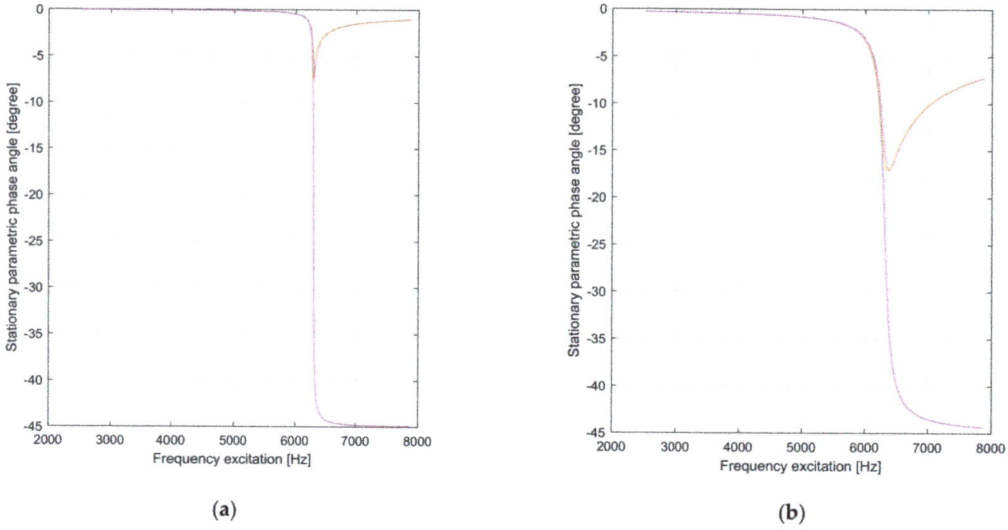

Figure 17. Stationary parametric phase angle Ψ of the bowl for $\chi_{nB} = 0.10$, where $\mu = 0.754 \times 10^{-4}$. (**a**) $\zeta = 0.00016$ (**b**) $\zeta = 0.00016$.

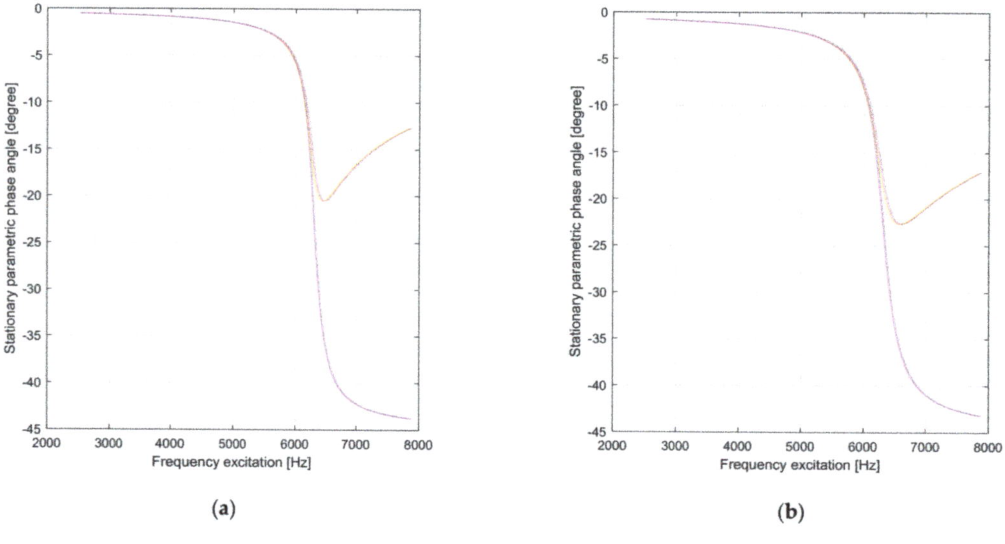

Figure 18. Stationary parametric phase angle Ψ of the bowl for $\chi_{nB} = 0.10$, where $\mu = 0.754 \times 10^{-4}$. (**a**) $\zeta = 0.0216$ (**b**) $\zeta = 0.0318$.

When analyzing the stationary parametric phase angle for the bowl presented in the graphs in Figures 17 and 18 for ζ = 0.0016–0.0318 and χ_{nB} = 0.10, the same kind of manifestation as that for the tulip, but in a different range frequency, can be seen, and for the cases, we had a manifestation of a "soft spring" with one branch for $\eta_3 \leq 2\Omega_3$, while for $\eta_3 \geq 2\Omega_3$, two branches of a "hard spring" existed, but with the increase in the damping ratio, the passing through principal parametric resonance was found for $\eta_3 > 2\Omega_3$, and in the region of the "hard" spring, the two branches were approaching one against the other. For the bowl, this aspect of the two branches closing one against the other was more accentuated. Additionally, the manifestation of increasing the value of the minimum of the second branch was sensitive to the increase in the value of the damping ratio in the range mentioned above (0.0016–0.0318). Figure 19a,b illustrates the stationary dynamic instability region of the tulip in the space (η_1, ξ, μ) using MATLAB software and Equation (81). When analyzing Figure 19a,b, it can be noticed that the two folded surfaces obtained were symmetrical concerning the plan given by (η_1, ξ) for the excitation frequency and the damping ratio. In contrast, the excitation coefficient μ could be positive or negative. The folded surface of the dynamic instability "kept" inside the two branches the region where it manifested the stationary instability. It can also be seen that increasing the damping ratio had a stabilizing effect on the dynamics of the tulip, as expected. Figure 20a,b illustrates the stationary dynamic instability region of the bowl in the space (η_3, ξ, μ) using MATLAB software and Equation (81).

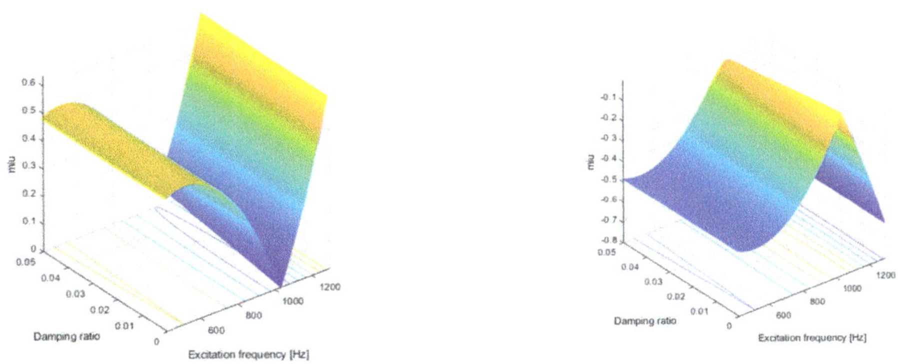

(a) (b)

Figure 19. Stationary dynamic instability surface frontier for the tulip. (**a**) $\mu > 0$. (**b**) $\mu < 0$.

By analyzing Figure 20a,b, it can be noticed that the two folded surfaces obtained were symmetrical concerning the plan given by (η_3, ξ) for the excitation frequency and the damping ratio, while the excitation coefficient μ could be positive or negative. The folded surface of the dynamic instability "kept" inside the two branches the region where it manifested the stationary instability for the bowl. It can also be seen that increasing the damping ratio had a stabilizing effect on the dynamic of the bowl, as expected. The only difference for these phenomena for the bowl was that the manifestation was in another range frequency than that of the tulip: the range frequency given by the natural frequency of the bowl in bending.

Figure 21 illustrates the nonstationary amplitude spectrum W of the FBVM (forced bending vibrating movement) for the tulip during transition through the PPR region.

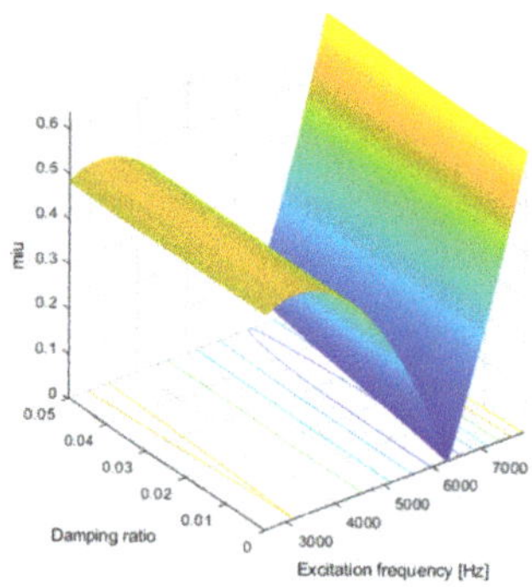

(a)

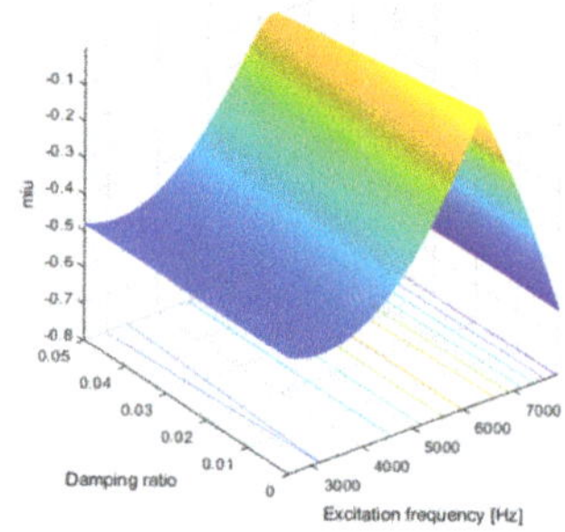

(b)

Figure 20. Stationary dynamic instability surface frontier for the bowl. (**a**) $\mu > 0$. (**b**) $\mu < 0$.

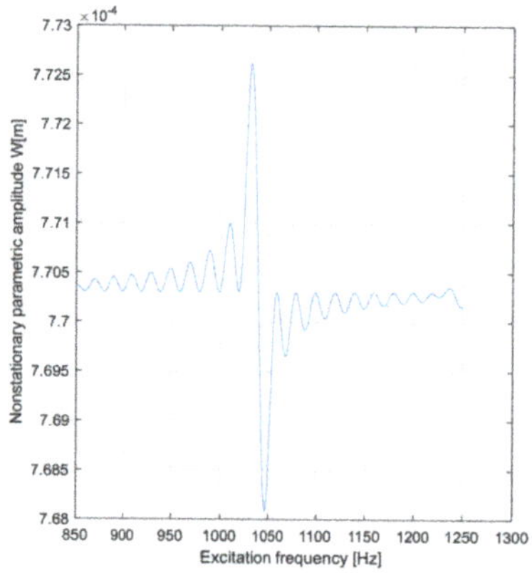

(a)

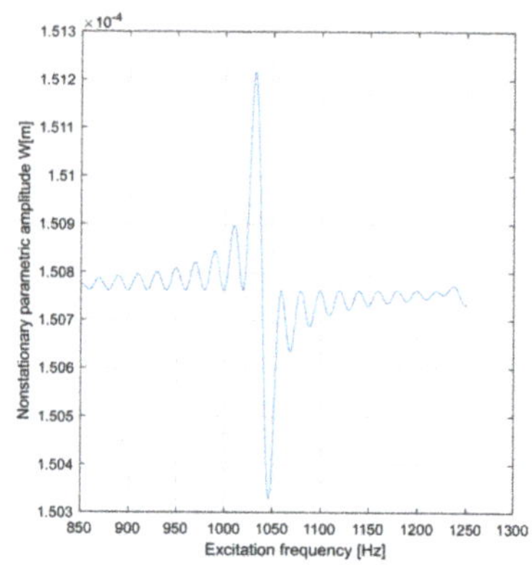

(b)

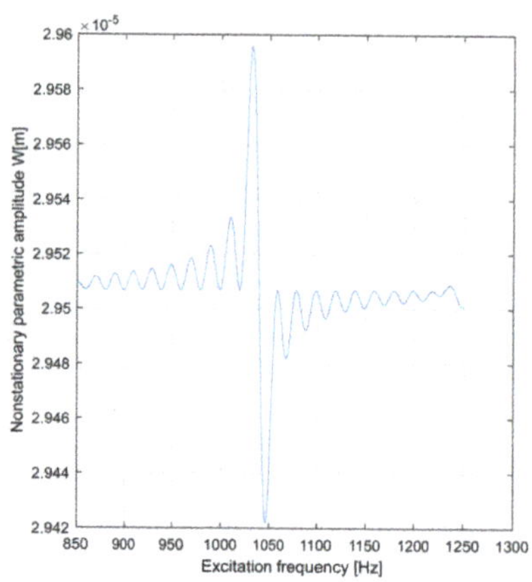

(c)

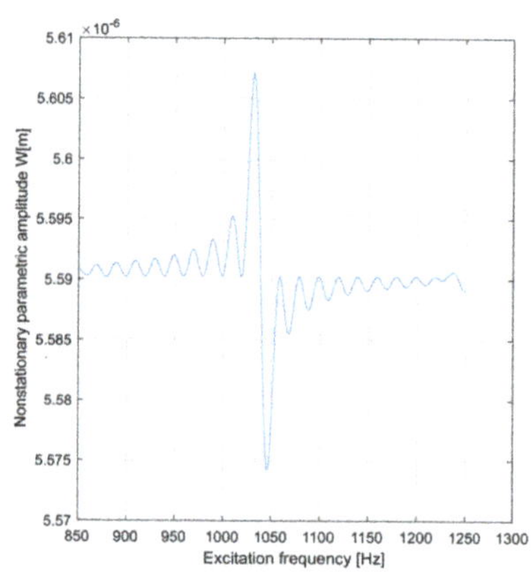

(b)

Figure 21. Nonstationary parametric amplitude spectrum W of the tulip for $\chi_{nT} = 0.25$, where $\mu = 0.623 \times 10^{-4}$. (**a**) $\zeta = 0.0016$. (**b**) $\zeta = 0.0116$. (**c**) $\zeta = 0.0216$. (**d**) $\zeta = 0.0318$. Here, $m = 8 \times 10^{-5}$ octaves/min.

It may be noted that while the excitation frequency was varied logarithmically through the resonant regime, the load parameter μ was kept constant. A vast number of numerical tests was performed to arrive at a reasonable value for the sweep rate of the excitation frequency m = 0.00008 octaves/min. When analyzing the nonstationary parametric amplitude spectrum presented in Figure 20 for $\zeta = (1.6–31.8) \times 10^{-3}$ and $\chi_{nT} = 0.25$, the manifestation of beating effects as well as a significant decrease in the amplitude peaks with the increase in the damping ratio value in the range mentioned above can be seen. The maximum peak of the amplitude was obtained when the excitation frequency was in the close vicinity of PPR 1038 Hz. Each time the damping ratio increased at a rate of 0.01, the maximum peak amplitude decreased by more than five times. Figure 22 illustrates the nonstationary velocity amplitude spectrum $\frac{dW}{dt}$ of the forced bending vibrating movement (FBVM) for the tulip during transition through the PPR region.

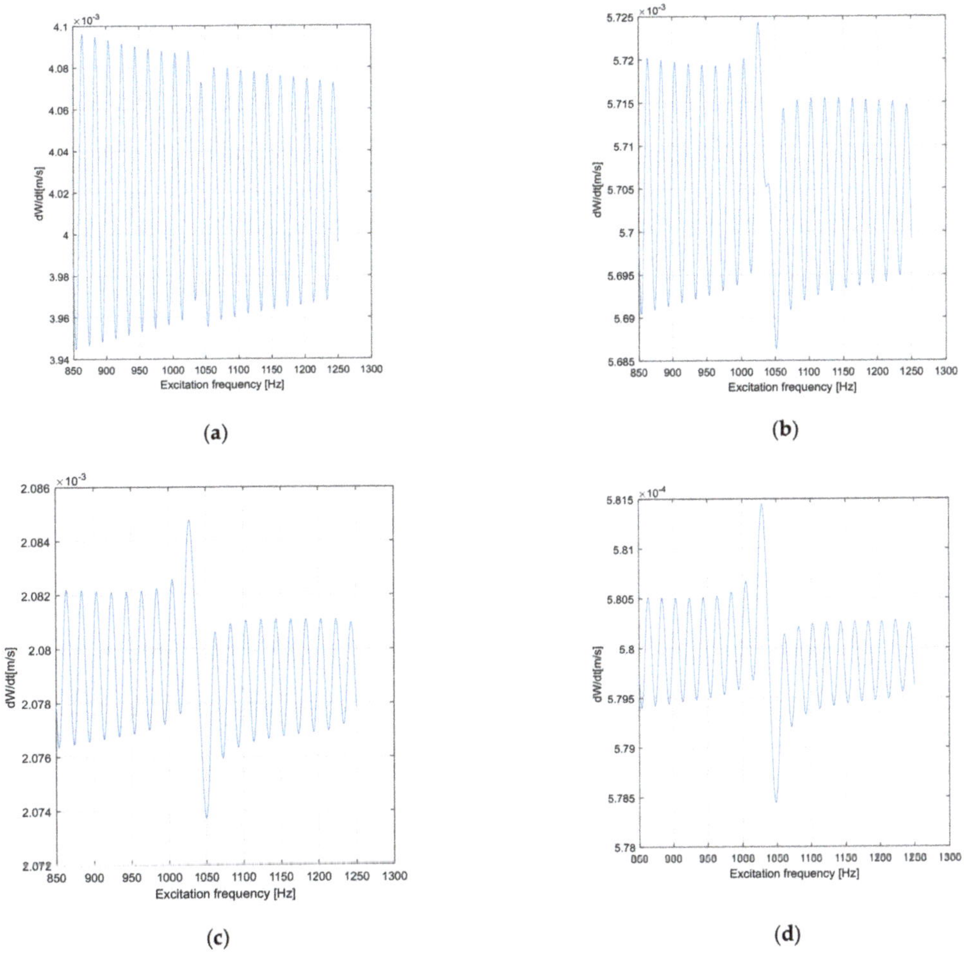

Figure 22. Nonstationary parametric velocity amplitude spectrum $\frac{dW}{dt}$ of the tulip for χ_{nT}= 0.25, where $\mu = 0.623 \times 10^{-4}$. (**a**) $\zeta = 0.0016$. (**b**) $\zeta = 0.0116$. (**c**) $\zeta = 0.0216$. (**d**) $\zeta = 0.0318$. Here, $m = 8 \times 10^{-5}$ octaves/min.

When analyzing the graphs in Figure 22, the moment of passing in close vicinity of the PPR by a sudden increase in the peak velocity amplitude can be seen, while the manifestation of beatings was like the nonstationary amplitude spectrum (see Figure 20). In the range of 0.0016–0.0116, the increase in the damping ratio had the effect of increasing the maximum peak of the nonstationary velocity amplitude by approximately 15%, while in the range of 0.0116–0.0318, the increase in the damping ratio had the beneficial effect of decreasing the maximum peak by 72%. It can be concluded that the growth of the damping ratio had a beneficial impact through a reduction of the nonstationary velocity amplitude only in the range of 0.0116–0.0318. Figure 23 illustrates the graphs of the phase space $\left(W, \frac{dW}{dt}\right)$ for the nonstationary FBVM of the tulip in the PPR region.

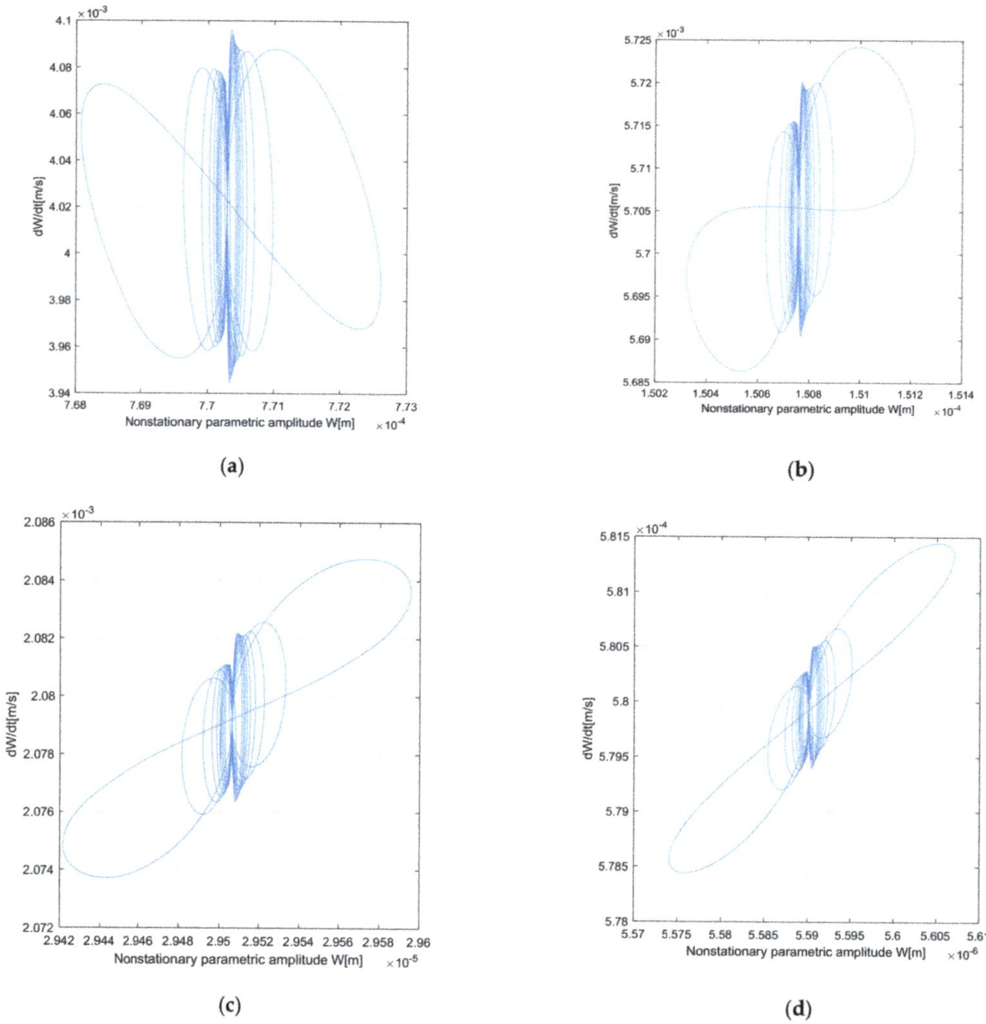

Figure 23. Phase space $\left(W, \frac{dW}{dt}\right)$ of the tulip for $\chi_{nT} = 0.25$, where $\mu = 0.623 \times 10^{-4}$. **(a)** $\zeta = 0.0016$. **(b)** $\zeta = 0.0116$. **(c)** $\zeta = 0.0216$. **(d)** $\zeta = 0.0318$. Here, $m = 8 \times 10^{-5}$ octaves/min.

By analyzing the graphs in Figure 23, a transition to chaotic behavior for the FBVM of the tulip due to the presence of limit cycles or even of strange attractors can be seen, but this last conclusion needs to be certified by detailed analysis using the methods of chaotic movements. One aspect that is evident is that the transition through the PPR region for the tulip was an unstable one. Figure 24 illustrates the nonstationary amplitude spectrum W of the FBVM for the bowl during the transition through the PPR region.

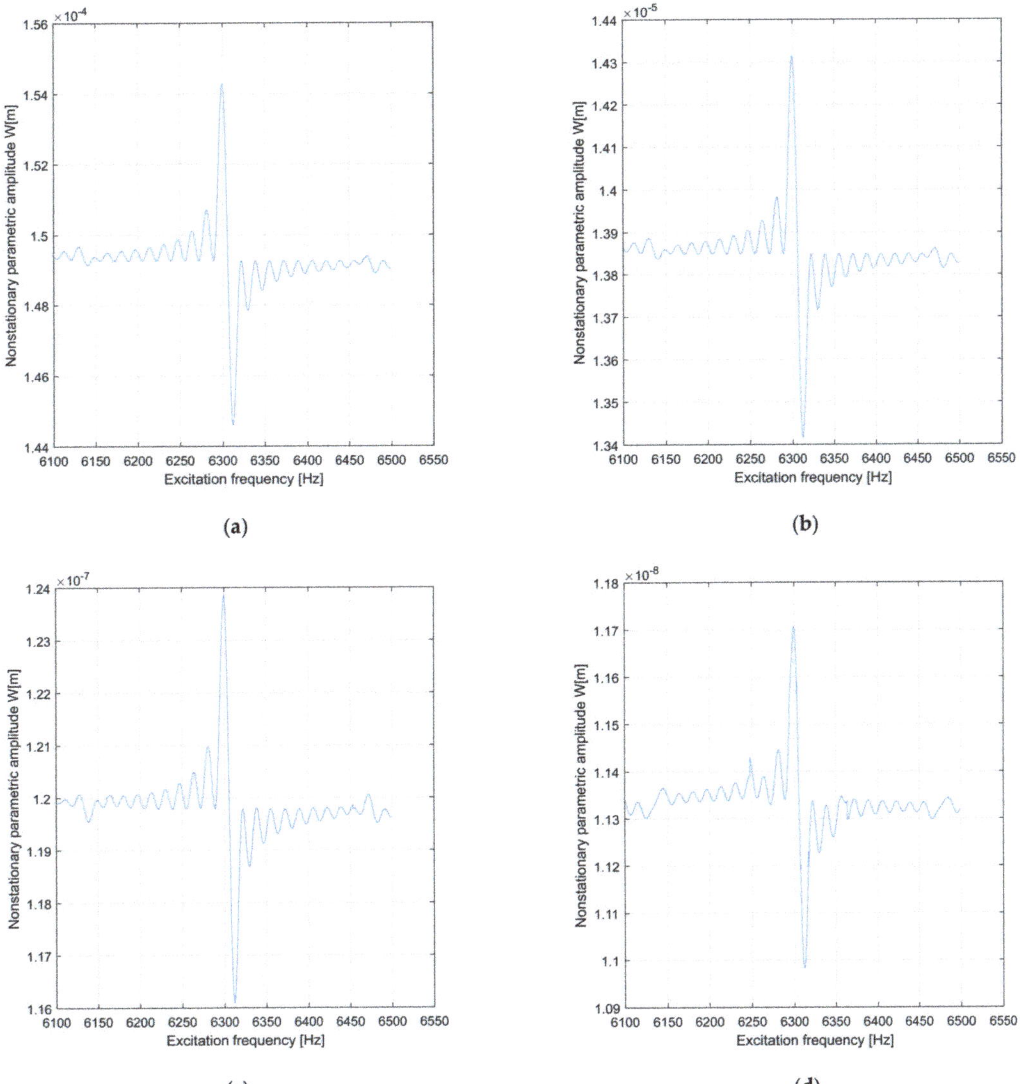

Figure 24. Nonstationary parametric amplitude spectrum W of the bowl for χ_{nB} = 0.10, where $\mu = 0.754 \times 10^{-4}$. **(a)** $\zeta = 0.0016$. **(b)** $\zeta = 0.0036$. **(c)** $\zeta = 0.0076$. **(d)** $\zeta = 0.0096$. Here, $m = 1.4 \times 10^{-5}$ octaves/min.

As for the tulip, a vast number of numerical tests was performed to arrive at a reasonable value for the sweep rate of the excitation frequency $m = 1.4 \times 10^{-5}$ octaves/min. By analyzing the nonstationary parametric amplitude spectrum presented in Figure 24 for $\zeta = (1.6–31.8) \times 10^{-3}$ and χ_{nB}= 0.10, the manifestation of beating effects as well as a significant decrease in the amplitude peaks with the increase in the damping ratio value in the range of (1.6–11.6) × 10^{-3} can be found, while in the range of (11.6–31.8) × 10^{-3}, the decrease was more significant. The maximum peak of the amplitude was obtained when the excitation frequency was in close vicinity of the PPR for the bowl (6306 Hz). Each time the damping ratio increased at a rate of 0.01, the maximum peak amplitude decreased by more than 10 times (5 times for the tulip), because the bowl had a more considerable rigidity than the tulip. Figure 25 illustrates the nonstationary velocity amplitude spectrum $\frac{dW}{dt}$ of the FBVM for the bowl during the transition through the PPR region.

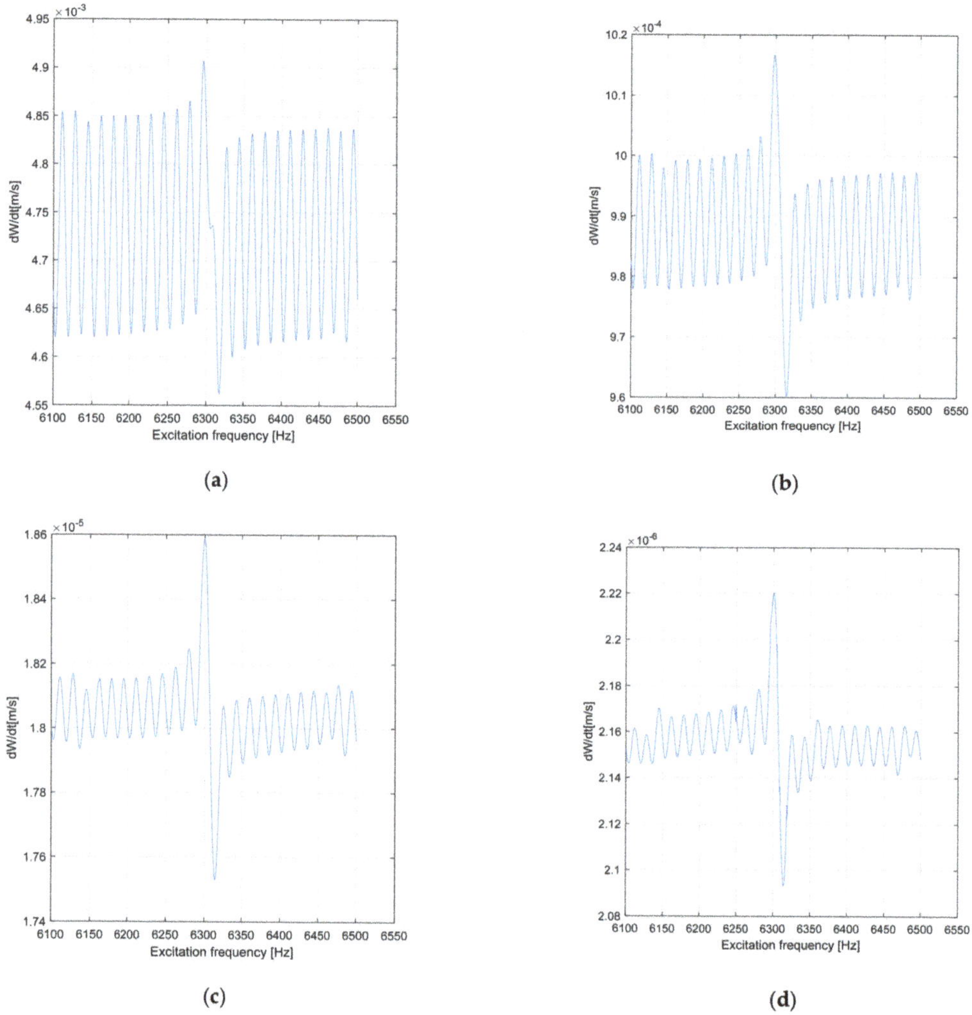

Figure 25. Nonstationary parametric velocity amplitude spectrum $\frac{dW}{dt}$ of the bowl for χ_{nT} = 0.25, where μ = 0.754 × 10^{-4}. (**a**) ζ = 0.0016. (**b**) ζ = 0.0036. (**c**) ζ = 0.0076. (**d**) ζ = 0.0096. Here, m = 1.4 × 10^{-5} octaves/min.

When analyzing the graphs in Figure 25, the moment of passing in close vicinity of the PPR was marked by a sudden increase in the peak of velocity amplitude, while the manifestation of beatings was like that of the nonstationary velocity amplitude spectrum for the tulip (see Figure 22). The manifestation for the bowl was different than that for the tulip (see Figure 22) because from the beginning, in the whole range of the damping ratio (0.0016–0.0318), the increase in the damping ratio had the effect of decreasing the maximum peak of the nonstationary velocity amplitude by approximately 5 times in the range of 0.0016–0.0116, while in the range of 0.0216–0.0318, the increase in the damping ratio had the effect of decreasing the maximum peak by more than 8 times. It can be concluded that the growth of the damping ratio had the beneficial impact of reducing the nonstationary velocity amplitude in the whole range (0.0016–0.0318) of the damping ratio. Figure 26 illustrates the graphs of phase space $\left(W, \frac{dW}{dt}\right)$ for the nonstationary FBVM of the bowl during the transition through the PPR region.

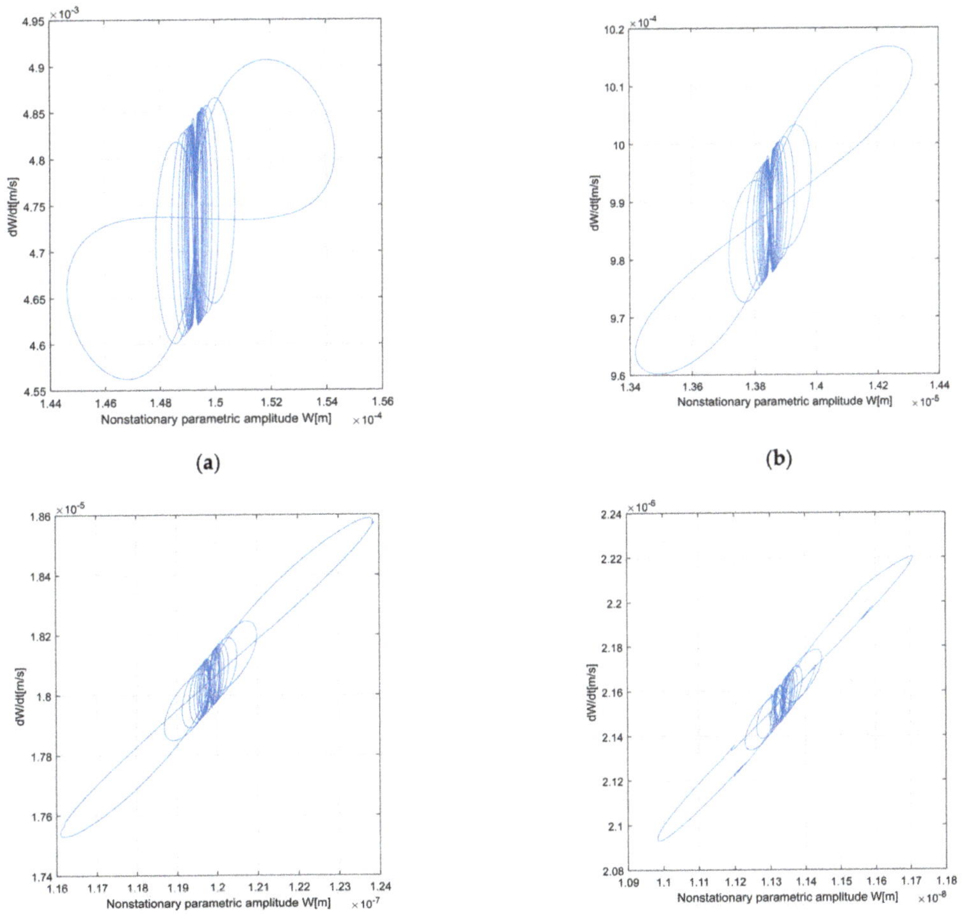

Figure 26. Phase space $\left(W, \frac{dW}{dt}\right)$ of the bowl for $\chi_{nB} = 0.10$, where $\mu = 0.754 \times 10^{-4}$. **(a)** $\zeta = 0.0016$. **(b)** $\zeta = 0.0036$. **(c)** $\zeta = 0.0076$. **(d)** $\zeta = 0.0096$. Here, $m = 1.4 \times 10^{-5}$ octaves/min.

When analyzing the graphs in Figure 26, the same manifestation for the bowl as for the tulip (see Figure 23) can be seen, that being a transition to chaotic behavior for the FBVM of the bowl due to the presence of limit cycles or even strange attractors. However, this last conclusion needs to be certified by detailed analysis using the methods of chaotic movements. This manifestation is valid only in the damping ratio range of 0.0016–0.0096.

One aspect that is evident is how the transition through the PPR region for the bowl was also an unstable one.

Unfortunately, there are no published studies which analyze in detail the dynamic behavior of each element of the AD for the FBVM (the tulip, bowl, and midshaft) apart from [4], as all the studies analyzed the global dynamic behavior of the automotive driveshaft. As can be seen from this paragraph, the use of AMA coupled with Hamilton's principle allowed the investigation of the stationary motion for the FBV movements of an AD's tulip and bowl in the PPR region by computing the amplitude (see Figures 10, 11, 14 and 15, the phase angle (see Figures 7, 12, 13 and 16), and the dynamic instability frontiers (see Figures 18 and 19) in the PPR region. In the meantime, the use of the AMA coupled with Hamilton's principle allowed the investigation of the nonstationary motion for the FBV movements of an AD's tulip and bowl in the transition through a PPR region by computing the amplitude spectrum (see Figures 21 and 24), the velocity amplitude spectrum (see Figures 22 and 25), and the velocity amplitude versus the amplitude in the phase space $\left(W, \frac{dW}{dt}\right)$ (see Figures 23 and 26). Figures 23 and 26 allowed the investigation of the dynamic instability in the transition through the PPR region. As is noted in Figures 21–26, the transition to the PPR region had an aspect of chaotic manifestation due to the manifestation of beating effects. To check whether this nonstationary dynamic behavior is deterministic chaos or a stochastic process, it would be necessary to use Lyapunov's exponents method coupled with the Poincare map method. All these aspects represent novelties in the dynamic instability investigation for the AD. Based on the phenomena presented above, the direction for future research involves the study of chaotic motion generated by the FBVM of the AD elements. Another future research direction is to investigate the dynamic instability in the proximity of each possible resonance type for the stationary and nonstationary FBVM of the AD elements, such as subharmonic resonance, superharmonic resonance, and combination and internal resonance.

8. Conclusions

The present work introduces a newly designed DMFFBVM for the AD, with the following phenomena being included for the first time: nonuniformity of the inertial characteristics of the AD's elements (see Equations (1)–(18)), serial stiffness and damping for the tulip and bowl (see Equations (19) and (20) as well as (23) and (24)), shock excitation due to the road geometry (see Equation (25)), and nonuniformity of the kinematic isometry (see Equation (61)). Based on this newly designed DMFFBVM, using Hamilton's principle coupled with the first-order approximation of the AMA, the stationary and nonstationary dynamic instability behavior of the AD elements were investigated in detail by computing the following:

- The amplitude, the phase angle, and the dynamic instability frontiers for the stationary FBV movements of the tulip and bowl in the PPR region;
- The amplitude spectrum, the velocity amplitude spectrum, and the velocity amplitude versus the amplitude for the FBV movements of the tulip and bowl in transition through the PPR region. As was remarked in the seventh paragraph, the growth of the damping ratio had a beneficial stabilizing effect for both the stationary and nonstationary FBV movements of the tulip and bowl. In the meantime, the dynamic instability frontiers were determined for the first time for the stationary FBV movements of the tulip and bowl in the parametric spaces (η_1, ξ, μ), (η_3, ξ, μ). In addition, for the first time, the velocity amplitude versus the amplitude was determined for the nonstationary FBV movements of the tulip and bowl in transition through the PPR region (see Figures 23 and 26). Agreements were found with the numerical and experimental data in the literature concerning the natural free frequency in bending and the manifestation of beating effects that conduct pitting and micro-cracking effects. Therefore, this DMFFBVM of the AD elements coupled with Hamilton's principle and first-order approximation of the AMA can be used in the early design stages to predict

an AD's durability. Moreover, the DMFFBVM must be added to the design algorithm for predicting the comfort elements of automobiles.

Author Contributions: Conceptualization, M.B.; methodology, M.B. and O.V.; software, M.B. and O.V.; validation, M.B.; formal analysis, M.B.; investigation, M.B.; resources, M.B. and O.V.; data curation M.B.; writing—original draft preparation, M.B.; writing—review and editing, M.B.; visualization, M.B.; supervision, M.B.; project administration, M.B. All authors have read and agreed to the published version of the manuscript.

Funding: This research received no external funding.

Institutional Review Board Statement: Not applicable.

Informed Consent Statement: Not applicable.

Data Availability Statement: Not applicable.

Acknowledgments: The authors of this article are thankful to the University POLITEHNICA of Bucharest for providing a serene environment and facilities for carrying out this research.

Conflicts of Interest: The authors declare no potential conflicts of interest with respect to the research, authorship, or publication of this article.

Appendix A. Expressions of the Coefficients Used in the Algorithm for Solving the SDEOFBVM for the DMFFBVM of the Tulip and the Bowl in the PPR Region

$$C_1 = \frac{a_1}{1+a_1}\chi_{nT}, \tag{A1}$$

$$C_2 = \frac{a_1 b_1}{J_{Z_2TA} + b_1(1+a_1)}\chi_{nT}, \tag{A2}$$

$$C_3 = \frac{a_2}{1+a_2}\chi_{nB}, \tag{A3}$$

$$C_4 = \frac{a_2 b_2}{J_{Z_2BA} + b_2(1+a_2)}\chi_{nB}, \tag{A4}$$

$$\lambda_1 = 1 - \frac{\eta_1}{2\Omega_1}, \lambda_2 = \frac{3}{8}\frac{\Gamma_1}{\Omega_1^2}, \tag{A5}$$

$$\lambda_3 = \frac{5}{16}\frac{\Gamma_2}{\Omega_1^2}, \lambda_4 = \xi^2 - \mu^2\left(\frac{\Omega_1}{\eta_1}\right)^2, \tag{A6}$$

$$\lambda_5 = \lambda_3^2, \lambda_6 = 2\lambda_2\lambda_3, \tag{A7}$$

$$\lambda_7 = \lambda_2^2 + 2\lambda_1\lambda_3, \lambda_8 = 2\lambda_1\lambda_2, \lambda_9 = \lambda_1^2 + \lambda_4, \tag{A8}$$

$$\lambda_{10} = 1 - \frac{\eta_3}{2\Omega_3}, \lambda_{20} = \frac{3}{8}\frac{\Gamma_3}{\Omega_3^2}, \tag{A9}$$

$$\lambda_{30} = \frac{5}{16}\frac{\Gamma_4}{\Omega_3^2}, \lambda_{40} = \xi^2 - \mu^2\left(\frac{\Omega 3}{\eta_3}\right)^2, \tag{A10}$$

$$\lambda_{50} = \lambda_{30}^2, \lambda_{60} = 2\lambda_{20}\lambda_{30}, \tag{A11}$$

$$\lambda_{70} = \lambda_{20}^2 + 2\lambda_{10}\lambda_{30}, \lambda_{80} = 2\lambda_{10}\lambda_{20}, \lambda_{90} = \lambda_{10}^2 + \lambda_{40}, \tag{A12}$$

References

1. Mitropolskii, Y.A. *Problems of the Asymptotic Theory of Nonstationary Vibrations*; Izdatel'stovo Nauka: Moscow, Russia, 1964; English Translation: D. Davey&Co.: New York, NY, USA, 1965.
2. Webber, H.; Kaczmarczyk, S.; Iwankiewicz, R. Non-linear Response of Cable-mass-Spring System in High-Rise Buildings under Stochastic Seismic Excitation. *Materials* **2021**, *14*, 6858. [CrossRef] [PubMed]
3. Sireteanu, T.; Gündisch, O.; Paraian, S. *Random Vibrations of Automotive*; Technical Publishing House: Bucharest, Romania, 1981. (In Romanian)
4. Steinwede, J. Design of a Homokinetic Joint for Use in Bent Axis Axial Piston Motors. Ph.D. Thesis, Aachen University, Aachen, Germany, 25 November 2020. Available online: https://www.google.com/search?client=firefox-b-d&q=%E2%80%9DDESIGN+OF+A+HOMOKINETIC+JOINT+FOR+USE+IN+BENT+AXIS+AXIAL+PISTON+MOTORS%E2%80%9D+J.+Steinwede+ (accessed on 19 December 2021).
5. Mazzei, A.J.; Scott, R.A. Principal Parametric Resonance Zones of a Rotating Rigid Shaft Driven through a Universal Joint. *J. Sound Vib.* **2001**, *244*, 555–562. [CrossRef]
6. Browne, M.; Palazzolo, A. Super harmonic nonlinear lateral vibrations of a segmented driveline incorporating a tuned damper excited by non-constant velocity joints. *J. Sound Vib.* **2008**, *323*, 334–351. [CrossRef]
7. Xia, Y.; Pang, J.; Zhou, C.; Li, H.; Li, W. *Study on the Bending Vibration of a Two-Piece Propeller Shaft for 4WD Driveline*; SAE International: Warrendale, PA, USA, 2015. [CrossRef]
8. Alugongo, A.A. Parametric Vibration of a Cardan Shaft and Sensitivity Analysis. In Proceedings of the World Congress on Engineering and Computer Science (WCECS), San Francisco, CA, USA, 23–25 October 2018; Volume II. Available online: https://www.google.com/search?client=firefox-b-d&q=Parametric+Vibration+of+a+Cardan+Shaft+and+Sensitivity+Analysis+Alfayo+A.+Alugong (accessed on 10 February 2022).
9. Wellmann, T.; Govindswamy, K.; Braun, E.; Wolff, K. *Aspects of Driveline Integration for Optimized Vehicle NVH Characteristics*; SAE International: Warrendale, PA, USA, 2007. [CrossRef]
10. Wellmann, T.; Govindswamy, K.; Braun, E.; Wolff, K. *Optimizing Vehicle NVH Characteristics for Driveline Integration*; SAE International: Warrendale, PA, USA, 2007; pp. 1–15. Available online: https://www.semanticscholar.org/paper/Optimizing-Vehicle-NVH-Characteristics-for/fc21c4c99a1bf0fe34a8a8848452876130cea8dd (accessed on 11 February 2022).
11. Yang, M.; Gui, L.; Hu, Y.; Ding, G.; Song, C. Dynamic analysis and vibration testing of CFRP drive-line system used in heavy-duty machine tool. *Results Phys.* **2018**, *8*, 1110–1118. [CrossRef]
12. Yao, W. Nonlinear Vibration Control of the Flexible Driveshaft System with Nonconstant Velocity Coupling via Torsional Input. Ph.D. Thesis, University of Tennessee, Knoxville, TN, USA, December 2019. Available online: https://trace.tennessee.edu/utk_graddiss/5699 (accessed on 28 February 2022).
13. Jadhav, M.M.; Jadhav, A.R. Vibration analysis of driveshaft with crack using experimental modal analysis and FEA. *Int. J. Eng. Res. Technol.* **2019**, *8*, 817–823. Available online: https://www.ijert.org/vibration-analysis-of-driveshaft-with-crack-using-experimental-modal-analysis-and-fea (accessed on 18 February 2022).
14. Qiu, B.; Shi, B.; Ding, C.; Fu, P.; Zhang, Y. *Influence of Dynamic Absorbers on Bending Vibration in Vehicle Propeller Shaft*; SAE International: Warrendale, PA, USA, 2018. [CrossRef]
15. Wu, Y.; Li, R.; Ding, W. The Application of Locally Resonant Photonic Crystals in Automotive Drive-Shaft Vibration Damper. 2016, pp. 2137–2143. Available online: https://past.isma-isaac.be/downloads/isma2016/papers/isma2016_0300.pdf (accessed on 25 January 2022).
16. Prakash, B.P.; Sinha, B.K. Analysis of drive shaft. *Int. J. Mech. Prod. Eng.* **2014**, *2*, 24–29. Available online: https://iraj.doionline.org/dx/IJMPE-IRAJ-DOIONLINE-505 (accessed on 24 January 2022).
17. Kumar, A.; Jain, R.; Patil, P.P. Dynamic analysis of heavy vehicle medium duty drive shaft using conventional and composite materials. *IOP Conf. Ser. Mater. Sci. Eng.* **2016**, *149*, 012156. Available online: https://iopscience.iop.org/article/10.1088/1757-899X/149/1/012156 (accessed on 26 February 2022). [CrossRef]
18. Alam, S.; Uniyal, A.; Bajaj, A. Evaluation of structural and vibrational characteristic of composite driveshaft using FEM. *Int. J. Eng. Sci. Res. Technol.* **2016**, *5*, 702–714. Available online: https://www.researchgate.net/publication/305394734_EVALUATION_OF_STRUCTURAL_AND_VIBRATIONAL_CHARACTERISTIC_OF_COMPOSITE_DRIVE_SHAFT_FOR_AUTOMOBILE_USING_FEM (accessed on 20 January 2022). [CrossRef]
19. Bugaru, M.; Chereches, T.; Trana, E.; Gheorghian, S.; Homotescu, T.N. Theoretical model of the dynamic interaction between wagon train and continuous rail. *WSEAS Trans. Math.* **2006**, *5*, 374–378. Available online: https://scholar.google.com/citations?view_op=view_citation&hl=en&user=tYI6MzwAAAAJ&citation_for_view=tYI6MzwAAAAJ:_FxGoFyzp5QC (accessed on 19 December 2021).
20. Deciu, E.; Bugaru, M.; Dragomirescu, C. *Nonlinear Vibrations with Applications in Mechanical Engineering*; Romanian Academy Publishing House: Bucharest, Romania, 2002; ISBN 973-27-0911-1.
21. Voinea, R.; Voiculescu, D.; Simion, F.P. Solid State Mechanics with Applications in Engineering. Romanian Academy Publishing House: Bucharest, Romania, 1989; ISBN 973-27-0000-9.
22. Rao, S.S. Transverse Vibrations of Beams. In *Vibration of Continuous Systems*; John Wiley & Sons: Hoboken, NJ, USA, 2007; pp. 317–392. ISBN 978-0-471-77171-5.
23. Rao, S.S. *Mechanical Vibrations*, 5th ed.; Prentice Hall: Upper Saddle River, NJ, USA, 2011; ISBN 978-0-13-212819-3.

24. Chan, K.T.; Lai, K.F.; Stephen, N.G.; Young, K. A new method to determine the shear coefficient of Timoshenko beam theory. *J. Sound Vib.* **2011**, *330*, 3488–3497. [CrossRef]
25. Duditza, F.; Diaconescu, D. Zur Kinematik und Dynamik von Tripode-Gelenkgetrieben. *Konstruction* **1975**, *27*, 335–341.
26. Bugaru, M.; Vasile, A. Nonuniformity of Isometric Properties of Automotive Driveshafts. *Computation* **2021**, *9*, 145. [CrossRef]
27. Seherr-Thoss, H.C.; Schmelz, F.; Aucktor, E. Designing Joints and Driveshafts. In *Universal Joints and Driveshafts*, 2nd ed.; Springer: Berlin, Germany, 2006; pp. 109–248.
28. Nayfeh, A.H.; Mook, D.T. *Nonlinear Oscillations*; John Wiley & Sons: New York, NY, USA, 1979.
29. Bugaru, M.; Vasile, A. A Physically Consistent Model for Forced Torsional Vibrations of Automotive Driveshafts. *Computation* **2022**, *10*, 10. [CrossRef]
30. Bolotin, V.V. *Dynamic Stability of Elastic Systems*, 2nd ed.; US Military Report; Aerospace Corporation: El Segundo, CA, USA, 1962.
31. Detroux, T.; Renson, L.; Masset, L.; Kerschen, G. The harmonic balance method for bifurcation analysis of large-scale nonlinear mechanical systems. *Comput. Methods Appl. Mech. Eng.* **2015**, *296*, 18–38. [CrossRef]
32. Bugaru, M. Dynamic Behavior of Geared System Transmission. Ph.D. Thesis, Auburn University, Auburn, AL, USA, University Politehnica of Bucharest, Bucharest, Romania, 13 October 2004. Available online: https://crescdi.pub.ro/#/profile/804 (accessed on 17 December 2021).

MDPI
St. Alban-Anlage 66
4052 Basel
Switzerland
Tel. +41 61 683 77 34
Fax +41 61 302 89 18
www.mdpi.com

Applied Sciences Editorial Office
E-mail: applsci@mdpi.com
www.mdpi.com/journal/applsci

www.ingramcontent.com/pod-product-compliance
Lightning Source LLC
Chambersburg PA
CBHW061521210926
51850CB00008B/1992
9783036577487